Multi-cultural Perspectives of the History of Science and Technology in China

Proceedings of the 12th International Conference on the History of Science in China

Edited by Liao Yuqun et al.

Science Press
Beijing

Responsible Editor: Guo Yongbin

Description

The proceedings contain more than 30 contributions made by researchers home and abroad at the 12th International Conference on the History of Science in China (ICHSC). Discussed are the following topics: ①Cross-cultural transmission and comparative studies in science and technology; ②Studies in ancient Chinese literature concerning science, technology and medicine; ③Traditional technology and non-material heritages in the world. A number of fields are covered, e.g. the history of science, technology, medicine, agriculture and traditional technology. The research perspectives include history, culture, philosophy, sociology, anthropology, archeology, and ecology.

图书在版编目(CIP)数据

多元文化视角下的中国科技史研究：第十二届国际中国科学史会议文集 = Multi-cultural Perspectives of the History of Science and Technology in China：英文／廖育群等主编．—北京：科学出版社．2012.8
　ISBN 978-7-03-034952-1

Ⅰ.①多… Ⅱ.①廖… Ⅲ.①科学史－中国－国际学术会议－文集－英文　Ⅳ.①N092-53

中国版本图书馆 CIP 数据核字（2012）第 133023 号

Multi-cultural Perspectives of the History of Science and Technology in China: Proceedings of the 12th International Conference on the History of Science in China

Copyright © 2012 by Science Press, Beijing

Published by Science Press
16 Donghuangchenggen North Street
Beijing 100717, China

Printed in Beijing

All right reserved. No part of this publication may be reproduced, stored in a retrieval system, or transmitted in any form or by any means, electronic, mechanical, photocopying, recording or otherwise, without the prior written permission of the copyright owner.

ISBN 978-7-03-034952-1
RMB 198.00

Editor-in-Chief
Liao Yuqun

Deputy Editor-in-Chief
Feng Lisheng
Hu Shenghua
Zhang Daqing

Preface

The 12th International Conference on the History of Science in China (12th ICHSC), hosted by the Chinese Society for the History of Science and Technology (CSHST) and organized by the Institute for the History of Natural Sciences (IHNS), Chinese Academy of Sciences (CAS) and Tsinghua University, was successfully held in Beijing on June 26-30, 2010 under the joint sponsorship of China Association for Science and Technology (CAST), CAS, National Science Foundation of China (NSFC), IHNS and the China-Portugal Center for the History of Sciences (CPCHS). 90 papers themed on the "Multi-Cultural Perspectives in the History of Science and Technology in China" were presented. More than 150 guests, speakers, and participants from 13 countries and regions attended the five-day conference. On behalf of the Organizing Committee of the 12th ICHSC, I would like to express my sincere gratitude to all the institutions and individual participants for their strong support for making the event a tremendous success. My thanks also go to the friends at home and abroad for their continuing interest in the history of science and technology in China.

The study on the development of science and technology is of tremendous value and significance to today's world. Essentially, it is the study of the history of how nature has been understood, explained, utilized and transformed, or in other words, what and how inventions, creations and discoveries were made through the wisdom and the accumulated experience of generations. Since HST became a new field of academic inquiry in the mid-20th century, specialists and scholars in science, technology and the humanities have gradually directed their research effort in the history of science and technology. There has been a growing consensus of opinion among them about the vital role of science and technology played in the progress of mankind.

China is a culturally rich nation with numerous well-known achievements in science and technology in ancient times. And knowledge of its history and intellectual roots enables an HST researcher to be more culturally sophisticated and avoid mistakes by learning from history. Studying the progress, laws and characteristics of the development of science and technology, and even addressing such issues as the development of contemporary intellectual frontiers from a historical perspective not only improve the structure of historiography, but also reveal the possible patterns and trends in the related fields by tracing the evolution of science and technology. The efforts may definitely provide strong references and deep inspiration for the future development of science and technology as well as the reform of its system in China. It is equally important to discuss issues concerning the history of science and technology from a cross-cultural perspective. Although originated in the West and later developed without much glaring geographical differences, the modern science and technology are still influenced substantially by different social, political, cultural and religious factors worldwide. And it is therefore of great significance to do comparative studies of the development, the dissemination and the schools of thought of science and technology between China and elsewhere in the world.

Recent years have witnessed a growing interest among the Chinese scholars of HST in interdisciplinary studies on cultural connotations of science and technology, the social environment of scientific research activities, and the possible cultural clashes between science, technology and the humanities. One more new trend in the field of HST is that a more open and culture-oriented attitude has been adopted on the part of the Chinese researchers. This has favorably led to a much closer cooperation with their overseas partners. China is actively pursuing self-reliant breakthroughs in science and technology as well as the cultural, social, scientific and technological growth and prosperity. It is no doubt that the multi perspectives on the study of the history of science and technology will serve the purposes well and help bring harmony to society.

Liao Yuqun

Contents

Preface

Standing Firm at Thirty: in Celebration to the 30th Anniversary of Chinese Society for the History of Science and Technology ················· Liu Dun（1）

Disciplinary Development of the History of Science and Technology in China ············ ················· Yuan Jiangyang（8）

Conference Address ················· Roshdi Rashed（12）

History of Science at the Beginning of the 21th Century ················· Roshdi Rashed（16）

Configurations Versus Equations: A Notational Difference ················· ················· Chen (Joseph) Cheng-Yih（22）

Zhu Shijie's Method of "Four Unknowns" as Inspiration for Wu Wen-Tsun ················· ················· Jiri Hudecek（49）

The Combination of Mathematics and Music—The Comparative Study of the Origin of the Calculation of Pitch in Ancient China and Greece ················· Liu Yaya（57）

Pythagoreanism in Edo—From ARAI Hakuseki to SAKUMA Shōzan ················· ················· Chikara SASAKI（65）

Algorithm and Principles of Division of Fractions in Chinese Ancient Literature ············ ················· Sun Xuhua（73）

An Exploration of the Original Sources of *Lvlv Zuanyao* ················· Wang Bing（83）

On Delisle's Correspondence to and from China Through the Archives of the Paris Observatory ················· Suzanne DÉBARBAT（97）

The Transmission of Western Astrolabe in Late Medieval China ······ Fung Kam-Wing（105）

Theories of Solar Motion in *Chongzhen Lishu*, *Yuzhi Lixiang Kaocheng* and *Lixiang Kaocheng Houbian* ················· Lu Dalong（124）

The Vacancy of Error Ideas about the Calculation of the Chinese Traditional Calendar ················· Wang Yumin（136）

Tentative Discussion on E. Diaz and the Influence of *Tianwenlue* on the Chinese Astronomy ················· Yao Licheng（147）

An New Explorations of the Origin of Chinese Alchemy ················· Han Jishao（153）

The Making of *Quanti Xinlun*	Chan Man Sing Law Yuen Mei	(163)
The Jesuit João de Loureiro (1717-1791) and the Medicinal Plants of China		
	Manuel S. Pinto Wang Bing Noël Golvers et al	(170)
The Feuds of the Medical Sects in Republic of China and Colonial Modernity		
	Xia Yuanyuan	(179)
Investigation on Traditional Spinning Wheels and Looms in Ze Zhou Region		
	Lu Wei Yang Xiaoming	(187)
Titles and Classifications of the Ancient Artisans in Dunhuang	Wang Jingyu	(195)
Guo Songtao and the Western Telegram Civilization	Xia Weiqi	(203)
On the Manufacturing Technology of Traditional Curved-Beam Plough in China		
	Feng Lisheng Huang Xing	(211)
Study on Indigenous Sugar-making Technology in Naman Tun of Daxin County		
	Liu Anding	(219)
British Iron and Steel Technology's Transfer in Early Modern East Asia: The Case of Qingxi Iron Works, China and Kamaishi Iron Works, Japan	Fang Yibing	(224)
How the Government Deal with the Drought from 989 AD to 992 AD in the Northern Song Dynasty	Dong Yuyu	(233)
Technologized Science: Representational Theories *vs.* Epistemological Engines		
	Byron Kaldis	(237)
A Shift in Interests to Science and Technology in the 11th China	Su Zhan	(248)
Traditional Chinese Science Among Vietnamese Minorities: Preliminary Results		
	Alexei VOLKOV	(258)
First Steps of Russian-Chinese Scientific Cooperation Contacts: Pyotr Kozlov's Visit to Beijing in 1925	Tatyana Yusupova	(268)
Negative Effects of Patent on Technological Development: The Perspective of the Theory of Modern Technological Process	Zhang Gaizhen	(273)
The Proceedings of the 12th International Conference on the History of Science in China		(281)

Standing Firm at Thirty: in Celebration to the 30th Anniversary of Chinese Society for the History of Science and Technology

Liu Dun

三十而立

——祝贺中国科学技术史学会成立三十周年*

刘 钝

女士们、先生们：

大家好！

今年是中国科学技术史学会的而立之年。30 年前的 10 月 6 日，在中国科学技术协会和中国科学院的大力支持下，来自全国各地的科学史工作者 247 名，聚集北京，成立了中国科学技术史学会（Chinese Society for the History of Science and Technology，以下简称"学会"）。那天的开幕式是一个高规格的学术集会，作为一只脚还在门槛外的 1978 级科学史专业研究生，我有幸窥见了当时的盛况。钱三强、茅以升、李昌、于光远、钱临照等科技界的重要人物亲临致辞；夏鼐、白寿彝等文史巨擘到场祝贺，时任中国史学会执行主席的周谷城与国内众多学术单位发来了贺电。这种热闹景象今日已难再现，它是在那个特殊年代里中国科学界、学术界对科学技术史这门学科在制度建设方面显示高度关注的象征。会议期间，与会者采取自愿申请的方式，每人填表一张，交入会费一元，经主席团审批通过，就构成了学会的第一批会员。

10 年前，在庆祝中国科学技术史学会成立 20 周年之际，时任学会理事长之一的席泽宗先生写过一篇纪念文章[1]；5 年之后，当时的学会秘书长韩健平等也发表了一文，对有

---

* 第十二届国际中国科学史会议暨中国科学技术史学会成立 30 周年纪念会开幕式报告，2010 年 6 月 27 日，北京。原为英文稿，题 Standing Firm at Thirty: in Celebration to the 30th Anniversary of Chinese Society for the History of Science and Technology。鲁大龙秘书长、自然科学史研究所分管学会工作的孙小淳副所长的鼓励与催促令我接受了这一任务；尽管做了一定的努力，但是没有韩健平研究员、王莹女士在提供原始资料和初稿撰写上提供的宝贵帮助，这份报告还是难以成文的。谨此一并致谢。

关进展作了补充[2]。今天，我受学会领导和第12届中国科学史国际会议组织者的委托，与在座的各位一道，回顾学会30年的光荣历史，分享它的成长历程。因为有上述两篇文章可供参阅，最近5年来的学会工作将是我报告的重点。

1. 学会简介

中国科学技术史学会是由中国科技史工作者自愿组织起来，依法登记的全国性、学术性的非政府组织，是中国科学技术协会下属的一级团体会员。它也是国际科学技术史学会的49个国家（或地区）会员之一，是中国科学史家在国际科学史界的正式代表。

学会的最高决策机构是全国代表大会，每三至四年召开一次。自1980年成立以来，中国科学技术史学会共召开过8次代表大会，先后在1980（北京）、1983（西安）、1986（北京）、1989（北京）、1994（北京）、2000（北京）、2004（哈尔滨）、2008（上海）等年份召开。

学会的首届理事长为钱临照，其后有柯俊（2届）、卢嘉锡、席泽宗/路甬祥（俩人共同承担两届）、刘钝、廖育群。可以说，2004年以前担任理事长的都是著名的科学家或中国科学界的领导。

学会的日常工作由常务理事会与秘书处主持，自成立以来秘书处及办公室一直挂靠在中国科学院自然科学史研究所。先后担任秘书长的有李佩珊、黄炜、范楚玉、周嘉华、王渝生、苏荣誉、韩健平，现任秘书长鲁大龙。他们是学会的总勤务。让我们对这些同事，以及所有担任过学会领导工作的老科学家、老前辈表示衷心的感谢。

学会的会员主要来自全国科研院所与高等院校，包括教师、研究人员和在读的研究生，也有一些热心科学史事业的业余研究者。现有注册会员1100人。下设16个专业委员会和2个研究分会，即：

数学史专业委员会
物理学史专业委员会
天文学史专业委员会
化学史专业委员会
地学史专业委员会
生物学史专业委员会
医学史专业委员会
农学史专业委员
技术史专业委员会
金属史专业委员会
建筑史专业委员会
综合史专业委员会
少数民族科技史专业委员会
咨询工作委员会
地方科技史志专业委员会
科技史教育专业委员会
计时仪器史研究分会

传统工艺研究分会

2. 组织高质量的学术会议

自成立之日起，学会就将举办学术会议当作主要工作来抓。早期经费困难，每年仍然举办学术会议达七八次之多。一些会议因其富有成效的组织形式和特殊的主题关注，在学术界一直获得好评。例如，全国青年科学技术史学术研讨会旨在为全国范围内从事科学史研究与教学机构的青年学者和研究生提供一个学术交流的平台。经过多年的运作，该会议已经积累了一些很好的经验。会议在全国范围内征集论文，组织专家初选，推荐大会报告；每场报告均有专家负责点评，并进行现场互动；又成立青年优秀论文奖评选委员会，严格遵照相关规则对所有的报告打分，再经专家讨论后决定奖项。学子们报告选题丰富、准备充分，专家点评鞭辟入里，异彩纷呈，会场互动气氛活跃。目前，全国青年科学技术史学术研讨会已经成为业内青年学子们自我展现和成长的最佳平台。

据不完全统计，学会成立30年来，共召开学术会议近200次。其中，中国少数民族科技史国际会议、国际中国科学史会议、全国数学史学术研讨会、中国地学史学术研讨会、中国技术史学术研讨会、中国天文学史研讨会、全国物理学史学术研讨会等，均已形成系列，正在向品牌学术会议的目标迈进。

这里值得提一下两个相关的系列国际会议，它们的源头是1982年在比利时召开的一次中国科学史会议。中国科学技术史学会成立不久，开始与国际科学史同行恢复往来，国内外都有人提出应该组织一次关于中国科学史的国际会议。按照何丙郁先生的说法：1978年他在北京饭店的一个座谈会上，提及自1956年竺可桢、李俨等人出席意大利第8届国际科学史大会以来，中国内地学者20多年来在国际舞台上几乎不见踪影，他也多次听到李约瑟对此表示遗憾。在场诸人莫不附和，并提议由何丙郁先生在海外谋划。不久何丙郁先生出任香港中文大学中文系主任，计划任内第一件大事就是举办中国科学史的国际会议。不过此时比利时的李倍始（Ulrich Libbrecht）已经筹到一笔专门的经费，遂建议他在鲁汶大学召开，是为第一届。当时有七八位中国学者获得邀请与会，在20世纪80年代初是相当引人注目的。何丙郁先生遂于次年在香港组织了第二届会议。出席会议的30人中，有14位来自中国内地，以考古学家夏鼐和科学史家席泽宗居首[3]。至此，被冠以"中国科学史国际研讨会"（International Conference on the Histoty of Science in China，ICHSC）的国际学术活动开始向系列化和品牌化的方向发展。

前六届ICHSC的时间和地点如下：

    1st ICHSC 1982 鲁汶
    2nd ICHSC 1983 香港
    3rd ICHSC 1984 北京
    4th ICHSC 1986 悉尼
    5th ICHSC 1988 圣迭戈（美国）
    6th ICHSC 1990 剑桥（英国）

剑桥会议的召开适逢李约瑟博士九十华诞。为老博士祝寿的同时，一些学者发起成立了一个新的学术组织国际东亚科学技术医学史学会（International Society for the History of

East Asian Science, Technology, and Medicine, ISHEASTM), 并决定延续先前的序列而将会名改为"国际东亚科学史会议"（International Conference on the History of Science in East Asia）。与此同时, 中国科学技术史学会则决定继续举办 ICHSC 系列, 同时鼓励其会员参加另一系列的活动。于是出现了两个系列并存的情况, 即

7th ICHSC	1994	深圳	7th ICHSEA	1993	京都
8th ICHSC	1998	柏林	8th ICHSEA	1996	汉城
9th ICHSC	2001	香港	9th ICHSEA	1999	新加坡
10th ICHSC	2004	哈尔滨	10th ICHSEA	2002	上海
11th ICHSC	2007	南宁	11th ICHSEA	2005	慕尼黑
12th ICHSC	2010	北京	12th ICHSEA	2008	巴尔的摩
13th ICHSC	2012[①]		13th ICHSEA	2011	合肥

3. 推动科学技术史教育在中国的发展

科技史教育近年来在中国有很大的发展。上海交通大学、中国科学技术大学、内蒙古师范大学三所高校创建了与科学技术史相关的系, 北京大学、清华大学、北京师范大学、北京科技大学、北京理工大学、华东师范大学、南京农业大学、西北农业大学、西北大学、天津师范大学、哈尔滨工业大学、东华大学等众多高校建立了一批科学史及其相关学科的研究与教学中心。科技史课程开始大规模进入高校的人文素质教育中。近年来, 学会将科技史教育作为工作的一个重点, 积极推动科学技术史教育在中国的发展。

2007 年 8 月, 学会主办了"首届全国科技史教学研讨会"。该次研讨会将科技史专业课程设置与研究生培养, 以及科技史课程与大学素质教育作为两大主题。会议对于深化国内一线科技史教育工作者的认识、推进科技史教育的发展、提升科技史学科的地位等方面, 都产生了积极的影响。一些专业委员会在推进科技史教学研究方面, 也付出了大量的努力。数学史专业委员会于 2005 年 5 月和 2007 年 4 月召开了第一届和第二届全国数学史与数学教育研讨会, 体现了专科史在高等院校相关专业中得到重视的现实。

2007 年底, 学会正式成立了科技史教学专业委员会, 标志着学会将促进科技史教学的发展当成自己的一项长期任务。

4. 创建学会网站, 优化和拓展学会的服务方式

在学会成立后的相当长一段时期里, 管理和服务工作都有赖于传统的书面作业方式。进入 21 世纪后, 网络信息技术在社会管理和服务领域中的应用日渐成熟, 也为学会的发展带来刺激和机遇。2004 年 10 月, 学会正式开通了官方网站。在中国科学技术协会所属的 150 余家学会中, 本学会是最早建立网站的学会之一。

网站的建立, 优化了学会的管理方式。例如, 过去学会每年都需编辑印发通讯等, 将一年来的工作情况和重要的学界消息报告给会员。但是这种方式时效性不强, 会员不能即时监督学会的工作, 也无法了解学界的最新情况。现在, 可以随时将学界资讯、工作动态、有关

---

① 本人报告之后, 从学会秘书长鲁大龙处获悉第 13 届 ICHSC 将于两年后在欧洲召开。

通知和信息，方便及时地上网，促进了会员及社会相关人士对学会工作的了解和监督。

网络信息技术的引入，拓展了学会的服务模式。会员数据库的建设，实现了会员对信息的分享。会员可以通过进入特定的会员社区访问数据库，了解同行的情况，并与各地志同道合的科技史工作者建立广泛的联系等。

在网站建设方面，一些专业委员会也走在了前列。例如技术史专业委员会、数学史专业委员会和少数民族科技史专业委员会等，都建立了自己独立的网站或网页，在加强专业委员会自身建设方面，迈出了重要的一步。

5. 推行事务公开，提高社会公信力

学会开展各项事业，都离不开社会各方面的支持与参与；反之，只有信任学会，社会力量才可能关注其发展。因此，学会的一个工作目标就是努力提高自身的社会公信力，从而赢得社会各方的信任和赞誉。社团提高社会公信力的一个有效途径，就是让公众知晓学会工作的开展情况，进而获得他们对学会工作的理解与认同。为此，我们在工作中积极推行学会事务公开，公布学会所组织的活动目标、运作方式及经费使用等情况，在公众中树立服务社会的学会形象。

常务理事会在会员代表大会闭幕期间负责领导学会的日常工作，制定规章和一些事务的暂行办法等。学会办公室即时将常务理事会的决议在网站上发布，让社会各方面了解学会的工作动态。近几年来，随着中国经济的发展，中国科学技术协会资助的项目日渐增多，学会总是在第一时间向全体会员群发邮件，通知学会项目申报组织事宜，欢迎各方踊跃申报。最近4年来，学会办公室及中国科学院自然科学史研究所、中国科学院研究生院、广西民族学院和北京科技大学等单位，先后在学会组织的项目申报中获得资助。

学会的经费主要来源于会费、学会申请的项目经费、挂靠单位的资助，以及相关科研教学单位对一些会议的资助等。虽然这些经费的额度大小不等，但学会都有义务和责任在合法的情况下使用好这些经费。为了便于社会监督，办公室建立了年报制度，不仅公布每年的主要工作，而且在年报中设有专门的部分，用来汇报当年度的经费使用情况。

6. 出版学术刊物

学会与挂靠单位中国科学院自然科学史研究所共同主办了两个科技史领域的综合性学术刊物。

《自然科学史研究》创刊于1982年。它在很长一段时间内以刊登中国古代科技史方面的论文为主，在学术界享有很高的声誉。近年来，刊物在继续发表学科史的考证研究论文的同时，较注意鼓励多学科多视角的综合性研究，倡导科学社会史、科学思想史，以及世界科技史和中国近现代科技史研究方面的广阔题材，密切关注国际科学史界、科学界和人文社会科学界的新问题、新方法和新理论，增加"研究讨论"、"书评"以及"学术信息"的分量，同时组建了一个由国际知名科学史家组成的顾问委员会，并邀请数位当今活跃在科学史前沿的年富力强的海外学者担任编委。

《中国科技史料》创刊于1980年，从1988年起交由学会和中国科学院自然科学史研究所合办。它侧重介绍清末以来的科学与工程技术方面的史料，特别是各个领域杰出科学

家的著述、传记、回忆录、创业史和治学方法等；同时，它也发表一些对中国科学技术事业有影响的外国科学家的生平及工作的介绍性文章。这些史料对于研究中国近现代科学技术的发展历程有着十分重要的学术价值，同时也面临散佚失传的问题，《中国科技史料》在抢救和整理这些文献方面做出了重要的贡献[4]。

《中国科技史料》2005年更名《中国科技史杂志》。新的办刊方针强调在搜集、抢救和整理史料的同时，加强对史料分析的解释工作，发表科学技术史领域的研究论文、综述评论、珍贵史料、学术信息、书评、教学研究等，主张以多元的视角开展科学技术史的研究，以展示科学发展的内在逻辑与社会文化特征，并以此推动与加强中国的科学技术史学科建设。

在长期的办刊过程中，《自然科学史研究》和《中国科技史杂志》形成了自己独特的风格，同时，随着时代的变化又不断进行一些调整，使得这两本学术刊物一直保持很高的学术水准。两本刊物连续入选中文核心期刊，并多次被中国科学技术协会评为优秀学术期刊。

7. 加强与台湾地区同行的交流

1980年，学会在成立之初就为台湾地区同行保留了两个理事名额。1982年，时任学会常务理事的席泽宗先生，在《中国科技史料》上发表了《台湾省的我国科技史研究》，向台湾地区同行发出了希望开展两岸同行合作交流的信号。席先生在文中写道："我们欢迎台湾的科学史工作者到大陆来参观访问和进行学术交流，并进行研究课题合作，为提高我国的科学史研究水平而共同努力。"此文在台湾地区同行中产生了良好的影响。从1985年起，两岸同行即在美国、澳大利亚等地的国际会议上频频会面。1991年，新竹清华大学历史研究所主编的《中国科学史通讯》出版，开始全面报道内地科学史界的学术资讯[5]。

1994年，借到台湾地区访问的机会，我们同台湾地区同行就参加学会事宜达成了共识：学会理事会为台湾地区学者留出三个名额，其中一名为常务理事。由此，在学会里形成了与台湾地区同行交流的友好局面，十余年来双边学术往来不断。这是两岸关系不断改善的结果，也是我们不断努力的结果。

8. 走向世界

新中国成立后，中国科学史界一直不断进行着同国际同行交流的努力，即使在相当严峻复杂的国际环境下。1956年，中国科学院副院长竺可桢等5人参加了在意大利佛罗伦萨召开的第8届国际科学史大会；在会上，中国被接纳为国际科学史学会（现更名为国际科学技术史学会）的国家成员。后因台湾问题和"文化大革命"，我们一度中断了同该组织的联系。

在学会成立后的第一次常务理事会上，大家就重返国际科学史组织及出席其重大活动一案进行了讨论。1981年，席泽宗等8人参加在罗马尼亚布加勒斯特召开的第16届国际科学史大会。第二届理事会成立后，柯俊理事长积极推动此事。1985年8月，在美国伯克利举行的第17届国际科学史大会上，中国科学技术史学会以国家成员的身份加入了国际科学史学会，李佩珊当选该组织理事，成为该组织建立以来首位进入理事会的女性学者。其后柯俊、陈美东和笔者本人等也先后被选进该组织领导机构，目前笔者担任着主席一职。

自1981年以来，中国科学技术史学会均组团出席国际科学史大会。2005年7月，学会与中国科学院自然科学史研究所在北京成功主办第22届国际科学史大会[6]。

女士们、先生们，中国科学技术史学会走过了 30 年艰辛的发展历程，取得了有目共睹的成绩。我们衷心祝愿她在未来有更大的发展，为推动中国的科技史研究和教育，做出更大的贡献。

参 考 文 献

［1］席泽宗. 中国科学技术史学会 20 年. 中国科技史料，2000，21（4）：289-296
［2］韩健平，李斌. 中国科学技术史学会的发展历程. 广西民族学院学报（自然科学版），2005，11（2）：17-19
［3］何丙郁. 学思历程的回忆：科学、人文、李约瑟. 新加坡：世界科技出版公司，2006.138-140
［4］林文照. 回顾与展望——纪念《中国科技史料》创刊 20 周年. 中国科技史料，2000，21（2）：95-101
［5］席泽宗. 中国科学技术史学会 20 年. 中国科技史料，2000，21（4）：289-296，292
［6］刘钝. 科学史的奥林匹克. 科技中国，2005，（7）：8-10

附　中国科学技术史学会在国际（非政府组织）科学大家庭中的位置①

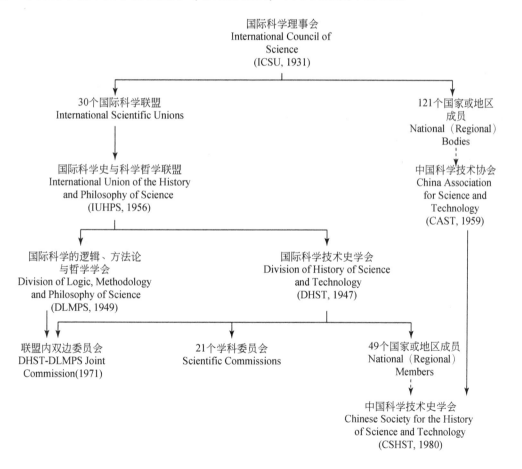

① 括号中为相应组织的英文缩写及创建年份，虚线箭头表示非直接属辖关系。表中数据来自两个国际组织的网站：http://www.icsu.org/；http://www.dhstweb.org/

Disciplinary Development of the History of Science and Technology in China

Yuan Jiangyang

(The Institute for the History of Natural Science, CAS)

I am glad to have this chance to talk about the disciplinary development of the history of science and technology in China in the past three years. We all know, the Chinese Society for the History of Science and Technology (CSHST) is 30 years old now. To celebrate this important anniversary, about 9 months ago, we started a research item supported by China Association for Science and Technology (CAST). Aiming at giving a systematic survey on the disciplinary development of our subject, lead by Prof. Liu Dun, the president of the International Association of the History of Science, and Prof. Liao Yuqun, the chairman of CSHST, and by co-operating between all the authors of chapters and other experts, this research produced a book named as "The Report on the Disciplinary Development of the History of Science and Technology" which was already published three months ago. What I would say is based upon the endeavor of all the co-authors of this book.

Chinese scholars, such as Li Yan and Qian Baocong, started their researches on HST in China in the early 20th century, and they focused on ancient Chinese literatures and relics related to science and technology. Since 1957, the institutional discipline of HST began to be established in China, and the starting-point is the foundation of a special research entity of HST known as "The Institute for the History of Natural Science, CAS". Nowadays, Chinese scholars' works have become an important part of the international research on HST since occidental HST theories and methods were introduced into China in 1980s.

In recent years, our discipline developed rapidly and experienced a process that Prof. Liu Dun and I called as "The re-institutionalization of the history of science in China". Now several universities have established the department named as "HST" or containing a unit of "HST". For instance, the first department of the history of science was found in Shanghai Jiaotong University in 1999. In 2006 and in 2008, the first and the second National Symposium on Teaching of HST came off in Beijing and Shanghai respectively. The Institute for the History of Natural Science, CAS, the Chinese Society for the History of Science and Technology, and the Inner Mongolia Normal University co-organized the first national seminar for the teaching of HST in Hohhot in 2009 to promote the teaching level of HST in Chinese universities.

Many top-level international academic conference of HST were held in China in recent years, such as the 22nd International Congress of History of Science (Beijing, 2005), the 11th International Conference on the History of Science and Technology in China (Nanning, 2007), and so on. Professor Liu Dun was elected as the President of the International Union of the History and Philosophy of Science/Division of the History of Science and Technology since 2009.

From 2007 to 2009, many important studies emerged on different fields of HST in China, especially on the history of science and technology in ancient China, in modern China, the oral history of science and technology in contemporary China, and the western history of science and technology.

The history of science and technology in ancient China remains as an important research field. Recently, some substantial research results were yielded in all these research areas.

In the research on the history of Chinese ancient astronomy, the archeological discovery of "Tao-si Observatory" (陶寺古观测台遗址) provided scholars an important, interesting and valuable subject-matter, which might give some substantial new clues to the origin of the Chinese astronomy. At the same time, the research on the history of mathematical astronomy in ancient China made great progress through new methods that used functions of modern celestial mechanics and computer program to examine the precision of the Chinese ancient calendars and re-construct the ancient calendars with the term of modern astronomy. In addition, the research direction involving the intercourse between ancient China and other nations in astronomy has also changed. Besides the study on the transplantation of Islamic astronomy in the Yuan and Ming Dynasties of China and European astronomy in the Ming and Qing Dynasties of China, which had been given much attention for a long time, the impact of Indian and Persian astronomy on the Tang and Pre-Tang Dynasties of China was studied as well in recent years.

In the field on the history of Chinese ancient mathematics, the discoveries of the *Suanshushu* (《算术书》) and *Shu* (《数》) from cultural relics of the Qin and Han Dynasties in the 1980s provided some important new materials for the scholars on the history of mathematics in early China. At the same time, the research on the history of the intercourse and the relationship between China and other nations in mathematics has also made great progress; and this kind of research is received supports from Wu Wenjuan Mathematics, Astronomy and Silk Road Fund (丝路基金). As achievements of the projects supported by the fund, two series of books have been published, one is *A Translated Series of Mathematics Masterpiece of Silk Road* (《丝路基金数学经典译丛》), and the other is *A Series of Comparative History of Mathematics* (《比较数学史丛书》).

The history of Chinese ancient technology has become a more noticeable field since the protection of intangible cultural heritage was regard as a very important issue in recent years. Two series of books have been published, the one is *Complete Works of the Chinese Traditional Crafts* (《传统工艺大全》) and the other is *Great Series of the History of*

Engineering and Technology in Ancient China(《古代中国工程技术史大系》) for the purpose of collecting, arranging, and conserving the knowledge of the ancient craft and technology. The Compass Project(指南针计划) is going along, which started in 2006 for the purpose of unearthing the value of Chinese ancient inventions.

As a new noticed field, the research on the history of science and technology in modern China is rapidly developing in these years. Scholars on this field studies the process of the institutionalization of the modern system of science and technology that was established in the late Qing Dynasty and the Minguo(民国) periods of China. The scientific societies and the modern industries built in the early 20th century China are the important topics. Recently, some researchers cleared up the history of the Science Society of China(中国科学社) founded in 1915, Academia Sinica of China(中央研究院) founded in 1928 and National Academy of Peiping(北平研究院) founded in 1929. And the others discussed the early history of Chinese industrialization since the Self-Strengthening Movement(洋务运动) from different angles.

The history of science and technology in contemporary China is also one of the fields which were received most attention in recent years. Firstly, studies on the oral history began to thrive. The *Chinese Journal for the History of Science and Technology*(《中国科技史杂志》,原《中国科技史料》) has offered a special column for the oral history since 1999 to collect and conserve the oral history materials from aged Chinese scientists. The publishing plan for *A Series of the Oral History of Science in the 20th Century China*(《20世纪中国科学口述史丛书》) has been executing smoothly since 2006. The first several books of the series have been published in recent years. By the way, I want to tell you a good news with great pleasure, that is, there comes an important project supported by China Association for Science and Technology and lead by Prof. Zhang Li. This project is aiming at surviving the information and the materials of the aged and excellent scientists and the sum of finance support comes up to 7 000 000 Yuan. Secondly, *The Disciplinary History Project* sponsored by the China Association for Science and Technology("中国科学学科史项目") has been on since 2008, and the books on the disciplinary development of chemistry, geology, communication engineering, and the combination of the Chinese/western medicine have been published since three months ago.

Compared with the above three fields, the western history of science and technology is still an underdeveloped field in China. However, It is the time for all of us to recognize the importance of this kind of research. Meanwhile, there are some obvious changes in the recent works coming from some young researchers. Firstly, studies on the famous scientists have enlarged to include more and more scientists as historical heroes, for instance, Faraday, Maxwell, Lavoisier, and so on, together with Newton and Einstein, have been studied in recent years. Secondly, new ideas and approaches were applied to the field of disciplinary histories, which cared about not only the processes of the accumulation of knowledge but also the processes of the formation and the development of disciplines. Thirdly, the research on the national history of science has made progress with some national characteristic problems as the key points. Finally, the foreign famous scientific and technological societies were given more and more attention to, for instance, the history of the Royal Society of London, the Lunar Society of Birmingham, and the Franklin Society in Philadelphia have been studied by the young researchers.

Naturally, there are some problems occurred in the development of our discipline at present in China. Firstly, the research on the history of science and technology in world is still given very little attention. Secondly, the history of science and technology in ancient China need to be re-examined with new methods and new points of view. Thirdly, the study on applied HST should be more noticed. I believe that our discipline, the history of science, will have a good future by improving all these three kinds of researches.

Conference Address

Roshdi Rashed

(Emeritus Director, National Center for Scientific Research, Paris;
Honorary Professor at the University of Tokyo)

Most esteemed colleagues, Ladies and Gentlemen:

It is both a great honour and a pleasure for me to be here, and to have this opportunity to address you. I am not intending to speak to you today about the scientific collaboration between France and China, which would require the expertise of a sociologist or a specialist in institutions, an expertise I do not myself possess. Nor shall I be talking about the exchanges between France and China, dating back to the 17th century. These are an intrinsic part of the history of the two countries, and have intensified and become widespread over the course of the past half-century. However, I shall keep to my own field, the history of the mathematical sciences, concentrating for the most part on those conceptions that have governed the dialogue between the two countries; that is to say, on the way in which Chinese science has been understood, and the position it occupies in the history of the sciences.

In fact, it was during the 18th century that French scholars first became interested in Chinese science, in the wake of ideas associated with the French Revolution, as well as through various accounts supplied by the Jesuit missionaries. It was at the time of the Enlightenment philosophy developed by the encyclopaedists that the history of the sciences first became an independent discipline. Indeed, the Enlightenment philosophers saw here the grounds of their main idea: that of a perpetual progress of the Enlightenment, and of the continuous eradication of error, the idea of the progress of the human mind and of humanity, which they understood in its totality as a single individual. Thus it was that the history of Chinese science—like that of Indian science, Arabic science, and so on—became part of the history of the sciences. In addition, the Jesuits made important material contributions. Indeed, it is to them that we owe some of the first accounts of the history of the Chinese sciences. To take astronomy, for example, we could mention Père Antoine Goubil's *Abridged History of Chinese Astronomy*, published in 1732, as well as his *Treatise on Chinese Astronomy*. From a later date, though still in the 18th century, we have de Mailla's translation of *The General History of China or the Annals of its Empire*, in 13 volumes (1777-1785). Moreover, a literature given over to the subject was becoming extensive in European languages during this period.

During that period, one of the best historians of the sciences in France, and indeed in Europe, was without doubt Jean-Baptiste Delambre (1749-1822). In 1817 he started publishing his monumental *History of Astronomy*, the first two volumes of which are given over to ancient and medieval astronomy. As Delambre himself said, the work was produced "for astronomers and mathematicians". Indeed, it is a technical history, in which Delambre debates the validity of the results and performs the calculations again himself. This aspect of the book is highlighted in the remark made by the eminent historian, Otto Neugebauer, that Delambre's work "remains much more than an impressive monument of the scholarship of the 19th century, it is a most valuable tool for the modern scholar attempting to understand the mechanism of theoretical astronomy of past periods. Whenever one tries to find one's way through an ancient treatise, it would be wise to look at what Delambre had to say about it"①. It is precisely in this monumental treatise that the Chinese astronomy makes an appearance as an integral part of the history of mathematical astronomy. Delambre dedicates a substantial chapter to it in the first volume, and revisits the theme once again in the second volume.

He starts by recalling that the Greeks only had knowledge of about six eclipses of the moon and none of the sun, whereas the Chinese were aware of 3858 consecutive years of eclipses, of which only one was lunar, all the others being eclipses of the sun. He states: "To predict eclipses reliably, besides having a knowledge of average movements, it was necessary to know about inequalities and parallaxes." Delambre undertakes a rigorous systematic study of the results of the Chinese astronomy, to which he had access via the transcriptions made by Goubil and de Mailla. He concludes the study with the following statement: "The Chinese knew about the movements of the planets. They only became aware of the movements of the stars 400 years after Jesus Christ, and nearly 600 years after Hipparchus. They understood nothing about stations or retrogradations, and held that all heavenly bodies revolve around the earth. Yet in individual works we find vestiges of a system that places the sun at the centre."② And he adds: "They made their astronomical discoveries after the Greeks made theirs, and their knowledge was quite inferior to that of the Greeks in every respect. The one thing the Chinese can point to are their eclipses of the sun, the most ancient, or at least the most numerous, known to us; it remains to be discovered whether these are authentic."③ If Delambre has a preference for the Greek astronomy, his judgment is based on the material knowledge he had of the Greek and Chinese astronomy. Perhaps that is a sufficient explanation. And yet, it should be noted that the philosophy of the Enlightenment, and of progress, implicitly contains this European *a priori*, which would find its clearest expression in the 19th century.

At a later date, although in attenuated form, one still comes across the same way of

---

① Jean Baptiste, Joseph Delambre. History de l'Astronomie Ancienne. Reprint from 1817 edition, Paris. With a new Preface by Otto Neugebauer. Johnson Reprint, N. Y. and Leiden, 1965.

② ibid. 363.

③ ibid.

writing history, where the Chinese contribution is put on apart with all the others, for example in B. Biot: Roman. *Studies on Indian Astronomy* and *Chinese Astronomy*, published in 1862.

However, throughout the 19th century German Romanticism maintained a firm hold over all the historical disciplines, including the history of the sciences. It was at that time that a preference for the Greeks became the "Greek miracle"; Chinese science, along with Indian science and Arabic science, now found itself pushed to the margins, far removed from the enlightened understanding of the Greeks and their successors. For about a century, one doctrine would dominate the history of the sciences: that of the occidentality of classical science. According to this doctrine, classical science is essentially European, and its origins are directly traceable to Greek philosophy and science; this tenet has survived intact through numerous conflicts of interpretation over the last two centuries. Almost without exception, the philosophers accepted it. Kant, as well as Comte, the neo-Kantians as well as the neo-positivists, Hegel as well as Husserl, the Hegelians and the phenomenologists as well as the Marxists, all acknowledge this postulate as the basis of their interpretation of Classical Modernity. Right down to our time, the names of Bacon, Descartes and Galileo (sometimes omitting the first, and sometimes adding a number of others), are cited as so many markers on the road to a revolutionary return to Greek science and philosophy. This return was understood by all to be both the search for a model and the rediscovery of an ideal. One might impute this unanimity to the philosophers' zeal to pass beyond the immediate data of history, to their wish for radical insight, or to their efforts to grasp what Husserl describes as "the original phenomenon (Urphänomen) which characterizes Europe from the spiritual point of view". One would expect that the position taken by those who have stuck more closely with the facts of the history of science would be quite different, but this is not the case. This same postulate is adopted by the historians of science as a point of departure for their work, and especially for their interpretations. Whether they interpret the advent of classical science as the product of a break with the Middle Ages, or defend the thesis of a continuity without a break or cut-off, or adopt an eclectic position, the majority of historians agree in accepting this postulate more or less implicitly.

Today, in spite of the works of many scholars on the history of Arabic, Indian and Chinese science, and notwithstanding the fact that non-western scientists are widely represented in the *Dictionary of Scientific Biography*, the works of the historians rest on the same fundamental concept: in its modern form as well as in its historical context, classical science is a work of European man alone. Furthermore, it is essentially the means by which this branch of humanity is defined. Occasionally the existence of a certain practical science in other cultures might be acknowledged; nevertheless, such a science remains outside history, or is integrated only to the extent that it contributes to what remain essentially European sciences. The contributions it makes are mere technical improvements, which do not modify the intellectual configuration or the spirit of European science in any way. An excellent illustration of this approach is the way in which Arabic science is depicted. In essence, Arabic science consists of a repository of the Greek tradition, transmitted intact or enriched thanks to the technical innovations made by the legitimate heirs of ancient science. The place accorded to Chinese science is no better. In all cases, scientific activity outside Europe is badly integrated into the history of the sciences; it is the product of an ethnography of science, which, translated into university studies, is nothing other than Orientalism.

Beginning in the 1950s, many historians of the sciences started to reject this doctrine of the occidentality of classical science. Among these historians were Joseph Needham and his collaborators. It is not by chance that Needham's *Science and Civilisation in China* opens with the following observation: "And still today the contribution of the East, and especially of its oldest and most central civilisation, that of the Chinese to science, scientific thought and technology, remains unrecognized and clouded in obscurity." It was this historical landscape that he sought to rectify; and yet, a better-documented knowledge of Arabic science would have enabled him to provide a more precise analysis of the formation of classical science.

In France too, at around the same time, research centres and research teams dedicated to Chinese civilisation began to multiply, whether in the National Centre for Scientific Research (Centre National de la Recherche Scientifique), in the Universities, at the Collège de France, or at the Practical College for Higher Studies (École Pratique des Hautes Études). As for the study of Chinese mathematics, this attracted a few individual researchers at first, but a real turning point occurred in 1982. In that year, two colleagues and I launched a research team that went under the name of Epistemological Research into the History of the Sciences and Scientific Institutions (Recherches Épistémologiques sur l'Histoire des Sciences et des Institutions Scientifiques, REHSEIS). In a radical departure from the old ideology of the occidentality of classical science, the members of this team developed multiple methods and fields of research. Chinese mathematics was studied at the same time as Greek, Arabic, modern and contemporary mathematics. A programme of collaboration with China furthered interaction between the two countries: researchers in our team were posted to China, and at the same time we hosted Chinese researchers. Indeed, the first of these was Professor Guo Shuchun, whose presence led to the critical edition and French translation of the *Nine Chapters*.

This collaboration with China, which has been constantly renewed, together with the presence of historians of Greek, Arabic and contemporary mathematics in our centre, has not only consummated the break with oblique notions of the history of the sciences. It has also made it possible to compare different styles of mathematics, and to discover the elements that unify these styles, as well as ways in which they differ. These comparisons will become richer and deeper as more and more critical editions of Chinese texts are produced, along with translations and analyses of them. Thus, as with all collaborations of this kind, Franco-Chinese collaboration is indispensable to the progress of the history of the sciences, whether by contributing to the development of methods in the discipline, or by enriching its content.

History of Science at the Beginning of the 21th Century

Roshdi Rashed

(French Society for History of Science and Technology)

Dear colleagues, Ladies and Gentlemen:

At the turn of the 19th century, and especially in the first decades of the 20th, historians of science discovered the full importance of textual research, and the necessity of retracing the textual tradition of each scientific writing examined. To a large extent, this new task imposed itself as an effect of the development of the historical and philological disciplines, which were themselves influenced by the German school of philology. This research on textual traditions brought to the history of science a host of auxiliary disciplines and historical techniques—paleography, codicology, philology, etc. —and it ended up as a definite achievement of the discipline, represented yesterday by Hultsch, Tannery, Heiberg... and attested today by the work on the translation of Archimedes by William of Moerbeke and those on the work of Newton, Leibniz, Euler and, more recently, by the studies devoted to the writings of Einstein, among others.

And yet, symmetrically, as it were, these achievements, to which we must add much other accumulated wealth which we will mention further on, did not take long to raise the problem of the gap between history and prehistory of sciences, which in turn gave rise to many other issues concerning scientific change. Such are the questions raised by the famous methodological debate which began in the sixties. In this debate, which was salutary, the goals went far beyond the questions raised. The historians who participated in it wished, in fact, to break with purely descriptive history, spontaneous history, the "history-novel" of scientists and their facts, and with history as the eclectic sum of people and facts. These were the first attempts to reflect on the discipline as such. In the life sciences, it was Georges Canguilhem who lead the reflection; In astronomy (minuscules), mechanics, and physics, it was G. Bachelard, A. Koyré, and above all T. Kuhn, among many others.

This debate interested those sociologists who, whether Weberians or Marxists, wished to give the history of science the social dimension it was lacking, by returning to institutions or social behaviors. In any case, this undertaking of methodological reflection, which in essence could not but remain unfinished, allowed the beginning of the first genuine effort to elucidate the discipline.

In addition to these methodological achievements, we must mention all the new fields to which we have laid claim. In the first place, the goal was to extend the history of sciences further back in time, thanks to the integration of Egypt and of Babylon. It was the work of Thureau-Dangin and Neugebauer, F. Thureau-Dangin and O. Neugebauer, in particular, that allowed this extension, at the same time as it incited historians to re-think the notion of "origin", and to situate Greek science differently. The other field was the rectification of the representation of Medieval science: research on Medieval Latin science was renewed, and scientific contributions which had until then been held to be peripheral were somewhat better integrated. Such was the case, in diverse degrees and in different modalities, with science in Arabic, Chinese, and Sanskrit; yet the view of these sciences as peripheral was to be long-lived, since it has not yet been abandoned. This task was accomplished by entire schools, some of which are associated with the names of J. Needham, A. Youschkevitch, P. Duhem, A. Maier, and M. Clagett.

Other fields were soon added to these newly-claimed domains, which opened the field of investigation of the discipline still further. The history of social sciences, the history of the diffusion of science from the centers of production towards the periphery, the history of institutions and of the great scientific laboratories, the history of applications, etc., today belong to the history of science, as is shown by the work presented at the various colloquia.

In these conditions, we will no doubt see the modification of the relations between the history of science and much other historical research, such as the history of philosophy and of technology.

In view of such diversity, not to say dispersion, we cannot avoid raising the question of the discipline itself, which was brought up in the course of the great methodological debate, only in order to be subsequently forgotten. Today, thanks to the work accomplished in the course of the preceding decades—Ch. Gillespie's D. S. B., and R. Cohen's immense collection of Boston Studies, among others—the question may be formulated in the following way: what is this discipline which, throughout its existence, and particularly from the 18th century on, when it was born as an independent discipline—deals both with epistemology and with history? Whether we think of Condorcet, in his Sketch or in his Academic Eulogies; or of August Comte and the role of the history of science in his *Course of Positive Philosophy*; or whether we come closer to our time, and bring up J. Needham, for instance: is the history of science really a discipline, and what is its true place between epistemology and social history?

The first part of the question—is it really a discipline? —can be disposed of quickly. As it presents itself today in the works of those who claim allegiance to it, the history of science is a domain of activity, and not by any means a discipline. Indeed, it lacks the unifying principle which could provide it with the means and the power to exclude; for a domain of activity does not exclude, but is indefinitely distended by successive additions. It is a heading designated by a label, not a discipline characterized by a genuine definition. Thus, in the history of science, the various doctrines are juxtaposed and opposed on the basis of dogmatic and exclusive options, and even *petitio principii*. According to some, the history of science presents itself as a history of ideas, or a history of mentalities; for others, by contrast, it is a history of scientific concepts; their formation, their development, and their rectification. For still others, who were originally historians, concepts and their nature had little importance, and the history of science is the history of a cultural production, in the same sense as that of painting or religion. Let us also cite those for whom it is a kind of social psychology of scientific actors, and those who make of the history of science an empirical sociology, such as has been developed particularly in the United States after the Second World War: a sociology of groups, laboratories, and institutions. This list is by no means closed, and this diversity continues to increase, not because of some internal necessity of research in the history of science, but rather as an effect of the successive importation of the views and methods of the social disciplines, and the trends that succeed one another therein.

This growing multiplicity has all the appearance of blind progress, which might spare us the examination of the second part of the question: what is the place of the history of science between epistemology and social history? But if we thus leave this question in the shadows, it will force us, whether we like it or not, to make up our minds about the object of the history of science. The difficulty—and it is considerable—is to be able to say in what sense the historian of science practices history, without formulating an arbitrary choice, and without imposing a methodology, be it empirical or transcendental. It was in order to avoid these shoals, upon which the methodological debate had run aground, that it seemed to me appropriate to start, in according with a well-known phrase, "from the things themselves"; that is, from works of science and the traditions within which they are integrated.

It will easily be granted that every work of science belongs to at least one tradition, and often to many—whether or not they are known to us—relatively to which it takes on its meaning. This means that we cannot understand anything about individual creations, however revolutionary they may be, unless we insert them within the traditions which witnessed their birth. If, by "work of science", we understand a result established in accordance with the precise norms of proof and consigned within a text, or realised within an object or an instrument, we shall for the moment give the word "tradition" the vague meaning attributed to this term, which has the advantage of not isolating the work of science from the community to which the scientist who conceives it belongs. Let us begin by considering this notion of tradition.

Historians of science, of whatever obedience, are quite willing to admit that one of their essential tasks is the reconstitution of scientific traditions. Yet the paths they follow in order to reach this goal are divergent and ramified. In fact, part of the methodological debate in the history of science refers to this diversity of conceptions of tradition and its nature. At first glance, the undertaking may seem easy and almost immediate: don't traditions most frequently present themselves under names, titles, institutions, and networks which ensure the exchange of information and human beings between poles, centers, places, and forms of learning? In this case, traditions would be immediately recognizable: we would speak of the tradition of the Euclidean theory of numbers, of the Japanese *Wasan*, of the tradition of the Italian algebraic school of the 16th century, of British quantum physics of the twenties, or of Bourbakian mathematics. To be sure, there are some exceptions, but they confirm the rule: I am thinking, for instance, of the Alexandrian tradition—or traditions—which finds its summit in the work of Diophantus, and of which, nevertheless, we are completely ignorant. How could we fail to be tempted to describe such easily-identifiable facts as men, titles, and institutions? In fact, it is this tendency which dominates a large part of historical writings, which present themselves under different names: history of ideas, social history of science, etc..

Nevertheless, if we are not satisfied with a simple empirical description, the status of a tradition is not easy either to delimit or to establish. How are we to isolate a tradition, assign to it a beginning and an end, and trace its borders, without proceeding by an arbitrary cut within the indefinitely mobile totality of living history? Who can found the unity of a tradition, which is constantly evolving through time? Why is it constituted, and why does it cease? What system of rules might its existence obey?

There is, it seems, no *a priori* response to these questions.

With simple description, the historian is only at the beginning of his labor. No sooner has he hitched himself to the task of reconstitution than the illusion is dissipated. The apparent simplicity evaporates, and all empirical data—names, titles, etc.—proves impotent to delimit a tradition while dominating all its ramifications. Let us try to be more precise, by describing the principal stages in a work of the history of science. At the first stage, the historian must restore a work of science—a mathematical theorem, a physical result, an astronomical observation, a biochemical experiment, etc.—in all its materiality. He must examine inscriptions, tablets, papyri, manuscript texts, and printed texts; he must redo experiments, and re-fabricate objects, if necessary... All these procedures contribute, in the first instance, to the reconstitution of the textual tradition, as well as of the technological tradition... In a word, of the "objectal" tradition (relatively to the notion of object in general). Although in many cases, this research is not independent of the very contents of the work of science, it requires competencies other than scientific knowledge: those which deal with the various historical disciplines, such as archaeology, codicology, paleography, philology, the history of technology, etc..

This level of analysis is indispensable, but it is not sufficient: in such a reconstitution we are still far from having exhausted the work of science. All that is known to us at this stage is its textual and technical authenticity, the networks along which it circulates, and the social context in which it was conceived and composed. All these elements are no doubt important, but they do not enlighten us with regard to its place within the science to which it belongs. Even more seriously, at this stage we would not be in a position to perceive the cleavages which may mark the work of one and the same scientist.

Other obstacles do not fail to arise along this way, which have their essential origin in the dialectic between increasing multiplicity and fundamental stability. After the study of numerous traditions, one general result imposes itself upon us: a work of science of some stature cannot be explained in terms of one single conceptual tradition, not even that to which this same work has contributed the most. Moreover, a conceptual tradition of some importance is distinguished by a certain stability, in spite of the diversity of authors and of contributions. Two somewhat paradoxical necessities seem to dominate the progress of a conceptual tradition: to exhaust all the logical possibilities inscribed within a given type of rationality, on the one hand; and, on the other, to reform this rationality and its means, in order to account for the new facts inexplicable within its framework. As examples, suffice it to reflect upon the Archimedean tradition in infinitesimal mathematics; the Euclidean tradition in the theory of parallels, etc..

These two terms: "objectal" tradition—of which textual tradition is a part—and conceptual tradition, seem concretely to translate the question of the place of the history of science between social history and epistemology. As an element of "objectal" tradition, the work of science is a material and cultural product; a product of men in a specific place and a time. As K. Marx would have advised, it is incumbent upon the historian to seek out the social and material conditions of this production. As a part of the conceptual tradition, however, this work also calls for an analysis of its conceptual structure, able to bring out its meaning, which will allow us to delimit the very notion of tradition. It might, of course, turn out that such a translation of our initial question may impoverish it. However it does seem susceptible of protecting us against two hazards: the reduction of history to pure epistemological analysis, such as it is for many of our eminent contemporaries; or else, even more, to a philosophy of history like that of A. Comte. The second risk is its assimilation to the history of any cultural domain, a practice which is current among historians. Yet the difficulty remains intact if we do not further specify what we mean by a conceptual tradition, to which a work of science belongs. Does this last question have the same meaning for all the scientific disciplines? Does the work of science belong to one conceptual tradition, or to several? These questions, among many others, emerge immediately, and they lead us necessarily to question ourselves about this notion of a work of science, and to ask ourselves what distinguishes it from all other social production of cultural works.

The differences between the objectal tradition and the conceptual tradition lie not only with objects and methods, but they are much more deeply rooted in the very nature of their necessity. The objectal tradition deals, in a word, with our actions: with psychological, social, and historical events, with things which are here and now; in brief, with contingent facts. The formation of an academy, the functioning of a great research center, the organisation of a laboratory, the modes of transmission of knowledge, the material support of texts, the allocation of resources, the scientist's social affiliations, his psychological profile, etc., are all so many contingent facts. Even if psychology, sociology, economics, and so on, can identify a kind of necessity in them, there is none in their relations to the facts of science. Conversely, it is to this very necessity that these facts owe their recognizability. Such is the case for a mathematical theorem, a physical law, etc. It is for this reason, moreover, that an objectal fact is not liable to be true or false, whereas for the conceptual fact its character of necessity is also a criterion of truth. We can thus understand that every globalizing vocation is condemned in advance to theoretical defeat. Conversely, to extend conceptual history to the objectal tradition leads to a "pure history", or a philosophy of history which is no longer a history of science. Yet the whole problem of the history of science, to which all its difficulty is reduced, consists in this: the production of the facts of science, quite determinate as the production of human beings and the results of their actions, transcends, as an effect, the contingent conditions of its advent; and this production transcends these conditions in order to distinguish itself from them by its necessary character. In a nutshell, the whole problem consists in knowing how necessity emerges from contingency. The historian of science then reveals himself to be what he has always tried to be: neither a "science critic", in the sense of an art critic; nor a historian, in the sense in which we understand a specialist in social history; nor a philosopher, like the philosophers of science; but simply a phenomenologist of conceptual structures, of their genesis and of their affiliations, in the midst of conceptual traditions which are always in transformation. Today more than ever, this self-awareness seems to me necessary if we want the history of science to be constituted as a genuine discipline, instead of being a mere domain of activity. Also today, we must construct a new discipline, as necessary as it is legitimate, simultaneously with the history of science, but independently from it: that of social research on the sciences. Such independence is the guarantee that both the history of science and social research on the sciences may be formed as true disciplines, which deal with the cultural phenomenon of science.

Configurations Versus Equations:
A Notational Difference*

Chen (Joseph) Cheng-Yih

(Department of Physics, University of California, La Jolla, California)

There are historians of mathematics and sinologists who viewed the presentation of early Chinese mathematic works in equations with signs (such as the equal " = ", plus " + ", minus " − ", etc.) and letter symbols (such as x, y, z, etc.) as equivalent to reading modern concepts of equations into old work. In this paper, I would like to suggest that misunderstandings such as these can be easily clarified, if one expands one's concept of notation from written mode to include mechanical mode.

In early Chinese work, mathematic problems were solved in Chousuan (筹算, computation with counting rods) not in Bisuan (笔算, computation with a pen). Consequently, the notational system developed is not in written mode but in mechanical mode. In Chousuan, one summarizes the mathematical information of the problem in configurations for solving the unknowns. Unlike in Bisuan, the mathematical information is summarized in equations for solving the unknowns. However, the equational representation in Bisuan and the configurational representation in Chousuan are equivalent because there is a one-to-one correspondence between the two representations. Consequently, one can transform between the two representations without changing the content of mathematics. History has shown that an important part of mathematical progresses is in the development of efficient calculating tools. A review of notation in mechanical mode is presented.

1. Introduction

In mathematics, there have been, since antiquity, two major modes for representing technical and quantitative mathematical concepts and relations, and for executing algorithms and derivations. In the mode known as the "written mode", one uses written symbols to compose equations for expressing mathematical thoughts and to execute algorithms in performing mathematical computations. In the "mechanical mode", on the other hand, one uses devised

---

* This is a talk presented at the 12th ICHSC (International Conference on the History of Science in China), Jun 26-30, 2010 at Beijing.

entities specifically related to the computational tools to compose configurations or patterns for expressing mathematical thoughts and to execute algorithms in performing mathematical computations. Therefore, the notational systems produced in the two modes are different in nature and intimately related to the mathematical tools involved. Obviously, cultural traditions have been one of the influential factors in one's choice, creation, and use of notations. The realization that mathematical notation exists in a number of different forms is one of the keys to appreciate the development of algorithms and derivation in early mathematics in different civilizations.

2. Numeral notation in written and mechanical modes

In most of the early civilizations, the written mode was favored for both performing computations and for keeping records in their early mathematic developments. In the early Chinese civilization, however, computations were carried out with a mathematical tool, known as the Suanchou (算筹, counting rods)① on a flat surface (or a counting board). In such a computational method, known in Chinese as the Chousuan, not only the numerals and mathematical relations are represented in terms of the configurational layouts of rods, but algorithms are also executed by manipulating the rods directly. Descriptions and records of the computations were kept in written form on a medium, such as shells, bones, bamboo strips, silk fabrics, or papers, depending on the different historical periods. As a consequence of such a practice②, two parallel systems of notation were developed in mathematics: one in mechanical mode for computation, and the other in written mode for writings.

An easy way to examine the difference in notational systems between the written and mechanical modes developed in the Chinese civilization is to compare the notational systems of the Chinese numeral systems. Two numeral systems were developed in early Chinese civilization. They are the positional numeral systems[2],[3]:37-44 and the grouping numeral systems.[4] In Tab. 1, the evolution of positional numerals from their rod form in mechanical mode to the ciphered form in written mode is illustrated for two specific examples: 35 648 and 10 933.

Tab. 1 Positional numerals in Chinese civilization

Notation	Example: 45 399	Example: 90 245
Chousuan notation		
Zhusuan (珠算) notation (Tang Dynasty)		

---

① Also known as Suance (算策), Chouce (筹策), or simply as Ce (策). Such counting rods were also found in archaeological excavations.[1]

② In fact, such a practice for performing computations, in a way, is similar to the present practice of carrying out computations electronically and recording them in printed written form.

续表

Notation	Example: 45 399	Example: 90 245
Northern Song written notation (Northern Song)	||||≡||⊥Ⅲ	Ⅲ○||≡||||
Sourthern Song written notation (Sourthern Song)	ㄨ᱑|||ㄨㄨ	ㄨ○||ㄨ᱑
Ming written notation (Ming Dynasty)	ㄨ᱑|||ㄨㄨ	ㄨ○||ㄨ᱑
Commercial notation	ㄨ᱑|||夊夊	夊○||ㄨ᱑
Character notation	四五三九九	九○二四五 九□二四五

The first two rows illustrate the positional numerals in mechanical mode. They are the rod form positional numerals in chousuan and the ball form in Zhusuan. The 3rd and 4th rows illustrate positional numerals in written form first appeared in the 13th century when the layout configurations of rods on computational board found their ways into the *Song books* as printed illustrations. The 5th and 6th rows illustrate the later written form of the positional rod numerals. The last entry in Tab. 1 illustrates the positional numerals expressed in terms of the character digits first appeared in grouping numerals.

The rod positional numerals shown in Tab. 1 are constructed in terms of numeral digits by alternating their horizontal and vertical tallying forms given in Tab. 2. The vertical form is used in the odd positions (i. e., the positions of units, hundreds, and so forth), and the horizontal form in the even positions (i. e., the positions of tens, thousands, and so forth). On the computational board, they are always displayed from left to right horizontally as illustrated and are aligned with respect to a designated position for the unit place. The demarcation of the successive powers of ten by alternating the two orientations of digits is not essential to the concept of positional value. Their adoption, however, makes it easier for calculations to be performed on flat surfaces without vertical line markings. Furthermore, such alterations also help to make the rod numerals easier to read.

The digit numerals listed in Tab. 2 illustrate that they are configurations composed of rods in different orientations.

Tab. 2 Tallying numeral digits formed by counting rods

Orientation	0	1	2	3	4	5	6	7	8	9
Horizontal form Hengshu (横数)		一	二	三	亖	亖	⊥	⊥	≛	≛
Vertical form Zongshu (纵数)		|	||	|||	||||	|||||	T	╥	╥	╥

The horizontal and vertical forms of these numeral digits are laid out by assigning orientational value in tallying. In the horizontal form of the numeral digits, each horizontal rod counts one and the vertical rod counts five. In the vertical form of the numeral digits, each vertical rod counts one and the horizontal rod counts five. ①

The positional rod numerals are constructed based on the base of 10, with groups (10^n, n = 0, 1, 2, etc.) denoted by their place values. These numerals are generated according to the multiplicative principles:

$$4 \times 10^4 + 5 \times 10^3 + 3 \times 10^2 + 9 \times 10 + 9$$

$$9 \times 10^4 + 0 \times 10^3 + 2 \times 10^2 + 4 \times 10 + 5$$

The zero digit of the system is always represented by a kongwei (空位), a vacant position with place value. These numeration principles found in the rod numerals suggest that the principles of positional numerals were already well developed by the Shang Dynasty (ca. -1600 to ca. -1046). With these numeration principles, the rod positional numerals can display all numbers, however large in principle, using only ten decimal digits.

The numerals invented in the written mode are the grouping numeral systems. Illustrated in Tab. 3 are the major stages in their evolution for two specific examples: 35 648 and 10 933.

Tab. 3 Grouping numerals developed in Chinese civilization②

Style	Thirty-five thousand six hundred and forty-eight (Example: 35 648)	Ten thousand nine hundred and thirty-three (Example: 10 933)
Shell-bone Inscriptions (-1400 to -1100)		
Present Form (since ca. -200)	三万五千六百四十八	一万九百三十三
Accounting Form (since ca. -200)	叁万伍仟陆佰肆拾捌	壹万玖佰叁拾叁

The entries for shell-bone inscriptions given in Tab. 3 are based on inscriptions found among the excavated shell-bone inscriptions listed in Tab. 4.

Tab. 4 The grouping numerals found in the shell bone inscriptions (ca. -1400 to ca. -1100)

Group of units									
	1	2	3	4	5	6	7	8	9

---

① These indicate that the positional rod numerals are constructed not only with place value but also with orientational value.
② There were also bronze inscriptions and variations in Han writing styles not included in Tab. 3.

续表

Group of tens	10	20	30	40	50	60	70	80
Group of hundreds	100	200	300	400	500	600		800	900
Group of thousands	1000	2000	3000	4000	5000		8000	
Group of tens of thousands	10000		30000					

It can be seen from Tab. 4 that these numerals are generated from the symbols for digits (see the shell-bone forms in Tab. 5) and for the basic groups (see the shell-bone forms in Tab. 6). The numeral for 8000, for example, is constructed by the combination of the symbol for the digit 8 with the symbol for the group 10^3 based on multiplicative principles. Hence, the value of the numeral is given by

$$= \text{)(} \times \text{{} = 8 \times 1000 = 8000$$

These numerals found in the shell-bone inscriptions suggest that the principles of grouping numerals were already well developed by the Yin period of the Shang Dynasty. The subsequent changes occurred only through the evolution of the digit and group symbols as illustrated in Tab. 5 and Tab. 6.

Tab. 5 The evolution of digits in the grouping numerals

Numeral types	1	2	3	4	5	6	7	8	9
Tallying form (ca. -3000)									
Shell-bone forms (ca. -1400 to -1100)									
Bronze and coin forms (ca. -1000 to -300)									
Han form (ca. -300 to +200)									
Present form (since ca. -200)	一	二	三	四	五	六	七	八	九

Configurations Versus Equations: A Notational Difference

Tab. 6　Evolution of the symbol for the basic groups in grouping numerals

Numeral types	10	10^2	10^3	10^4
Shell-bone form (ca. −1400 to −1100)	⌣ ∣ ⊥	⊎	↑	(symbol)
Bronze and coin forms (ca. −1000 to −300)	↑ 十	(symbols)	↑	万
Present form (since ca. −200)	十	百	千	万 萬

By examining the digit components in the ciphered grouping numerals listed in Tab. 5, it is seen that the digits in the unit group are in the horizontal form, but in the group of tens they are in the vertical form. Then, in the group of hundreds, the digits are again in the horizontal form. Not until the group of thousands, did the alternation in the two forms of the numeral digits stop. Since the practice of alternating the two forms of digits is a characteristic of the positional numerals in rod form, not needed at all for the ciphered grouping numerals, it suggests that the practice must have been carried over from the rod positional numerals. Thus, the residual alternation found among the ciphered grouping numerals provides the epigraphic evidence that the rod numerals existed before the ciphered grouping numerals. This confirms that the rod positional numeral systems were developed before or in parallel with the ciphered grouping numerals and their numeration principles were established no later than the Yin period of the Shang Dynasty.

In the modern world, there are at least two numeral systems needed in one's daily usage. In writing a check for $10 933.00, for example, one uses both the place value numeral (10 933), and the ciphered grouping numeral (ten thousand nine hundred and thirty three). The numeration principles of these two numeral systems turned out to be identical to those of the rod positional numerals and ciphered grouping numerals, respectively. This illustrates that although the notation in the Chinese numeral systems is very different from the notation of the current numeral systems adopted worldwide, behind the different notation their mathematical contents and properties are identical.

• **The zero dilemmas: concepts and symbols**

The absence of an explicit written symbol for zero digits in the counting rod positional numerals has been a source for misunderstanding. As noted earlier, the zero digit in positional numerals is left empty in the layout configurations in both Chousuan and Zhusuan. An illustration of such layout configurations is given in Tab. 1 (see Rows 1 and 2 for number 90 245). There are historians of mathematics, who viewed that, without a written zero symbol, the Chinese rod positional numeral system was not fully developed. This view is erroneous, resulting from a failure to appreciate the differences in notation between the written and mechanical modes. It is certainly correct that in written presentation of positional numerals, a written symbol for zero digits is essential. However, a written symbol is not the only way that the zero digits can be represented. The empty space left for zero digits in positional numerals is in itself a notation for zero in mechanical mode known in Chinese as Kongwei (literally a blank position with a place value) .

A Kongwei so displayed in chousuan has all the function of a symbol for zero digit since its aligned position provides the correct place value and its distinct kongwei configuration is uniquely identifiable with the zero digit. The use of absence of counting rod to denote zero in Chousuan is similar in concept to the use of "signal off" to denote zero in electronic calculations.① It should be also noted that the use of a blank position for denoting zero digits was for mechanical convenience, since this would allow calculations to be performed simply with a bag of rods. No later than the −3rd century, colored rods were used for displaying negative numbers in Chousuan.② This analysis reveals that if one expands one's concept of notation to include mechanical mode, one would recognize that a Kongwei is a configurational symbol denoting zero digit. Not only the absence of a written symbol for zero digits does not imply the absence of zero-digit concept but also the absence of ciphering in rod numerals does not imply an arrested development of the rod positional numeral system.

The Chinese rod positional numeral system is a fully functional positional system. Its numeration principles with decimal multiplicative place value were fully developed long before the emergence of our current decimal place-value numeral system.③ In studying the origin of our current place-value numeral system, it is necessary to consider the following four necessary characteristic properties of the numeral system: ① a decimal base, ② the multiplicative principle, ③ place value, and ④ the concept of a zero digit. This consideration would eliminate a number of ancient numeral systems as possible candidate. The Mesopotamian cuneiform numerals, for example, constitute probably the earliest extant positional numeral system④, but because it is a sexagesimal system with 60 digits, it is likely played no role in the development of our current place-value numeral system. The Mayan numeral system, on the other hand, has probably the earliest existing written symbol for zero, but because it is a mixed-base system with 20 as its principal base, it likely played no role in the development of our current place-value numerals. Among the ancient numeral systems, the Chinese rod numerals constitute the only

---

① The digit zero and digit one of the binary positional numeral system used in electronic calculations are expressed by the "off" and "on" signals.

② See Chapter 8, Problem 3 of the *Jiuzhang Suanshu* (《九章算术》, *Nine Chapters of Mathematics*)[5]. Here, in speaking of negative numbers, it is of course distinguished from simple subtrahends. In ancient Chinese mathematics, the positive and negative numbers are called Zhengshu (正数) and Fushu (负数), respectively, while additions and subtractions are called Bing (并) (or Jia, 加) and Jian (减), respectively. The concept of negative number is found in India in the 17th century work of mathematician Brahmagupta and in Europe in the 17th century work of the Flemish mathematician Albert Girard (1590-1633) and the French mathematician René Descartes (1596-1650).

③ For a comparison of the major numerals systems with the Chinese numeral systems see, for example, Joseph Cheng-Yih Chen, *History of Mathematics in Chinese Civilization*[6]:45-97.

④ Based on current extant records, it is the Mesopotamian cuneiform numerals (ca. −2400) constitute the earliest positional numeral system. The Cuneiform numerals were print-written on tablets and used for both computation and description. By the 3rd century, special symbols appeared in the tablets for zero.

place-value numeral system which possesses all the four necessary characteristic mathematical properties of our current place-value numeral system.

Since the written symbols in our current place-value numerals probably originated from Hindu numerals, recent studies on the origin of our current place-value numeral system centered on Hindu numerals. Although the early Hindu Brahmin numerals, found among the Nana Ghat and Nasik cave inscriptions and dated to approximately 100 years after the Asoka Dynasty (ca. −200)[7], are not place-value numerals and the multiplicative principles were not used consistently, but place-value Hindu numerals dated to 595 were found on the Gurjara grant plate from Sankheda. On the plate, the date Samvat 346 (+595) is written in decimal positional notation.[8]:52 Depending on the point of view, two different hypotheses were developed in searching for the origin our current place-value numerals. One hypothesis holds that the Hindu place-value numerals evolved directly within Hindu civilization without outside influence; the other holds that Hindu place-value numerals were developed under the influence of the Chinese rod positional numerals.

The earliest datable evidence for written zero symbols is found in ancient Indo-China.[8] Two inscriptions, one found at Palemberg in Sumatra and the other in Cambodia, are dated as "the 605th year of Saka era". In both cases the zero in numeral 605 is inscribed as a dot (see Fig. 1). A circle symbol for zero is found in the inscription "the 608th year of Saka era" on the island of Banka (see Fig. 1). Taking the date for the beginning of the Saka era to be −78, we obtain the dates 683 and 686 for the epigraphic evidence for the dot and circle symbols of zero, respectively.[9]:323

605   608

Fig. 1   The dot and circle written symbols for zero found in stone inscriptions in Sumatra, Cambodia, and the Island of Banka, dated to the year of 683 and 686, respectively

Epigraphic evidence for zero found in India are also from the Bhojadeva inscriptions at Gwalior.[8]:52 In one of these inscriptions dated Vikrama Samvat 933 (the year 876), the circle symbol for zero is found in numerals for 50 and 270. A zero symbol is also found in an undated Bhojadeva inscription, which according to Datta and Singh[7]:25-32 is probably from the year 870.

Based on the fact that the instances of epigraphic evidence for zero found in the borderline cultural areas of India and China are earlier than those found in India by more than a century①, David Eugene Smith (1860-1944)[11] and Dirk Jan Struik (1894-2000)[12]:71-72 suggested

---

① The exact date for the beginning of the Saka era is not certain. Instead of +78, it could be, although unlikely, as late as +128[10]; Even if this is the case, the date for the epigraphic evidence for zero would still antedate those found in India by more than a century, being 733 and 736 instead of 683 and 686.

independently that the Hindu place-value numerals probably originated from Chinese civilization. A comparison of early Hindu Brahmi numerals with the written form of digit numerals found in the shell-bone inscriptions (see shell-bone form in Tab. 5) are given in Tab. 7.

Tab. 7 A comparison of early Hindu Brahmi decimal digits with those found in the Chinese shell bone inscriptions of the Shang Dynasty

Digit numerals	0	1	2	3	4	5	6	7	8	9
Shell bone inscriptions		一	=	≡	≣	㐅	个	十	八	㇉
Hindu Brahmi numerals		一	=	≡	ᛉ	۴	۴	7	ら	?

Smith further suggested that the current 2 is merely a cursive form of = and the current 3 is similarly derived from ≡ . [11]:67-68 In this connection it is of interest to note that, in the vertical form of the Chinese rod numerals, the digit for 1 is represented by a vertical rod (see Tab. 2). Joseph Needham (1900-1995) and Wang Ling (王铃) (1917-1994) suggested that the symbol for zero was probably created from marking the blank space left for zero in Chousuan. [13]:11-12 The zero symbols found among the east Arabic numerals are also a dot. No written symbols of zero have been found among the early west Arabic numerals.

In addition to the datable epigraphic evidence for the written zero symbols discussed above, there is also the dot zero symbols from the Bakhshâlî manuscript discovered in 1881. It is an undated mathematical work written on birch bark. Based on its contents, early historians of mathematics placed the manuscript between the third and fourth century. From 1927 to 1933, the Bakhshâlî manuscript was edited by G. R. Kaye and published with an English translation and a transliteration together with facsimiles of the text. The manuscript was dated by Kaye to the twelfth century. This dating has given rise to much controversy which will not be discussed here. Recent scholars, adding linguistic analyses of the manuscript, placed the date between the second and fourth century, and claimed the zero symbols were an indigenous development. Indeed, this dating would make the written symbols for zero found in Hindu numerals earlier than the zero symbol in the Mayan numerals.

It is should be noted that being the first to create a written zero symbol does not necessarily imply being the first to create the concept of zero, since a written symbol is only one of many ways that the zero concept can be expressed or denoted. The zero concept is first found among the Chinese rod numerals used in Chousuan. In fact, the Chinese rod numeral system is the only place-value numeral system that is earlier than the Hindu numeral system, and also possesses all the four necessary characteristic mathematic properties of our current place-value numeral system. It is reasonable to claim that our current system of place-value numerals is a multi-cultural product. It was produced from the convenient Hindu written symbols and the efficient numeration principles of the Chinese numeral systems and transmitted through the Arabic civilizations to the West and then the world.

3. Notation in Algebra

The difference in notation between written and mechanical modes also contributed to misunderstandings of early Chinese work of algebra. An analysis of notation used for expressing mathematics is made in this section for both the written and mechanical modes.

3.1 Representation of linear equations

A practical way to examine the difference in notation between written and mechanical modes for linear equations in ancient China is to examine an actual problem from an early Chinese mathematics classic. Let us consider as an example, Problem 1 of Chapter 8 Fangcheng (方程, Square Array) from the Jiuzhang Suanshu:

Given: Shanghe (top grade paddy) 3 Bing (bundles), Zhonghe (medium grade paddy) 2 bundles, and Xiahe (low grade paddy) 1 bundle yield a Shi of 39 Dou of grain.

Top grade paddy 2 bundles, medium grade paddy 3 bundles, and low grade paddy 1 bundle yield a shi of 34 Dou of grain.

Top grade paddy 1 bundle, medium grade paddy 2 bundles, and low grade paddy 3 bundles yield a Shi of 26 Dou of grain.

Q.: How much paddy does one bundle of each grade yield?

A.: Top grade paddy: $9\frac{1}{4}$ Dou per bundle; medium grade paddy: $4\frac{1}{4}$ Dou per bundle; and low grade paddy: $2\frac{3}{4}$ Dou per bundle.

今有：上禾三秉、中禾二秉、下禾一秉、实三十九斗；
上禾二秉、中禾三秉、下禾一秉、实三十四斗；
上禾一秉、中禾二秉、下禾三秉、实二十六斗。
问：上中下禾实一秉各几何。
答曰：上禾一秉九斗四分斗之一，中禾一秉四斗四分斗之一，下禾一秉二斗四分斗之三。

This problem given in written mode deals with three linear algebraic relations for three unknowns: Shanghe, Zhonghe, and Xiahe, corresponding to the three different grades of the paddy. The constants of the equations are given under the technical term Shi.

On the counting board in Chousuan, this problem is represented by a rectangular array configuration, expressed in terms of the rod numerals as shown in Fig. 2.

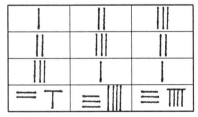

Fig. 2 The rectangular array configuration for a set of linear equations expressed in terms of rod numerals with their units Bing and Dou (a standard measurement in volume) omitted. For the convenience of the reader, the rectangular array configuration is also displayed in terms of our current numerals on the right.

The rectangular array (Juzhen) is composed of a square array of coefficients at the top and a row of the constants (Shi) at the bottom, taking from the set of three linear relations. The coefficients in the square array are arranged in accordance with the traditional writing order, from the top to bottom and from right to left. In the rectangular array configuration, the three unknowns in the set of linear relations are identified by their row positions.

In current written notation adopted universally, this set of linear algebraic relations would be expressed in a set of equations as follows:

$$\begin{aligned} 3x + 2y + z &= 39 \\ 2x + 3y + z &= 34 \\ x + 2y + 3z &= 26 \end{aligned} \qquad (1)$$

where the three unknowns Shanghe, Zhonghe, and Xiahe are denoted by letter symbols: x, y, and z.

By comparing the current written notation with the early Chinese mechanical notation, one observes that the basic layouts are similar; the layout in current written notation is horizontal but the array configuration layout is vertical. By rotating the written linear equations 90° clockwise, one would then have the coefficients and constants of the linear equations arranged exactly as those given in the array configuration of the linear equations. The major differences between the two systems of notation lie in denoting the unknowns. In the Chinese mechanical mode, the unknowns are denoted by their corresponding row positions, not by letter symbols.

3.2 Higher-degree equations with one variable

The representation of higher-degree equations in mechanical mode follows a scheme similar to that of linear equations, using layout configurations on a counting board to represent the problem. In this scheme, each equation with one variable would be represented by a single columnar configuration. The column would be composed of the coefficients of this variable in different degrees, listed in ascending order with the constant at the bottom. Such a columnar configuration layout is termed a Kaifangshi in Chinese, meaning a configuration for root extraction.

3.2.1 The 2nd-degree equations

To examine the difference in notation between written and mechanical modes for equations of higher degrees, let us begin with the problem of square root extraction using again an actual problem from the Chinese mathematics classics. An earliest extant problem, which is general in nature, is Problem 20, Chapter 9 of the *Nine Chapters in Mathematics*. Reproduced below is the problem:

> Given: A square city of unknown dimensions has a gate in the middle of each of its four sides. Twenty Bu (paces) north from the north gate there is a tree, which is visible when one walks 14 Bu south from the south gate and then turns westward walking 1775 Bu more.
>
> Q.: What is the length of its sides?
>
> A.: 250 Bu
>
> Method: By multiplying the distance from the north gate (i.e., 20 bu) with the westward

distance (i. e. , 1775 Bu) and double the product, one obtains the constant term (Shi) [of the 2nd-degree equation]. By adding the distance from the south gate (i. e. , 14 Bu) with the distance from the north gate (i. e. , 20 Bu), one obtains the tag-on (Congfa) term (i. e. the linear term) [of the Kaipingfangshi (2nd-degree equation)]. The root extraction [of the 2nddegree equation] yields the side of the city.

今有：今有邑方不知大小，各中开门。出北门二十步有木，出南门十四步，折而西行一千七百七十五步见木；

问：邑方几何。

答曰：二百五十步。

术曰：以北门步数乘西行步数，倍之为实，并出南门步数为从法。门方除之，即邑方。

In this example, the equation is given in terms of the coefficient of the linear (Congfa) term and the constant (Shi) term, without explicit explanation as to how these quantities were derived from the geometric relation described in the problem (see Fig. 3).

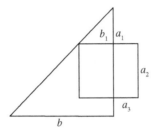

$a_1 = 20$ Bu, $a_2 = x$, $a_3 = 14$ Bu, $b = 1775$ Bu, and $b_1 = \frac{1}{2}x$.

Fig. 3   An illustration of the geometric relation of a square city of unknown dimension x and the sighting of a tree on the north side[①] from the south side of the city

Liu Hui (fl. +263), the well-known commentator of the *Nine Chapters in Mathematics*, provided the following instructions for determining these quantities:

此以折而西行为股，自木至邑南十四步为句；以出北门二十步为句，率北门至西隅为股，率即半广数。故以出北门句乘西行股，得半广股率乘句之幂。然此幂居半以西，故又倍之，合半以东西也。

Let the westward distance be the Gu b ($b = 1775$ Bu, see Fig. 3) and the distance from the tree to the 14-Bu south of the south gate be the Gou a ($a = a_1 + x + a_3 = x + 34$ Bu); let the northward distance 20 Bu from the north gate to the west corner be the Gou a_1 ($a_1 = 20$ Bu) and let the ratio $\frac{x}{2}$ from north gate to the western corner be the Gu b_1 which is half the side of the city ($b_1 = \frac{1}{2}x$, see Fig. 3). Therefore, the

---

① The directions used in the *Nine Chapters in Mathematics* are based on the traditional Chinese convention which is directly opposite to the current convention. Thus, the top of a page is south, the bottom north, the right-hand side west, and the left-hand side east. In this illustration, the directions are converted to the current convention, for the convenience of the reader.

product of the north-gate Gou a_1 with the westward-distance Gu b (i.e., $a_1 \times b =$ 35500 Bu^2) gives the area obtained from the product of Gu b_1 (i.e., x) and the Gou a (i.e., $x + 34$ Bu). However, the area so obtained is only the western half side of the area. By doubling it, one then obtains the combined area of the western and eastern halves.

In terms of symbols, one gets from Liu Hui's instructions the following relation:
$$(20 \text{ Bu}) \times (1775 \text{ Bu}) = (x/2) \times (x + 34 \text{ Bu}) \tag{2}$$
This gives the desired 2nd-degree equation for root extraction:
$$x^2 + (34 \text{ Bu})x = 71000 \text{ (Bu)}^2. \tag{3}$$
From the illustration given in Fig. 3 for the problem, it is obvious that the relation given in Eq. 2 is derived from the proportionality laws of similar right triangles:
$$\frac{1775 \text{ Bu}}{x + 34 \text{ Bu}} = \frac{x/2}{20 \text{ Bu}} \tag{4}$$
where the identification of the Gou a and the Gu b of the two similar right triangles are given in Liu Hui's explanation (see the above translation of the text and Fig. 3).

From the derivation of the equation, it is clear that the early Chinese mathematicians viewed the 2nd-degree equations in terms of their corresponding geometrical relations in areas. An illustration of this interpretation for the above example is given in Fig. 4.

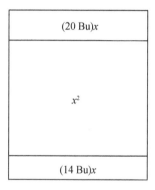

Fig. 4   An interpretation of the equation $x^2 + (34 \text{ Bu}) x = 71000 \text{ (Bu)}^2$ in terms of the area of the city x^2, the additional areas from the tag-on (Congfa terms), and the total area for the constant Shi (see Fig. 3)

Setting the coefficient of the quadratic term to be 1, and using the coefficient 34 Bu of the linear term[①] and the constant 71 000 [square] Bu of the equation[②] given in the method, the columnar configuration for root extraction can be displayed on the counting board in Chousuan

---

① That is 20 Bu plus 14 Bu.
② That is 2 × (20 Bu) × (1775 Bu).

as illustrated in Fig. 5 below:①

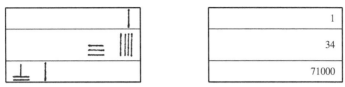

Fig. 5 The columnar configurational presentation of a 2nd-degree equation expressed in terms of rod numerals. For the convenience of the reader, the columnar configuration is also displayed on the right in terms of our current numerals

In modern written notation, this columnar configuration would be expressed as a 2nd-degree equation as follows:

$$x^2 + 34x = 71000. \tag{5}$$

where the linear $34x$ term is the congfa term. An apparent difference between this written representation and the columnar configurational representation in Chousuan is in the denotation of the variable of different orders. In the current written notation, the different degrees of the variable are denoted by the different powers of the letter x for the variable. On the counting board, the different degrees of the variable are denoted by the corresponding orders in the column.

3.2.2 The higher-degree equations

The columnar configurational representation of square root extraction in mechanical mode can be easily extended to higher-order extractions. The order is specified by the highest entry in the column and the coefficients coming from the higher-order terms can be accommodated simply by adding them in the column in ascending order. The situation is similar to the equational representation in which extension to higher degrees can be made just by adding terms with the higher powers of x. Thus, in principle, the columnar configurational representation of root extractions can be extended to any orders, just like the equational representation can be extended to any degrees.

By the Song Dynasty (960-1279), the columnar configuration also began to appear in mathematical books as printed illustrations. In such printed illustrations, the numerals in the columnar configuration are presented by rod positional numerals in written mode instead of in mechanical mode (see entries for positional numerals for the Northern Song and Southern Song periods in Tab. 1)②. The color scheme for negative number is replaced by drawing an extra rod diagonally across the last non-zero digit of the numeral. In addition, the character tai is adopted to mark the position of the Shi (constant) term specified in the columnar configuration

---

① It should be noted that the convention of order, ascending or descending, was not universally fixed. It depends on the mathematical schools and time period.

② It should be noted, on the counting board in Chousuan, the circle zero symbol is not used. The zero digits are still represented by kongwei in the mechanical mode.

for the root extraction.

As an example, consider the illustration given in Fig. 6 taken from the book Ceyuan Haijing (《测圆海镜》, *Reflections on Circle Measurements*) by Li Ye (1192-1279).[14]

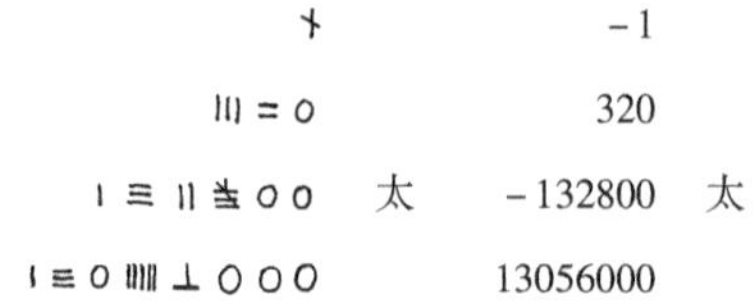

Fig. 6 An illustration of columnar configuration representation of a equation with a negative power term in the written mode taken from the book Ceyuan Haijing by Li Ye (1192-1279)

The columnar configuration given in Fig. 6 represents, in equational representation, the following equation:

$$-x^2 + 320x - 132800 + 13056000x^{-1} = 0 \qquad (6)$$

This is an equation that has a negative power term. From Fig. 6, it is seen that the constant term marked by character symbol Tai is -132800. Hence, the term 13056000 below the constant term corresponds to the first negative power of the unknown, i.e. $13056000\,x^{-1}$.

For the description and recording mathematics in written form, written symbols are also needed for the coefficients of the higher-order terms in addition to the variables. A unified written system was developed by the Song algebraists. In this system, the technical terms in the equations are systemized and variables are denoted by designated characters such as Tian, Di, Ren ... in a way much like the current use of letters $x, y, z, \ldots$ The term Tianyuan would mean the variable Tian, corresponding exactly to the variable x in modern usage.

To examine the written symbols for technical terms in the unified written system developed by the Song algebraists, let us compare them with those in the currently adopted notation, exhibited in the following polynomial equation:

$$a_n x^n + a_{n-1} x^{n-1} + \cdots + a_2 x^2 + a_1 x + a_0 = 0, \qquad (7)$$

A comparative listing of these symbols is tabulated in Tab. 8.

Tab. 8 A comparison between current notation and the Song notation for higher-degree equations with one variable

Degree	Term	Coefficient (current)	Coefficient (Chinese)	+ Positive	− Negative
(n) th	x^n	a_n	Yu (隅)	Zheng	Yi
$(n-1)$ th	x^{n-1}	a_{n-1}	$(n-1)$ th-lian (下廉)	Cong	Yi
$(n-2)$ th	x^{n-2}	a_{n-2}	$(n-2)$ th-lian (廉)	Cong	Yi
...	...	...	...	...	...
4th	x^4	a_4	3rd-lian (三廉)	Cong	Yi
cubic	x^3	a_3	2nd-lian (二廉)	Cong	Yi

Degree	Term	Coefficient (current)	Coefficient (Chinese)	+ Positive	− Negative
quadratic	x^2	a_2	1st-lian (上廉)	Cong	Yi
linear	x^1	a_1	Fang (方)	Cong	Yi
	x^0	a_0	Shi (实)	Zheng	Yi

In terms of these systemized character symbols for the technical terms, the Song algebraists would be able to express a higher-degree equation precisely and explicitly in written form. A major difference in the current written notation as opposed to the early Chinese written notations lies in the use of symbols + and − to denote positive and negative coefficients, instead of using characters Cong① and yi to denote positive and negative coefficients.

For an explicit example, consider, for the following 6th-degree equation② in columnar configuration given in the first column in Tab. 9 with the rod positional numerals replaced by current numerals.

Tab. 9 A 6th-degree equation in columnar configuration with the written description given in Chinese and in English translation

Configurational Form	Written Description (Chinese)	Written Description (English Translation)
1 15 85 225 274 120 − 665 280	得六十六万五千二百八十为益实，一百二十为从方，二百七十四为从上廉，二百二十五为从二廉，八十五为从三廉，一十五为从四廉，一为正隅，五乘方开之	one has negative 665280 for Shi, positive 120 for Fang, positive 274 for 1st-lian, positive 225 2nd-lian, positive 85 for 3rd-lian, positive 15 for 4th-lian, and positive 1 for Yu for the 5th-order (i.e. 6th-degree)③ root extraction

In the current equational representation, the configuration for root extraction can be presented in follows:

$$x^6 + 15x^5 + 85x^4 + 225x^3 + 274x^2 + 120x - 665280 = 0. \quad (8)$$

With such systems of notation, the higher-degree equations with one variable can be represented systematically and conveniently both in computation and in written description.

3.3 Higher-degree equations with multiple variables

With the increase in the number of variables, the logistics in implementing the positional notation in chousuan increase in complexity. For one variable, just a single column would be

---

① For the first and last terms Shi and Yu (corresponding to the a_0 and a_n terms), the character Zheng is used instead of the character Cong.

② The example is taken from Problem 6, Section 1, Volume 3 of the Siyuan Yujian[15] of 1303. by Zhu Shijie.

③ The counting on root extraction began with the solving of a 2nd-degree equation, $a_2x^2 + a_1x + a_0 = 0$, since the solving of a linear equation, $a_1x + a_0 = 0$, is not counted in the ordering root extraction.

adequate since the column denotes the variable and the positions in the column denote the different powers of the variable as discussed in Section 3.2.2 above. However, for two variables, the designation of two columns, one for each variable, would not be adequate because there could be cross terms of the two variables in different powers, in addition to the different powers of each variable. Consequently, it requires a column and a row to account for the two variables so that the cross terms of the two variables could be denoted by the positions defined with respect to both the column and the row. An illustration of such a scheme is shown in Fig. 7.

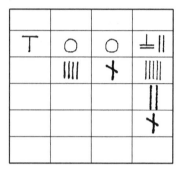 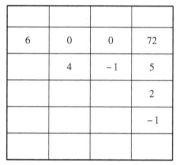

Fig. 7 A written illustration of a configuration for an
equation with Tian and Di, two variables denoted by the column and the row, respectively

This configuration represents the following equation in current written notation:
$$-x^3 + 2x^2 + 5x - xy + 4xy^2 + 6y^3 + 72 = 0 \qquad (9)$$
It is seen from the coefficients -1 and 4 of the cross terms xy and xy^2, respectively, that these terms are denoted by their positions specified with respect to both the column for Tian (x) and row for Di (y) variables. To extend this notational scheme from two variables to three variables, it would be equivalent to change the representation from a two-dimensional configuration to a three-dimensional configuration. Obviously, it would be difficult and impractical, if this scheme were to be followed rigorously to higher degrees.

A modified scheme capable of accommodating up to four variables was adopted by the well-known mathematician Zhu Shijie (1249-1314).[①] In this modified scheme, the character Tai fixes the position of the Shi term and, in addition, divides simultaneously the column into the upper and lower segments and the row into two left and right segments. These four segments are then used to represent the four variables: the lower column segment for variable Tian (x), the

---

① Zhu Shijie lived during the war-torn period in which the Mongols rose to power and established the Yuan Dynasty. Many of the mathematical works of this period were lost. Zhu's earlier book namely *Suanxue Qimeng*[16] of 1299 was later recovered from Korea. However, many books cited by Ji Xianfu in his Postface for the *Siyuan Yujian* were lost. Among these books were the *Yiguji* (《益古集》) by Jiang Zhou (蒋周) (1060-1140), *Zhaodan* (《照胆》) by Li Weiyi (李文一), *Lingjing*(《铃经》) by Shi Xindao(石信道), and *Rujishisuo*(《如积释锁》) by Liu Ruxie(刘如谐). In addition, there were also lost books on two-variable higher equations such as *Liangyi Qunying Jizhen* (《两仪群英集臻》) and three-variable higher equations such as *Qianqun Kuonang* (《乾坤括囊》).

Configurations Versus Equations: A Notational Difference

left row segment for variable Di (y), the right row segment for variable Ren (z) and the upper column segment for variable Wu (w). In specifying positions for the degrees of the variables, the character Tai acts as if it is the origin of a two-dimensional coordinate system with the column and row acting as the two axes. In this modified notational system, Zhu Shijie was able to accommodate most of the cross terms systematically. For cross terms between variables "Di" (y) and "Ren" (z), between variables Tian (x) and Wu (w), and cross terms involving more than two variables, special designations would be specified.

In his Introductory Volume of the Siyuan Yujian of 1303, Zhu Shijie provided diagrams to illustrate how mathematical expressions such as the sum of the four variables and the square of this sum can be represented in this modified notional system. These diagrams are reproduced in Fig. 8.

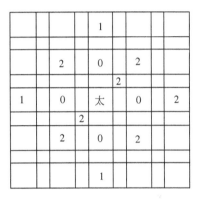

Fig. 8 An illustration of the mathematical expressions in Chousuan: Left, the sum of Tian, Di, Ren, and Wu four variables; Right, the square of the sum. For the convenience of the reader, the rod numerals are replaced by current numerals

It is seen from the left-hand diagram of Fig. 8 that coefficients for the four variables in the sum $x + y + z + w$ are placed one unit position away from the cross point of the column and row. The right-hand diagram of Fig. 9 illustrates how the 11 terms in the square of the sum $(x + y + z + w)^2$:

$$(x + y + z + w)^2 = x^2 + y^2 + z^2 + w^2 + 2xy + 2xz + 2xw + 2yz + 2yw + 2zw, \quad (10)$$

would be positioned. Terms, such as x^n, are located n-units from the character Tai on its corresponding axis segments. The cross terms, such as $x^n y^m$, are located n-units and m-units from their corresponding axis segments, respectively, if the segments are adjacent to each other. Otherwise, the cross terms, such as xw and yz, are located along the axis xw or axis yz at the vertex of the corresponding units.

An illustration of a configuration of an equation is given in Fig. 9.
This configuration represents the following equation in current written notation:
$$- x^2 + 2x - xy^2 + xz + 4y + 4z = 0 \quad (11)$$

It is seen from the coefficients of the cross terms xy and xy^2 that these terms are specified with respect to their corresponding segments of the column and row.

0	4	太	4
-1	0	2	1
0	0	-1	0

Fig. 9 A written illustration of a configuration for an equation with Tian, Di and Ren three variables in the modified positional notation system

3.4 Configurations versus equations

The difference in notation between written and mechanical modes also contributed to misunderstandings of early Chinese work in algebra. There are historians of mathematics and sinologists who view the invention of equations with mathematical signs (such as the equal =, plus +, and minus − signs) and with letter symbols (such as x, y, and z) for the unknowns as the essence of algebra development. From their perspective, early Chinese work in algebra should not be viewed as such because in this early work the "unknowns" were not identified with designated symbols and mathematical relations are not set in equations with signs. All these indicate to them that the concept of algebra was not properly developed in ancient China. Not until the Song Dynasty (960-1279) when characters (such as Tian, Di, Ren, etc.) were adopted for designating unknowns, did algebra emerged in China. In their views, to present the early Chinese work in equations with signs and symbols is equivalent to reading modern concept of equations into old work. Misunderstandings such as these can be easily clarified, as soon as one expands one's concept of notation from written mode to include mechanical mode.

It is certainly true that equations with symbols and signs are essential in written mode, for it is in terms of such equations that one summarizes the mathematical information of an algebra problem for solving the unknowns. However, equations with signs and symbols are not the only way that algebraic concepts can be expressed. In Chousuan, the mathematical information of an algebra problem would be summarized in terms of configurations for solving the unknowns, not in terms of equations as in equational representation. Nor would one specify the variable in symbols. In configurational representations, variables are specified by their positions configurations. To clarify this point, it is instructive to put the two representations side by side for a direct comparison.

For the algebraic problem having three variables with linear relations discussed in section 3.1 above, one has the following comparison:

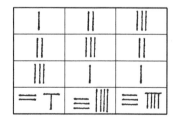

$$3x + 2y + z = 39$$
$$2x + 3y + z = 34$$
$$x + 2y + 3z = 26$$

Configurations Versus Equations: A Notational Difference

It is seen that the algebraic information represented by the set of linear equations is also well represented by the rectangular array (Juzhen) configuration. Instead of the letter symbols x, y, and z, the three variables are identified by their corresponding rows in the rectangular array configuration. The first row is identified with the unknown x, the second row with the unknown y, and so on until the row before the last row reserved for the constant terms. By construction, each constant in the last row equals the sum of all the entries above within the same column of the constant. The plus + and minus − signs are specified by the color of the rods for the coefficients and the constant occupying the configuration. ①

For algebraic problems having one variable with second-degree relations discussed in Section 3.2 above, one has the following comparison:

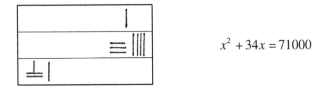

$$x^2 + 34x = 71000$$

Again one can see that the algebraic information represented by the second-degree equation is also well represented by the columnar (Hanglie, 行列) configuration. By construction, each of the columnar configurations is designed to represent only one variable. The different degrees of this variable are specified by their positional orders in the column with respect to the constant term of the problem at the bottom of the column.

These comparisons reveal that configurations and equations have the same mathematical function in their respective representations. In addition, these comparisons also reveal that there is a one-to-one correspondence between the two representations. This implies that the configurational representation in mechanical mode and equational representation in written mode are equivalent. Consequently, one should be able to transform early Chinese algebraic work in configurational representation to equational representation without changing the content of the mathematics. From the perspective of world mathematics, this choice of written presentation is preferred since the equational representation is now universally adopted and easily understood by readers not familiar with Chousuan.

It should be noted that the adoption of characters such as Tian, Di, Ren, etc., for designating unknowns by the Song Dynasty was for written description of the configurational layout in Chousuan when such configurations first appeared in mathematical books as illustrations. In such printed illustrations, the rod numerals in the configuration are ciphered into new written symbols (see Tab. 1 and Fig. 6 or Fig. 7) with the Kongwei for zero denoted by the circle symbol. Thus, the place value in the rod numerals is preserved in the illustrations. The

---

① By the −3rd century, a negative number in Chousuan was represented in terms of colored rods. Often, the red (or black) colored rods were used for displaying negative numbers.

practice of using colored rods for the negative numbers in Chousuan is replaced by a diagonal line drawing across the number in the illustrations. In addition, the character tai is used in the printed configuration to mark the fix position of reference in Chousuan for the constant term (see Fig. 5). All these changes were designed to preserve the original algebraic information contained in the configurations as they found their way from the Chousuan computational board into mathematical books as printed illustrations.

When the word "equation" in algebra first appeared in China during the seventeenth century, it was automatically identified with "Fangchengshi" (the "configuration of a square-array") in Chinese①. This is natural since equation represents the same mathematical concepts as those represented by Fangcheng configurations. In fact, the word equation, in fourteenth-century usage, means literally "the equality or equivalence of mathematical expressions"②. It is the algebraic concepts behind it and the functions it represents that make the word meaningful and important in algebra. The identification of equation with Fangchengshi led to the terminology based on the association of different degrees of equations with the different orders of Kaifangshi (configuration for root extraction). In this terminology, one names the basic columnar configuration for linear extraction as the 1st-degree equation (Yici fangchengshi), the 1st-order columnar configuration for square root extraction (Kaipingfang) as the 2nd-degree equation (Erci fangchengshi), the 2nd-order columnar configuration for cubic root extraction (Kailifang) as the 3rd-degree equation (Sanci fangchengshi), etc..

4. Notations and mathematical tools

Pens and counting rods were among the earliest mathematical tools devised for computations. Although the use of a pen or counting rods appeared similar in nature at the early stage of computational development, but the notational systems created with these tools are rather different in nature. In discussing mathematical notations and tools, it is convenient to distinguish notation in written mode from notation in mechanical mode. Mathematical notation in the written mode is made by the use of a tool, generally known as the pen, and is found among almost all major written languages in a variety of media since antiquity. It appeared, for example, among writings on papyrus, bamboo strips (Zhujian), parchment, silk fabrics, papers, etc. This includes also the cuneiform writings on tablets and the shell-bone inscriptions[③] (Jiaquwen) on turtle shells and ox bones discovered archaeologically at the turn of the 20th century.

---

① It is often simply called Fangcheng.
② Its usage was, of course, not confined to mathematics. It was extended, for example, to logic as well as to the sciences.
③ The translation "shell-bone inscriptions" is advised based on the technical terms "Jiaguwen" coined by Chinese linguistic communities. The common translation "oracle-bone inscriptions" in the West initiated by sinologists is misleading, depicting only one aspect of the inscriptions. The turtle shells and ox bones were the favored writing media during the Shang Dynasty and their use probably begun in antiquity.

Configurations Versus Equations: A Notational Difference

Mathematical notation in the mechanical mode, on the other hand, is found in association with calculating tools other than a pen. Such a notation is composed of the devised entities specifically related to the computational tools. The early example offered by Chousuan in the Chinese civilization reveals that the notation and the tool are intimately related. Not only is the notation expressed in terms of the calculating rods but mathematical computations are also executed by manipulating the rods. This specific relation between notation and the calculating tool requires the notation in mechanical mode to evolve with the calculating tools.

History has shown that an important part of mathematical progresses is in the development of efficient calculating tools. To meet the growing needs in commerce for faster calculations in transactions during the Tang Dynasty (618-907) in Chinese civilization, an improved version of the Suanpan (abacus) with axial-sliding balls (or beats) was developed. ① With a Suanpan, one computes with sliding balls on columnar axes, permitting faster arithmetic operations. The Suanpan became an important calculating tool primarily in commerce. The notation in Zhusuan (calculation with balls) is therefore expressed in terms of layouts of balls displayed on columnar axes of the Suanpan. With only one row for display, Zhusuan are usually carried out with the assistance of algorithmic rhymes and verses. ② These rhymes and verses were often recited subconsciously during calculations to facilitate the recall of algorithms and to speed up repetitive calculations. In a way, such algorithmic rhymes and verses can be viewed as instructions for operating mechanical devises for performing repetitive computations.

The next generation of calculating tools involves wheels and gears instead of rods and balls. To use gears and wheels for assisting calculations, one needs to relate wheel rotations with numbers and to create rotational patterns for representing mathematical relations. Such a practice first appeared in astronomy as one used gear-wheel rotations for keeping time in relating the apparent celestial rotational motion to the social activities of man. In Chinese civilization, such a practice has a long history and probably began with the armillary sphere (Hunxiang) invented by the Han astronomer Zhang Heng (78-139). ③ However, it is in the clock of astronomer Su Song (1020-1101) that one finds that the clockwork mechanism is designed to compute time

---

① For a description of early Chinese computational schemes and mathematical tools see the book Shushu Jiyi[17] of ca. 190 by Xu Yue.

② One has, for example, the Qiuyige (求一歌), Hualingge (化零歌), Guichu Gejue (归除歌诀) and Jiugui Gejue (九归歌诀). Such rhymes and verses are efficient algorithms, often originated from Chousuan.

③ According to the *Jinshu*[18] of 635 by Fang Xuanling, Zhang Heng astronomer (78-139) of the Eastern Han (+25 to +220) constructed a water-powered armillary sphere. The account stated that the rotation of the armillary sphere is operated by falling water (Yilou Shuizhuan) but provides little descriptions as to the mechanism involved. One could argue that gears were probably used in Zhang's armillary sphere, because for maintaining an acceptable accuracy it would be necessary to use gears to control the rotation and bronze gears were available in Chinese civilization no later than the Zhanguo (Warring States, −475 to −221) long before the time of Zhang Heng. The water-powered bronze armillary sphere with bronze gears constructed by astronomer Zhang Sui (684-727, also known as Monk Yi Xing) in the Tang Dynasty (618-907) is well documented. For further discussions, see for example, Lu Jingyan 《中国机械史》[19]:179-182.

intervals for reporting automatically.① In addition, the clockwork mechanism is powered by a water-wheel and regulated by an escapement mechanism. Even though a clock is not a calculating tool, it is however constructive to examine the clockwork mechanism since it contains important features which are of relevance to calculating tools.

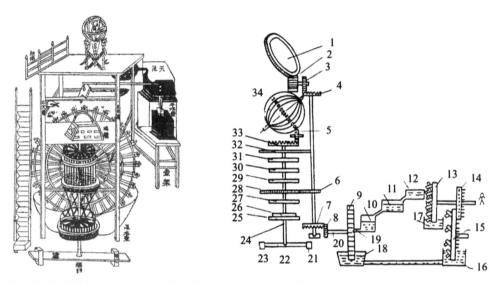

Fig. 10   Left diagram: An illustration of an armillary Clock (viewed from the south) housed in the Water-Powered Observatory (Shuidong Yixiangtai)② invented by Su Song and constructed by Han Gonglian. The project took four years (1088-1092) to complete. Right diagram: A modern diagram of the clockwork mechanism (viewed from the east) by Liu Xianzhou

1. Armillary diurnal motion gear-ring (yunyitianyunhuan); 2. Front gear-wheel (Qiangu); 3. Back gear-wheel (hougu); 4. Upper wheel (Shanglun); 5. Celestial axle (Tianzhou); 6. Middle wheel (Zhonglun); 7. Lower wheel (Xialun); 8. lower gear-wheel (Digu); 9. main driving wheel (Shulun); 10. Constant-level tank (Pingshuihu); 11. Upper reservoir tank (Tianci); 12. Upper flume (Tianhe); 13. Upper water-raising wheel (Shengshuishanglun); 14. Stream paddle-wheel (Heche); 15. Lower water-raising wheel (Shengshuixialun); 16. Lower water-raising tank (Shengshuixiahu); 17. Upper water-raising tank (Shengshuishanghu); 18. Water withdrawing tank (Tuishuihu); 19. Sphon (Kewu); 20. Main driving shaft (Shuzhou); 21. Base stand Dizu; 22. Mortar-shaped end-bearing (Shujiu); 23. Lower bearing beam (Diji); 24. Time-keeping shaft (Jilunzhou); 25. Night-watches jacks-reporting wheel (Yelougeng Chousichenlun); 26. Night-watches rod-indicator wheel (Yelou Jianlun); 27. Night-watches gong-striking wheel (Yelou Jinzhenglun); 28. Time-keeping gear-wheel (Boya Jilun); 29. "Quarter"-reporting jacks wheel (Baoke Sichenlun); 30. On-hour jacks wheel (Shichuzheng Sichenlun); 31. "Quarter"-hour bell-drum wheel (Shike Zhonggulun); 32. Upper bearing beam (Tianshu); 33. Upper gear-wheel (Tianlun); 34. Celestial globe equatorial gear-ring (Chidao Yaju).

Two illustrations of the Su Song armillary clock is reproduced in Figure 10. The left diagram is the Su Song's original illustration taken from his book *Xin Yixiang Fayao* of 1094. The illustration reproduced at right is a modern diagram of the clockwork mechanism

---

① See *Xin Yixiang Fayao*[20] of +1094 by Su Song.
② Reproduced from Page 4a, Chapter 3 of the book *Xin Yixiang Fayao* of +1094 by Su Song.

(viewed from the east) by Liu Xianzhou.① It is seen that the Su Song clock is a composite system of instruments housed in a observatory called Shuidong Yixiangtai (Water-powered Armillary Observatory). The major components of the system is an armillary sphere (Hunyi) located at the top of the observatory, below which at the second level are various instruments including a celestial globe (Yixiang) mounted inside of the earth horizon box (Digui), behind which is the operating gear-wheel system (Shuji Lunzhou) supported in the mortar-shaped end-bearing (Shujiu) below. The water power is supplied by a reservoir tank (Tianci) at the upper right and beneath which is the constant-level tank (Pingshuihu).②

The clockwork mechanism illustrated in the right diagram of Fig. 10 reveals that the transmission-shaft drives both the armillary and the time-keeping shaft. Since the early Chinese divided a fully day (midnight to midnight) into 100 Ke (quarters) as well as into 12 Shi (double-hours)③, the clockwork mechanism is designed to determine and report time at the beginning and the middle of the Shi as well as at the Ke intervals. These intervals are calculated by setting the appropriate gear ratios at the transmitting connections which led to the wheels for keeping and reporting time as shown in the clockwork mechanism.

The rate at which the main transmission- shaft (Shuzhou, 20) of the driving wheel (Shulun, 9) drives the vertical celestial axle (Tianzhou, 5) is given by the gear ratio between the gear-wheel (Digu, 8) and the pinion (Xialun, 7). The vertical celestial axle in term drives the vertical time-keeping shaft (Julunzhou, 24) in accordance with the gear ratio between middle wheel (Zhonglun, 6) and the time-keeping gear-wheel (Boya Jilun, 28). In accordance to the transmission ratios, the time-keeping shaft rotates the following six time-keeping or time-reporting wheels: the night-watches jacks-reporting wheel (Yelougeng Chousichenlun, 25), the night-watches rod-indicator wheel (Yelou Jianlun, 26), the night-watches gong-striking wheel (Yelou Jinzhenglun, 27), the "quarter"-reporting jacks wheel (Baoke Sichenlun, 29), the on-hour jacks wheel (Shichuzheng Sichenlun, 30), and the "quarter"-hour bell-drum wheel (Shike Zhonggulun, 31).

The clockwork mechanism of Susong contains several innovative features which were of important in the development of better mechanical calculating tools. In the armillary clock, the tasks of determining and reporting time are built in the operating mechanism by design, eliminating the need for reading the notation and the need of following the processes step-by-step mentally. These innovations were revolutionary in tool making since they change the nature of interfacing with machine. In addition, the operation of the clockwork mechanism is powered by

---

① Liu Xianzhou《中国机械工程发明史》[21]:111-113.

② For further details, see Lu Jingyan and Hua Jueming《中国科学技术史·机械卷》[22]:297-312. In English, see J. Needham, Wang Ling, and D. J. Price, Heavenly Clock Work: The Great Astronomical Clock of the Medieval China[23], and Joseph Needham, Science and Civilization in China[13], vol. 4, part 2.

③ In modern reckoning, a Shi is 120 min. and a Ke is 14 min. 24 Sec. The hour (60 min.) of modern reckoning is termed Xiaoshi referring to a half of the Shi.

water, greatly reduced the need of labor. These innovations once incorporated in the design of calculating tools would eliminate the need for one to follow the intermediate calculating process visually and would allow, in principle, the speed of future calculating tools to exceed the speed limit set by the physical and mental human factors.

It was Wilhelm Shickard (1592-1635), a professor of mathematics at the University of Tübingen, who first recognized the importance of clockwork mechanism and devised in 1623 the first mechanical tools for addition in Europe[①] which he called a calculating clock. The adding function was based on the movement of six dented wheels, geared through a wheel which with every full turn allowed the wheel located at the right to rotate 1/10th of a full turn. By design, it performed manually and was able to handle up to six-digit numbers. The adding feature was devised to help performing multiplication with a set of Napier's cylinders included in the upper half of the machine. An overflow mechanism rang a bell. Unfortunately, one of the two prototypes of this machine was destroyed by a fire and the other lost. In addition, Shickard and his family did not survive the calamity of a bubonic plague. However, brief descriptions of the design for the calculating clock were discovered among the 1623-1624 correspondence between Shickard and Johanes Kepler (1571-1630) by historian Franz Hammer in 1935-1956. Based on the descriptions, mathematician Bruno Von Freytag reconstructed in 1960 Shickard's calculating clock. It is now preserved at the Deutsches Museum in München.

Unaware of Shickard's work, a mechanical adding device called "Pascaline" was built in 1643 by the French mathematician Blaise Pascal (1623-1662). It was intended to assist his father, a tax collector, with the tedious activity of adding and subtracting large sequences of numbers. Built on a brass rectangular box, the Pascaline made use of a set of notched dials to move internal 5 wheels clockwise in a way that a full rotation of a wheel caused the wheel at the left to advance one 10th. Later units were built with 6 and 8 wheels. Subtraction was done by adding the nine's complement. However, the Pascaline was difficult to use and, in principle, less versatile than the calculating clock of Schickard. By 1652 about 50 units had been made but less than 15 had been sold.

In 1672, German mathematician Gottfried Wilhelm Von Leibniz (1646-1716) improved the Pascaline and built the "Stepped Reckoner", which is capable of perform multiplications and divisions. Leibniz made use of a cylinder drum with nine bar-shaped teeth of incrementing length parallel to the cylinder's axis. When the drum is set to rotate by cranking, it also rotates a regular ten-tooth wheel, fixed over a sliding axis, from zero to nine positions depending on its relative position to the drum. This function allows the user to slide the mobile axis so that when the drum

---

① By the 17th century, Europe began to develop innovative devises for assisting calculations. The first practical calculating tool in Europe was built in 1617 by the Scottish mathematician John Napier (1550-1617). It was known as the Napier's bones consisting of multiplication tables written on strips of wood or bones for performing arithmetic operations. Through this device, he also developed logarithms to convert multiplication into addition. Later in 1633, the slide rule was invented based on logarithms by the English mathematician William Oughtred (1574-1660).

is rotating it generates in the regular wheels a movement proportional to their relative position. This movement is then translated by the device into multiplication or division depending on which direction the stepped drum is rotated. For each digit, it requires one such set of wheels.

In 1674, Leibniz completed his design and commissioned the building of a prototype to a craftsman from Paris named Olivier. Probably only two prototypes of this machine were made. These machines remained in the attic of the University of Göttingen until it was found in 1879 while fixing a leak in the roof. One of the machines is now preserved in the State Museum of Hanover and the other in the Deutsches Museum in München. Not until the 19th century did such a calculating tool received proper attention. Based on Leibniz's designs, Charles Xavier Thomas de Colmar (1785-1870) obtained in 1820 the patent right in France and produced approximately a total of 1500 sets of "arithmometer" between 1820 and 1878.

Calculations with wheels and gears, since they first conceived and applied in the computation of time for keeping time in a clock in 1092 and in an adding tool in 1623 for addition in general, took on a number of designs and improvements before they were evolved into an electrically driven calculating machine with input keyboard in the early sixty's of the 20th century, but only to be replaced by the revolutionary pocket-sized electronic calculators in the seventy's of the same century. Even though wheel-gear calculators never succeeded in attaining the desired speed and versatility to dominate calculations[①], their evolution process revealed, however, that the ability of recognizing mathematical information in mechanical notation and of executing computation can be incorporated in the machine by properly designed interfacing procedures.

With the development of micro-electronics for calculators, one entered modern computations. In electronic calculators, one uses the binary numeral system for implement calculations, the "zero" and "one" digits are respectively designated by the "off" and "on" signals. Numerals and mathematical relations are expressed by the appropriate combinational patterns of these signals. Capabilities of interpreting such signal patterns are built in the integrated circuitries of chips in electronic calculators and microprocessor in computers for expressing mathematical relations. With greater versatility and processing speed available in electronic calculations, programs for interfacing with the machines for calculations can now be incorporated in the electronic calculators and computers. Furthermore, there were computer languages developed for converting programs into the binary machine codes. With these innovations, mechanical notation and calculating tools are now well integrated in modern electronic machines.

---

① When electric calculators such as Monroe, Friden, and Marchant first appeared in the market, there were international competitions between electric calculators and abacus. Often there were categories that abacus came out ahead. In Japan, such competitions were not abolished until the appearance of electronic calculators in the second half of the 20th century.

References

［1］ 卢连成，时协中，梅荣照. 千阳县西汉墓中出土算筹考古，1976，2：85-88，108

［2］ 钱宝琮. 古算考源. 北京：商务出版社，1930

［3］ 钱宝琮. 钱宝琮科学史论文选集. 北京：科学出版社，1933

［4］ 李俨. 中国数学史道言. 北京：商务印书馆，1932

［5］ 九章算术（Nine Chapters in Mathematics），Zhou Dynasty. Restored by Zhang Cang（张苍）(fl. -210 to -162, d. -152) in ca. -165 based on Zhou Dynasty fragments and edited by Geng Shouchang（耿寿昌）(fl. -75 to -49). The earliest extant commentary is that by Liu Hui（刘徽）of the 3rd century. Author（s）or original compiler（s）unknown

［6］ Chen C-Y. History of Mathematics in Chinese Civilization（Chinese Studies 170 Lecture Notes，University of California, San Diego, 1980）

［7］ Datta B, Singh A N. History of Hindu Mathematics（Vols I and II. New York，1962）Eves, Howard. An Introduction to the History of Mathematics. New York, 1953

［8］ Smith D S, Karpinski L C. Hindu-Arabic Numerals. Boston, 1911. 52

［9］ Goedes G. À propos l'origin des chiffres Arabes. Bulletin of the London School of Oriental and African Studies，1931（6）：323

［10］ Tarn W W. The Greeks in Bactria and India. Cambridge, 1951

［11］ Smith D S. A History of Mathematics. Vol 2 ［1923-1925］. Boston, 1953

［12］ Struik D J. A Concise History of Mathematics ［1914］. New York, 1967. 71-72

［13］ Needham J. Science and Civilisation in China. Vol III, Sec 19. Cambridge, 1959

［14］ 孔国平. 测圆海镜导读. 武汉：湖北教育出版社，1996

［15］〔元〕朱世杰. 四元玉鉴，1303

［16］〔元〕朱世杰. 算学启蒙，1299.（fl. +1299 to +1303）

［17］ 徐岳. 数术记遗. 公元 190

［18］〔唐〕房玄龄. 晋书

［19］ 陆敬严. 中国机械史. 台湾：越吟出版社，2003

［20］〔宋〕苏颂. 新仪象法要，1094

［21］ 刘仙洲. 中国机械工程发明史. 北京：科学出版社，1962

［22］ 陆敬严，华觉明. 中国科学技术史·机械卷. 北京：科学出版社，2000

［23］ Needham J, Wang L, Price D J. Heavenly Clock Work：The Great Astronomical Clock of the Medieval China. Cambridge：Cambridge University Press, 1960

Zhu Shijie's Method of "Four Unknowns" as Inspiration for Wu Wen-Tsun

Jiri Hudecek

(Needham Research Institute / Department of History & Philosophy of Science, University of Cambridge)

1. Introduction

In 1977, Wu Wen-Tsun (吴文俊) (Pinyin: Wu Wenjun) designed a method of automated proving of geometric theorems, now called "mathematics mechanisation" or "Wu's method". Wu famously announced that his method received inspiration from ancient Chinese mathematics—from its "basic characteristics" (i.e. its overall spirit), and more directly from specific algorithms for the reduction of polynomials. The focus of this paper is to identify these specific algorithms and show their relationship to Wu Wen-Tsun's work; the influence of the spirit of Chinese mathematics on his own mathematical thought will not be discussed here.

I will first briefly review what Wu Wen-Tsun has written about his inspiration in Chinese mathematics; then compare the central algebraic technique of his method with one part of the method of "Four Unknowns" (Siyuanshu, 四元术), recorded by Zhu Shijie in 1303, and finally argue that Qian Baocong's *History of Chinese Mathematics* (1964) played a crucial role in making Wu aware of the method of "Four Unknowns" and its possible applicability in his work. I will conclude, however, that although this is a very probable connection between the method of "Four Unknowns" and Wu's method, it is of limited significance from the mathematical perspective. The inspiration from traditional Chinese mathematics was a psychologically important driving force and trigger of Wu's method, but it did not supply concepts or techniques lacking in mainstream Western mathematics.

2. Wu Wen-Tsun and ancient Chinese mathematics

Wu Wen-Tsun's life, work and inspiration in ancient Chinese mathematics have been summarised by several authors.[1-4] He was born in Shanghai in 1919. He studied topology in France, but returned to China in 1951 and became one of the leading researchers of the Institute of Mathematics of the Chinese Academy of Sciences. He received a First Class State Prize for

Natural Sciences in 1956 and was elected to the Academic Department of Mathematics, Physics, and Chemistry in 1957. During the Campaign against Lin Biao and Confucius in 1974, he became interested in ancient Chinese mathematics. After the fall of the Gang of Four, he resumed research work interrupted by the Cultural Revolution, but in a new field of mathematics mechanisation. The method he has been developing since the late 1970s enables automated proving and discovery of theorems in various types of geometries, but has also been widely applied in e. g. computer-aided drawing systems. Wu received another major prize, The Highest State Award for Science and Technology, in 2000 for his work on mathematics mechanisation, as well as international awards (Herbrand Prize, Shaw Prize and others).

Wu Wen-Tsun repeated on several occasions that his method was inspired by traditional Chinese mathematics. His clearest pronouncement on this subject, quoted by Li Wenlin[5], is however of unclear date and provenance:

> My method of solving equations is basically derived from Zhu Shijie. He used elimination to eliminate variables one by one, which provided me with a basic model for the method. Of course, Zhu Shijie had no theory, it was very crude, only calculations. I developed it and put it on a truly modern mathematical ground, i. e. algebraic geometry.

Wu Wen-Tsun's earlier remarks were less clear. He published his first article on mechanisation of theorem-proving in elementary geometry in 1977. Already in this article, he mentioned the connection between his method and algebraic techniques used in China in earlier times:

> The algorithm we use for mechanical proofs of theorems in elementary geometry involves mainly some applied techniques for polynomials, such as arithmetic operations and simple eliminations of unknowns. It should be pointed out that these were all created by Chinese mathematicians in the 12-14 century Song and Yuan period, and already reached considerable development then. The work of Qian Baocong can be consulted for detailed introduction.[6]

The passage shows that Wu considered Qian Baocong's book[7] the best place to understand these ancient techniques, instead of any particular original works by 12-14 century Chinese mathematicians. It is worth noting that Wu Wen-Tsun did not at this point explicitly claim inspiration from the ancient techniques, although the note clearly suggests that awareness of them was useful for the creation of Wu's method. Wu had been arguing for the relevance of Chinese traditional mathematics for modern research since 1975[8, 9], but this was the first time he referred to it in a serious mathematical article.

Wu first used the word "inspiration" in a 1980 article in the popular magazine *Baike Zhishi*.[10] Here the case was made much more strongly than in 1977. Wu emphasised that even though David Hilbert in 1899 and Alfred Tarski in 1950 have already come close to his method, he developed it unaware of their results:

> We set out the question and came up with a method of solution under inspiration from Chinese ancient algebra.

But even then Wu did not pinpoint any single work or method that inspired him. The article includes his summary of the history and achievements of Chinese algebra, which discusses in one breath all mathematicians of Song and Yuan periods:

> Concepts corresponding to modern polynomials were introduced, calculation rules for polynomials and algebraic tools for elimination were established, which brought systematic development to the method of algebraisation of geometry, as can be seen in many works of Yang Hui, Li Ye, and Zhu Shijie, fortunately preserved until today.

The first evidence of Wu Wen-Tsun's direct reading of Zhu Shijie's works appeared in his article "On the Double-Difference Theory in Mensuration Studies of Ancient China, with Comments on Certain Methodological Questions in the History of Mathematics", published in 1982 but probably written somewhat earlier.[11] Wu referred to problem 4 of the Gougu Cewang (勾股测望) chapter, including the working (Cao, 草) from Luo Shilin's (罗士琳) edition of 1834.[12] Later in 1987, Wu Wen-Tsun wrote explicitly about Zhu Shijie and the method of "Four Unknowns"[13], saying that his own method "is only a modern generalisation of Zhu Shijie's method". We should now compare the two methods in detail to see their relations to each other.

3. Wu Wen-Tsun's method of mathematics mechanisation

Although Wu Wen-Tsun started his work in mathematics mechanisation with the aim to mechanise geometric proofs, we will only analyse the algebraic method he created, which can be used in other contexts, too. Its input is a system of polynomial equations, and its output a reduction of one of these equations (the conclusion) with respect to all the other equations (premises) by a process of "pseudo-division".

Polynomials generally describe relations between unknown, mutually dependent variables $x_1, \cdots, x_r$. The order of these variables is arbitrary but important for the definition of "reduction": One polynomial is reduced with respect to another when its leading variable is of a lower index, or at least of a lower power.

This can be demonstrated on the example of two polynomials $F(x,y)$ and $G(x,y)$. Initially F is reduced with respect to G: the highest power of y in F is m, whereas the highest power of y in G is higher, M:

$$F(x,y) = I_F(x)y^m + f(x,y), \\ G(x,y) = I_G(x)y^M + g(x,y) \tag{3-1}$$

Here I_F and I_G are called *initials* of the two polynomials. The initials, as well as the remainders f and g, are already reduced with respect to their polynomial.

If we multiply both polynomials crosswise by the initial of the other one, and also raise F to the same power in y as G, that is to M, we get two polynomials in the same power:

$$I_G y^{M-m} F = I_G I_F y^M + I_G y^{M-m} f, \\ I_F G = I_F I_G y^M + I_F g \tag{3-2}$$

Then we subtract the multiplied F (the "divisor") from the multiplied G (the "dividend") and get a remainder:

$$R = I_F G - I_G y^{M-m} F = I_R y^{M-1} + r. \tag{3-3}$$

The remainder has lower power of y than the original conclusion G. This process can be repeated with the remainder taking the place of G, and so on, until the final polynomial will be reduced not only with respect to G, but also to F.

The procedure is more complex for sets of several hypotheses. They have to be first replaced with an equivalent well-ordered basic set, where each hypothesis introduces one variable in the correct order $1, \cdots, r$, and their initials are reduced with respect to the preceding hypotheses. The original set of hypotheses is thus 'triangulated' into a set $A_1 \cdots A_r$. Any polynomial $H(x_1 \cdots x_r)$ can then be algorithmically decomposed by Wu's theorem

$$I_1^{S_1} I_2^{S_2} \mathrm{K} I_r^{S_r} H = Q_1 A_1 + Q_2 A_2 + \cdots + Q_r A_r + R, \tag{0-4}$$

where R is the last remainder, reduced with respect to all A_i. This decomposition is achieved by repeated polynomial division by the A_i. If the decomposition succeeded in eliminating all dependent variables from the remainder, which is finally identical to zero, the polynomial H is implied by the basic set, and therefore by the premises.

This technique of decomposition has another application in the construction of the basic set. The basic set is obtained by replacing the original polynomials with their simplest possible equivalents. From the decomposition $I_{i-1} A_i = Q A_{i-1} + R$, it is clear that given $A_{i-1} = 0$, $A_i = 0$ is equivalent to $R = 0$ (initials have to be non-zero). Remainders can thus take place of the original premises. Another decomposition can be attempted with the new set, until the basic set has been derived.

As can be seen from equation (3-2), polynomial division consists of repeated reciprocal multiplication and mutual subtraction. This is in principle a very common and ancient algebraic technique. In China, it goes back as far as the method of "Excess and Deficit" (盈不足术, Ying Buzu Shu), which is already recorded in the earliest 2nd century BC collection *Writings on Reckoning* (Suanshushu). What justifies focusing on Zhu Shijie's method of "Four Unknowns" as a source of inspiration? Firstly, the parallels in this case are much closer and more direct; secondly, Qian Baocong highlighted the possible universality of Zhu Shijie's method, and this had major impact on Wu Wen-Tsun. I will now discuss these two issues separately.

4. Polynomial reduction in the method of "Four Unknowns"

The method of "Four Unknowns" (literally 'four elements', referring to Heaven, Earth, Man, and Matter, Tian Di Ren Wu, 天地人物), is recorded in the book *Jade Mirror of the Four Unknowns* (Siyuan Yujian, 四元玉鉴), published in 1303 by Zhu Shijie (朱世杰). A new critical Chinese edition[14] as well as an English translation[15] of the text have recently been published. The reader is referred to these publications for details about its history and influence.

What should be mentioned is that the method of Four Unknowns was developed by gradual expansion of the method of two unknowns, introduced by a certain Li Dezai (李德载), according to the prefaces to Zhu Shijie's book. We will see that this method of two unknowns is the core technique similar to Wu's method. "Zhu Shijie" and "Four Unknowns" are therefore conventional labels for the source and whole complex of techniques rather than the actual author and technique.

The crucial method of two unknowns, i. e. reduction of two polynomials with two unknowns into a single polynomial with a single unknown, is explained by Zhu Shijie only in very terse slogans:

(1) Huyin Tongfen (互隐通分)

(2) Liangwei Xiangxiao (两位相消)

Only the final step of the technique shows what Huyin Tongfen actually means:

(3) Neiwaihang Xiangcheng (内外行相乘)

Let us illustrate it on the third exemplary problem in the introductory chapter (*Juanshou*) of the *Jade Mirror*. We will use the tabular notation used for recording what we would call coefficients of polynomial equations. In this notation, each term is placed in the table as if the powers of the unknowns within it were its coordinates. Substituting Hindu-Arabic numerals for the original Chinese numerals, we can observe the principle on the simplest binomial $x + y$, or "Heaven + Earth", and its squaring, which would be represented by these tables:

1	0C
	1

1	0	0C
	2	0
		1

The letter "C" denotes the constant member. The final table has been obtained by joining two copies of the original one, shifted down and left respectively. These shifts correspond to the multiplications by "Heaven" and "Earth".

Getting back to the third exemplary problem of the Jade Mirror, we start with two tables obtained after some previous manipulations:

Former (Qianshi)

1	1	-2C
-1	1	-1
	1	-2

Latter (Houshi)

1	-2	2	0C
	-2	4	-2
		1	-2

Now the method prescribes the step Huyin Tongfen, literally "reciprocally hidden [terms] are equalised in parts". This means bringing both equations to the same initial, in terms of Wu Wen-Tsun's method. The former form has to be multiplied by Earth, the latter form by (1-Heaven)—the leftmost columns in both cases:

Former (Qianshi)

1	1	-2	0C
-1	1	-1	0
	1	-2	0

Latter (Houshi)

1	-2	2	0C
-1	0	2	-2
	2	-3	0
		-1	2

Then the second step is executed—Liangwei Xiangxiao, "the two positions eliminate each other". When subtracting the latter form from the former, we get a result with only three columns:

3	−4	0C
1	−3	2
−1	1	0
	1	−2

This is quite clearly the very same process that is going on in Wu's method. What we have got here is the remainder of polynomial division in Wu Wen-Tsun's terms. If we go back to the calculation procedure for Wu's method

$$R = I_F G - I_G y^{M-m} F,$$

We can identify F with the former form, G with the latter form, multiplication by y with the left shift of the former form, and the leftmost columns with the initials. The mutual elimination (Xiangxiao) is subtraction.

The two steps are repeated until we get two remainders with the same minimal degree of Earth, the "left form" and the "right form":

Left (Zuoshi)

7	−6
3	−7
−1	−3
	1

Right (Youshi)

13	−14
11	−13
5	−15
−2	−5
	2

(Note that the remainder has at one point become the new divisor, which has a parallel in the process of formation of basic sets, but not in the actual theorem-proving algorithm.)

Now the last step occurs, Neiwaihang Xiangcheng. The inner columns and the outer columns mutually multiply—this crosswise multiplication has in fact been going on also in previous cases, but with two or more columns multiplied at the same time. These two multiplications create two single columns, which can be again subtracted, finally forming a single polynomial in one unknown (Heaven) only.

There can be no doubt that polynomial manipulations induced by Wu's algorithm are structurally the same as those occurring in Zhu Shijie's book. But a more direct evidence of actual influence is needed. In this case, it is Wu's acknowledged use of Qian Baocong's book.

5. Secondary inspiration

Wu Wen-Tsun mentioned repeatedly that he found original Chinese mathematical texts, written in classical Chinese, too difficult at the outset and started from "secondary sources".[16, 17] We have already seen that he explicitly mentioned Qian Baocong's 1964 classic History of Chinese Mathematics as an entry point.

Qian's book includes a clear modern symbolic algebraic explanation of the crucial step of Huyin Tongfen Xiangxiao:

"Reciprocal elimination by equalisation of mutually hidden fractions" is a method of elimination for general systems of equations in two unknowns, which means not only with two columns. For example tables with three columns (or polynomials with a square of y) would in general be denoted like this:

$$A_2 y^2 + A_1 y + A_0 = 0 \cdots \quad (5\text{-}1)$$

$$B_2 y^2 + B_1 y + B_0 = 0 \cdots \quad (5\text{-}2)$$

The upper-case letters represent polynomials without y.

If we want to eliminate the term with y^2, we can multiply all terms in (5-2) apart from $B_2 y^2$ by A_2, then multiply all terms in (5-1) apart from $A_2 y^2$ by B_2, subtract them from each other, and we get

$$C_1 y + C_0 = 0 \cdots \quad (5\text{-}3)$$

Then (5-3) is multiplied by y and set again against (1) or (2), and by the same method of elimination we get

$$D_1 y + D_0 = 0 \cdots \quad (5\text{-}4)$$

(5-3) and (5-4) are then "two-column forms with two unknowns", from which all members with y can be eliminated by the preceding method [i. e. Neiwaihang Xiangcheng].

Of course, a mathematician like Wu Wen-Tsun would not need this explanation to understand the original text, especially not with the detailed numerical workings in Luo Shilin's edition, which he had clearly consulted by 1982. But it seems likely that Wu's only knowledge of the method of "Four Unknowns" at the earliest period of his studies of traditional Chinese mathematics was through Qian Baocong. The fact that Qian's book explained the technique so clearly must have greatly facilitated Wu's decision to use it in his own method.

6. What kind of inspiration

Wherever Wu Wen-Tsun learned the method of 'Four (or actually two) Unknowns' from, it seems at the first sight too elementary to warrant so much excitement about his inspiration from Chinese mathematics. Closely related techniques have been used in algebra before, although it seems that their application to multivariate polynomials, and to geometrical questions, is rather peculiar if not unique.

In other words, Wu Wen-Tsun could probably have created his method even without the knowledge of the method of "Four Unknowns". But he truly derived great inspiration from its existence in traditional Chinese mathematics: the knowledge that the method of "Four Unknowns" was traditional, proven and efficient, was itself a source of comfort and motivation. More crucial, however was the ability to claim Chinese intellectual ancestry for his research direction, and to use it as a proof of the value of traditional Chinese mathematics in modern era. The strength of this cultural factor is a subject, which I discuss in my PhD dissertation, is however a separate issue.

References

[1] 纪志刚. 吴文俊与数学机械化. 上海交通大学学报, 2001, (3)
[2] 胡作玄, 石赫. 吴文俊之路. 北京: 科学出版社, 2002
[3] 柯林娟. 吴文俊传: 让数学回归中国. 南京: 江苏人民出版社, 2009
[4] 李文林. 古为今用、自主创新的典范——吴文俊院士的数学史研究. 载: 姜伯驹等. 吴文俊与中国数学. 新加坡: 世界科技出版公司, 2010. 27-41
[5] 李文林. 古为今用的典范——吴文俊教授的数学史研究. 载: 林东岱等. 数学与数学机械化. 济南: 山东教育出版社, 2001. 49-60
[6] 吴文俊. 初等几何判定问题与机械化证明. 中国科学, 1977, (6): 507-519
[7] 钱宝琮. 中国数学史. 北京: 科学出版社, 1964
[8] 顾今用. 中国古代数学对世界文化的伟大贡献. 数学学报, 1975, 18 (1): 18-23
[9] 吴文俊. 文化大革命为数学研究开辟了广阔的前途. 应用数学学报, 1976, 1 (2): 13-16
[10] 吴文俊. 数学的机械化. 百科知识, 1980, (3)
[11] 吴文俊. 我国古代测望之学重差理论评介——兼评数学史研究中某些方法问题. 科技史文集 (8), 1982. 载: 吴文俊论数学机械化, 济南: 山东教育出版社, 1996: 112-150
[12] 罗士琳. 四元玉鉴细草. 1834
[13] 吴文俊. 对中国传统数学的再认识. 百科知识, 1987, (7-8)
[14] 朱世杰. 四元玉鉴校证. 李兆华校. 北京: 科学出版社, 2007
[15] Zhu S J. The Jade Mirror of the Four Unknowns. Transl. Hoe J. Christchurch: Mingming Bookroom, 2008
[16] 吴文俊. 在教育部主办的全国高校中外数学史讲习班开学典礼上的讲话. 中国数学史论文集, 1984, 2 (1986): 1-8
[17] 李向东, 张涛. 大家: 我的不等式. 中央电视台, 2006

The Combination of Mathematics and Music—The Comparative Study of the Origin of the Calculation of Pitch in Ancient China and Greece

Liu Yaya

(Center for History of Mathematics and Sciences, Northwest University)

According to the analysis and comparison, this paper tries to analyze the mutual relationship between mathematics and ancient temperament theory in both East and West at the initial period of its creation, point out how to use mathematics method to explore the rules of the temperament theory. Through the methods of mathematical logic to comparing with the musical forms, come to a result that the calculation of pitch between ancient China and Greece are significant difference. Music systems between different civilizations have different value positions. Although their aims are different and they experience different ways, all of them create calculation method of pitch and analysis result, they mirrored modernization the "correctness"[1] of mathematics and music. Because of this, the calculation of pitch by ancient Chinese and ancient Greek from different civilizations can be forming and development independently.

1. The nature of musical acoustics: understanding from two ways

1.1 The physical nature of pitch

The theory of pitch starts from the rules of vibration of sounding body, uses mathematics calculation to define the pitch standardization and intervals of musical sound[2], which is an important subdiscipline of musicology.

1.2 The perception of pitch

Much of our acoustical experience comes naturally and spontaneously. The early students of acoustics nevertheless made penetrating analyses based on their perception of sound. Furthermore, the early peoples found the consonant interval. This find of music temperament impels the mutual intermingle between mathematics calculation and the theory of music, and stimulates the early peoples to make an impressive improvement on the theory of music temperament.

The explanation above shows that, no matter what kind of ways of expression, the music consonance as an objectively existed thing, the nature should be in accord.

Thus, there are some issues regarding to weather the origin of calculating the pitch in both of the East and West are comparatively individual. According to the analysis and comparison, this paper tries to point out there is a preferable way to further discuss weather the origin of pitch in early China and ancient Greece are comparatively individual, that is doing the analysis through the aims of making pitches and the calculating method of pitches.

2. The origin and the process of the calculation of early Chinese pitch

2.1 The aim of making pitch

Yinzhou (殷周) Dynasty is the time of slavery. The noble families have high level music like bell, chime stone, Qin, Se, Sheng, Yu (琴, 瑟, 笙, 竽) and have other enjoyment. The high-level enjoyment was monopolized by the high class, however the low class and the slave can't share the enjoyment.[3] So the ruling class must enjoy the pitch system by themselves like they enact the calender, so as to achieve the classification and consolidation of rank. It also promotes the development of the calculation of pitch in early China.

The Chinese term for pitch is Yinlv (音律) or simply Lv (律). The earliest extant mention of pitch is found in the *Shundian* (《舜典》) *The Canon of Shun* of the *Yushu* (《虞书》), *The Book of Yu* which survived through its incorporation in the *Shangshu* (《尚书》), The Book of Documents. We have:

Ode expresses thought in lyric, Song puts lyric in singing, Sound is produced to facilitate singing, Pitch is used to harmonize sound… (诗言志, 歌永言, 声依永, 律和声……[4])

Among the descriptions of the activities and polices implemented by Shun after taking over the responsibilities of the king of Yao (尧), it is stated in the *Shundian*:

Standardize pitched, measures and weights.[5] (同律度量衡)

So we can see that, the aim of making pitch in early China is the need of stabilizing the social economy. The accordance of music can stabilize the politics, otherwise, the ruined music may represent the declining even the collapse of the society. Thus, making of pitch is the way to find harmonious music and achieve it. And the aim of making the pitch is safeguarding the harmony of the social life.

2.2 Computational process

At this motive of making pitch, along with the pursuit of social harmony, the anceint Chinese found the consonant interval based on their perception of sound, further details on the calculation of the research, they created a series of related theory and method, the specific contents of the following several parts:

2.2.1 A mathematical procedure for the generation of pentatonic scale

A mathematical procedure for the generation of pentatonic scale has come down to us in the *Guanzi* (《管子》) of year 480-422 BC. The relevent passage runs as follows:

As for the generation of the five tones [by a mathematical procedure], one begins first by

The Combination of Mathematics and Music—The Comparative Study of the Origin of the Calculation of Pitch in Ancient China and Greece

taking the principal to be unity and then multiplying it by the number 3. The fourth power of this number gives 9×9 which is the smallest leading number for the generation [of the scale] from [the tone] Gong in the Huangzhong pitch. By increasing the number by one-third of third of its value, one obtains the number 108 for [the tone] Zhi. By decreasing the number so obtained by nothing more than one-third of its value, one obtains [the number 72] which is sufficient to generate [the tone] Shang. By giving back to [the number 72] one-third of its value, one finds [the number 96] for the generation of [the tone] Yu. By decreasing the number so obtained by one third of its value one finds [the number 64] for the generation of [the tone] Jue (凡将起五音：先主一而三，四开以合九九。以是生黄钟小素之首，以成宫；三分而益之以一，为百有八，为徵；不无有三分去其乘，适足以是生商；有三分而复于其所，是以生羽；有三分而去其乘，是以生角)[6].

In current mathematical notation, the procedure for the generation of a set of five notes can be summarized as follows:

$1 \times 3 = 3$

$3 \times 3 \times 3 \times 3 = (3 \times 3) \times (3 \times 3) = 9 \times 9 = 81$ ················· For the note Gong (宫)

$81 \times (1 + 1/3) = 81 \times (2 \times 2/3) = 108$ ························ for the note Zhi (徵)

$108 \times (1 - 1/3) = 108 \times 2/3 = 72$ ····························· for the note Shang (商)

$72 \times (1 + 1/3) = 72 \times (2 \times 2/3) = 96$ ······················· for the note Yu (羽)

$96 \times (1 - 1/3) = 96 \times 2/3 = 64$ ································ for the note Jue (角)

These formulas show a mathematical procedure for the generation of a set of five notes, as Sanfen Sunyi (三分损益) method. There are two factors 2/3 and 4/3: multiply 2/3, means decreasing the number by one-third of its value, multiply $2 \times 2/3$, means increasing the number by one-third of its value. Since the intervals between notes are given by the ratio of their corresponding numbers, the scales generated are independent of the initial number 81. This seems to imply that number 81 is assigned as the initial number for the purpose of avoiding non-integers because, in generating the rest of four notes in the scale, the procedure requires the number to be divided by 3, four times, and the number 81 is the smallest integer that can be evenly divided by 3 four times.

2.2.2 The generation of chromatic scale

A description of the up-and-down procedure for the generation of chromatic scales is given in the *Lvshi Chunqiu*. We have from Chapter Yinlv (Tonal Pitch) the passage:

The Huang Zhong pitch generates the Linzhong pitch. The Linzhong pitch generates the Taicu pitch. The Taicu pitch generates the Nanlv pitch. The Nanlv pitch generates the Guxian pitch. The Guxian pitch generates the Yingzhong pitch. The Yingzhong pitch generates the Ruibin pitch. The Ruibin pitch generates the Dalv pitch. The Dalv pitch generates the Yize pitch. The Yize pitch generates the Jiazhong pitch. The Jiazhong pitch generates the Wuyi pitch. The Wuyi pitch generates the Zhonglv pitch. The addition of 1/3 of its value gives rise to the up-generation. The subtraction of 1/3 of its value gives rise to the down-generation.

Huangzhong, Dalv, Taicu, Jiazhong, Guxian, Zhonglv, and Ruibin are obtained by up-generation. Linzhong, Yize, Nanlv, Wuyi and Yingzhong are obtained by downgeneration. (黄钟生林种，林钟生太簇，太簇生南吕，南吕生姑洗，姑洗生应钟，应钟生蕤宾，蕤宾生大吕，大吕生夷则，夷则生夹钟，夹钟生无射，无射生仲吕。三分所生，益之一分以上生；三分所生，去其一分以下生。黄钟、大吕、太簇、夹钟、姑洗、仲吕、蕤宾为上，林钟、夷则、南吕、无射、应钟为下)[7].

Guanzi used Sanfen Sunyi (三分损益) method to calculate pentatonic scale as we mentioned above. By using the Huangzhong pitch $3^4 = 81$, *Guanzi* got the integer numbers of pentatonic scales. Here we are looking for twelve chromatic scales, in order to avoid getting the non-integral, so we use $3^{11} = 177147$ as the large number. The mathematics procedure for the generation of twelve pitches as mentioned in Yinlv are as follows:

$3^{11} \times 1 = 177147$ Hangzhong (up)

$3^{11} \times \frac{4}{3} = 3^{10} \times 2^2 = 236196$ Linzhong (down)

$3^{11} \times \frac{4}{3} \times \frac{2}{3} = 3^{11} \times \frac{2^3}{3^2} = 3^9 \times 2^3 = 157464$ Taicu (up)

$3^{11} \times \frac{4}{3} \times \frac{2}{3} \times \frac{4}{3} = 3^{11} \times \frac{2^5}{3^3} = 3^8 \times 2^5 = 209952$ Nanlv (down)

$3^{11} \times \frac{4}{3} \times \frac{2}{3} \times \frac{4}{3} \times \frac{2}{3} = 3^{11} \times \frac{2^6}{3^4} = 3^7 \times 2^6 = 139968$ Guxian (up)

$3^{11} \times \frac{4}{3} \times \frac{2}{3} \times \frac{4}{3} \times \frac{2}{3} \times \frac{4}{3} = 3^{11} \times \frac{2^8}{3^5} = 3^6 \times 2^8 = 186624$ Yingzhong (down)

$3^{11} \times \frac{4}{3} \times \frac{2}{3} \times \frac{4}{3} \times \frac{2}{3} \times \frac{4}{3} \times \frac{2}{3} = 3^{11} \times \frac{2^9}{3^6} = 3^5 \times 2^9 = 124416$ Ruibin (up)

$3^{11} \times \frac{4}{3} \times \frac{2}{3} \times \frac{4}{3} \times \frac{2}{3} \times \frac{4}{3} \times \frac{2}{3} \times \frac{4}{3} = 3^{11} \times \frac{2^{11}}{3^7} = 3^4 \times 2^{11} = 165888$ Dalv (up)

$3^{11} \times \frac{4}{3} \times \frac{2}{3} \times \frac{4}{3} \times \frac{2}{3} \times \frac{4}{3} \times \frac{2}{3} \times \frac{4}{3} \times \frac{4}{3} = 3^{11} \times \frac{2^{13}}{3^8} = 3^3 \times 2^{13} = 221184$ Yize (down)

$3^{11} \times \frac{4}{3} \times \frac{2}{3} \times \frac{4}{3} \times \frac{2}{3} \times \frac{4}{3} \times \frac{2}{3} \times \frac{4}{3} \times \frac{4}{3} \times \frac{2}{3} = 3^{11} \times \frac{2^{14}}{3^9} = 3^2 \times 2^{14} = 147456$ Jiazhong (up)

$3^{11} \times \frac{4}{3} \times \frac{2}{3} \times \frac{4}{3} \times \frac{2}{3} \times \frac{4}{3} \times \frac{2}{3} \times \frac{4}{3} \times \frac{4}{3} \times \frac{2}{3} \times \frac{4}{3} = 3^{11} \times \frac{2^{16}}{3^{10}} = 3^1 \times 2^{16} = 196608$ Wuyi (down)

$3^{11} \times \frac{4}{3} \times \frac{2}{3} \times \frac{4}{3} \times \frac{2}{3} \times \frac{4}{3} \times \frac{2}{3} \times \frac{4}{3} \times \frac{4}{3} \times \frac{2}{3} \times \frac{4}{3} \times \frac{2}{3} = 3^{11} \times \frac{2^{17}}{3^{11}} = 3^0 \times 2^{17} = 131072$ Zhonglv (down)

Above formula shows that, take $3^{11} = 177147$ as the Huangzhong large number to generate the other eleven pitches according to the Sanfen Sunyi method. The Huangzhong large number, namely $3^{11} = 177147$, was multiplied by 2/3 and 4/3 for eleven times according to the sequence of up-and-down generation. Then we get the largest denominator which makes sure that all the number of the twelve pitches are integers.

Obviously, this is similar to the pentatonic scale, which takes $3^4 = 81$ gong pitch given in *Guanzi*. And the procedure of generating the pentatonic scale is also made of two steps, namely up-and-down generating. This further improves the result of theory of the generation by the fifth (Wudu Xiangsheng, 五度相生). From the above analysis, it is obvious that the *Guanzi* procedure, also known in the Chinese as the Sanfen Sunyi method, is actually a general method which can be used to generate scales in different modes or with different number of notes.

2.2.3 Up-and-down generating

These five notes are generated either by the multiplication of the factor 4/3 or by the factor 2/3. The down-generation with the ratio 2/3 generates a perfect fifth since it corresponds to decrease the length of the vibrating string by 1/3 (1-1/32/3). The up-generation with the ratio 4/3 generates also a perfect fifth but an octave lower, since the operation to add 1/3 the length of the vibrating string is equivalent to shorten the length by 1/3 and then double the remaining length[8].

From the above analysis, it is obvious that *Guan Zi* procedure, also known as in Chinese as the Sanfen Sunyi method, is actually a general method which can be used to generate scales in different modes or with different number of notes.

The analysis of generation of Chinese pitch system shows that the up-and down-procedure is a common procedure. Although the process of making the pitch is simply and it is a "simplest mathematical procedure", it should be applicable to the mathematical analysis.

2.3 Brief summary

Rites and music is the main aspect of education and politics. If the rites and music is upright, it can promote the smoothness of education and politics. Rites can control people's heart, while music can smooth people's character[9]. The success of rites and music can achieve a harmonious society.

Pitch and calendar are made from numbers. The ancient Chinese mathematics depend on the production practice, and than they give services to the production practice. Before the calculation of pitch, the mathematics methods of calculating pitch are used in the area of calendar and weights and measures. So the point is mathematics come from practical count. When the abstract of this practical count reaches a certain degree, it will be used for the rites and music on spiritual level. The discussions about pitch, weights and measures as well as the practical count and etc. form a special pattern of the combination of music mathematics in early Chinese.

3. The origin and the process of the calculation of pitch in Ancient Greece

The school of Pythagoras (570-496 years BC) regard number as a principle of everything. They tried to explore the eternal rules of the astronomy by revealing the mystery of numbers. They made lots of deep research on number, and got many results. They paid attention to the harmonious relationship between numbers and music, the relationship between numbers and geometrical figure as well as the relationship between numbers and the movement of celestial bodies.

The music on the earth is just the imitation of the music on the heaven. However, the concord of the music on the heaven is based on numbers, so there is a harmonious relationship between music and numbers.

3.1 The first to find the mathematical ratios of the harmonious sound and the harmonic sequence

Pythagoras was regarded as the first person who finds the mathematical ratios of the harmonious sounds [10].

Pythagoras found that if two strings that have same stretching force, and the length ratio of these two strings are the simplest ratio between two integers like 1/2, 2/3 or 1, so the music of these two strings made is concordant and pleasant. This issue regarding to the concordance of music was brought into his model that is "number is the origin of all things". After the further study of the simplest ratio between two integers, Pythagoras found that the reciprocal of them are 2, 3/2 and 1 can make an arithmetic sequence. And these sequences are called "harmonic sequence" [11].

3.2 Tetraktys and the determination of the tone relations

According to the theory of "simplest ratio between two integers", Pythagorean built a musical theory. The sum of the first four natural numbers is 10, and these four numbers conduct a triangle with four points each sides, so it was called Tetraktys. 4:3, 3:2, 2:1 make up the main tonalities. These tone relations are the foundation of the ancient Greek music system, and be regarded as "the eternal source of the world" [12].

Thus, Tetraktys can be reproduced by 6, 8, 9, 12. Divede the chord into twelve share, twanged the chord which constitute by the number of 6, 8, 9, 12 separately, and the harmonic sequence can be palyed. So that, the octave, fifth, fourth can be arranged in 2:1, 3:2 and 4:3. Diatonic scale stereotype was based on gradual dividing of the Tetraktys musical interval. Within two fourth-chord, they can detect two major second like 9:8 individually: (9:8) · (9:8) · (256:243) =4:3

3.3 The establishment of the generation by the fifth

The musical structure made by Pythagoras and Plato (427- 327years BC) was greatly influenced by the inner-relationship between music and Mathematics. After Pythagoras and Pythagorean believed 3/2, 4/3and 2 are Consonant interval, they use the intervals of fifth as 3/2 to make the generation by the fifth method.

Tab. 1 The scales of Chinese and Greek[13]

Generation	Ancient China	Ancient Greece
C	1	1
#C	2187/2048	
D	9/8	9/8

续表

Generation	Ancient China	Ancient Greece
#D	19683/16384	
E	81/64	81/64
F	177147/131072	4/3
#F	729/512	
G	3/2	3/2
#G	6561/4096	
A	27/16	27/16
#A	59049/32768	
B	243/128	243/128
C	2/1	2/1

The musical interval of Pythagorean scale also has only two style: major tone, the frequency ratio is 9:8, 204 cent; Minor semitone, the frequency ratio is 256:243, 90 cent, its size is same as the *Sanfensunyi* mathod. Though it has only one combine factor, there is only one direction to downward.

4. Conclusion

In ancient Chinese, they believe that "the numbers of the pitch and calendar are the truth of the cosmos"[14], music and make temperament are the expression of Yuehe Zhengping (乐和政平). They creating the special computing method along the way to find the social harmony, continue to perfect and improve the method to agree with the demand of practical music.

According to the model of "the number as a principle of everything", ancient Greek Pythagorean transform the music to the relation of numbers, on the way of finding cosmic harmony, though the research of numbers to go deeper to reveal the essence of music.

References

[1] Wilie R. Musik und Mathematik. Beijing: Science Press, 1986. 23
[2] 戴念祖. 中国物理学大系. 长沙: 湖南教育出版社, 2001. 188
[3] 人民音乐出版社编辑部.《乐记》论辩. 北京: 人民音乐出版社, 1983. 24
[4] 尚书. 郑玄注. 北京: 商务印书馆, 1937
[5] 郭沫若. 青铜时代. 上海: 上海古籍出版社, 1951. 4
[6] 夏纬英. 管子·地员篇校释. 北京: 中国农业出版社, 1981. 35
[7] 〔战国〕吕不韦. 吕氏春秋. 上海: 上海古籍出版社, 1989. 25
[8] Chen C-Y. Early Chinese Work in Natural Science. Hong Kong: Hong Kong University Press, 1996. 49-51
[9] 同 [5]. 12
[10] Cohen M R, Drabkin I E, A Source Book in GREEK Science. Cambridge: Harvard University Press, 1958. 294

[11] 梁宗巨,王青建,损宏安.世界数学史(上、下).沈阳:辽宁教育出版社,2001.34
[12] 同[1].25
[13] 同[2].188
[14] 〔西汉〕刘安.淮南子·天文训卷三.北京:中华书局,1936

Pythagoreanism in Edo

—From ARAI Hakuseki to SAKUMA Shōzan

Chikara SASAKI

(The Euler Institute in Japan & The University of Tokyo)

1. Introduction: Pythagoreanism and the study of mathematics in early modern Japan

During the Edo Period (the 17th through the mid-19th centuries), Neo-Confucianism became the official learning for the ruling warrior class by the authority of the Tokugawa central government. As a part of this system, the difficult classic *I Ching* or *Yi Jing* (《易经》, *Book of Changes*) was seriously studied and became influential. A kind of Pythagoreanism in *I Ching* made a basis of the thought emphasizing mathematical understanding of the structure of the world. In this essay, Pythagoreanism or, precisely speaking, scientific Pythagoreanism means a philosophical thought which considers that numbers consist of the first principles, by nature, of all things, thus, implying that mathematics is an essential discipline.

As Joseph Needham states in *Science and Civilisation in China*, Vol. 1: "Introductory Orientations" (1954): "While the Pythagorean School flourished (600BCE-300BCE) the scholars and divines in China were developing the *I Ching* (*Book of Changes*) into a universal repository of concepts which included tables of antinomies (Yin and Yang) and cosmic numerology; all this was systematised in the Han." (p. 228) The text of *I Ching* seems to have encouraged worrior-intellectuals to study mathematics from the middle of the 17th century on. Thus a kind of Pythagoreanism in *I Ching* provided the mathematico-philosophical foundations of Wasan (和算), traditional mathematics in Tokugawa Japan.

Then, I try to characterize mathematics in the history of Japan. Before the 16th century, mathematical classics in Japan were just a subset of ancient Chinese classics such as *Jiuzhang Suanshu* (《九章算术》, *Nine Chapters on the Mathematical Art*). Even this most important canon of mathematical disciplines cannot be said to have been studied by Japanese intellectuals. In the second half of the 16th century, two important events occurred in the history of Japanese mathematics. First, Jesuit missionaries who equipped themselves with mathematical knowledge visited Japan and transmitted it to the Japanese youth, even though we know almost nothing about what actually happened. It is said that Carlo Spinola (1564-1622) taught mathematics at a Jesuit school in Kyoto in the early 17th century. Second, at the wars of invasion to Korea attempted by Toyotomi Hideyoshi some Chinese books were looted and brought into Japan. For example, Korean editions of Zhu Shijie's (朱世杰) *Suanxue Qimeng* (《算学启蒙》, *Introduction to Mathemaics*, 1299), and of Chen Dawei's (程大位) *Suanfa Tongzong* (《算法统宗》, *Systematic Treatise on Arithmetic*, 1593) became essential sources of Japanese mathematics.

With the establishment of the strong Tokugawa central government in the 17th century, a rather peaceful era emerged and the population of the Japanese nation increased drastically. Thus the Wasan began to flourish.

With the arrival in the Edo Bay of Perry in the United States of America in 1853 a new era of Japanese mathematics started, namely Western or European mathematics began to be introduced. The new imperial government adopted western arithmetic rather than traditional Japanese arithmetic for its school system in 1872, the 5th year after the Meiji Restoration. After that event, the style of mathematics in modern Japan became western mathematics, with an exception of the abacus calculation.

The most influential explanation as to why and how mathematical studies flowered in Tokugawa Japan was presented by Mikami Yoshio (三上义夫, 1875-1950), eminent historian of Japanese mathematics with his "Japanese Mathematics from the Viewpoint of Cultural History" of 1922. According to Mikami, traditional Japanese mathematics was essentially not a science in the western sense of the word but just an art, or "geido" like the art of tea ceremony or the art of flower arrangement in addition to several utilities such as ordinary and calendrical calculations and the art of land surveying.

Now I am going to present a new thesis that in the 17th century mathematics became to be thought as an important element of the Neo-Confucian system of learning which was introduced deliberately by the Tokugawa government in 1600 on. In this juncture, the Chinese classic *I Ching* with the reading by Shao Yong (邵雍, 1011-1077) and Zhu Xi (朱熹, 1130-1200) played a crucial role.

2. Neo-Confucianism as "Tokugawa ideology" in the 17th century and Arai Hakuseki's policy of money making

With the conversion from Zen Buddhism to Confucianism of Fujiwara Seika (藤原惺窝, 1561-1619), Neo-Confucianism became an official system of learning in Tokugawa Japan. He is said to have appeared before Tokugawa Ieyasu in 1600 with a Korean costume of Confucianists. It is known that this conversion became possible with his communication with Confucian scholars and books looted or imported from Korea. He and his talented disciple Hayashi Razan (林罗山, 1583-1657) began to study seriously Neo-Confucianism with these Chinese books in Korean editions which were brought into Japan in the late 16th century. Also

Yamazaki Ansai (山崎闇斎, 1619-1682) is considered to have contributed to the introduction of Neo-Confucianism to Tokugawa Japan.

The Neo-Confucianist samurai in the late 17th and early 18th centuries named Arai Hakuseki (新井白石, 1657 – 1725) was a scholar belonging to the genealogical line of Fujiwara Seika. He became a high officer of the central government and an important political and intellectual adviser of the 6th Tokugawa shogun Ienobu (徳川家宣). Thus, Hakuseki made an advice of the money making to the Shogun Ienobu, actually concerning the production of oval golden coins called "koban". He mobilized his knowledge of *I Ching* to insist to maintain a high percentage of gold in the "koban". According to him, the earth and the heaven are full of the great mystery of numbers which cannot be seen in real abacus. He stated in his proposal to the Shogun:

> Small numbers are visible numbers which appear in abacus and which can be known through counting, while large numbers are numbers of which great arithmetic exists between the heaven and the earth even though they cannot be seen in abacus. Hence, those who are familiar with arithmetic can count small numbers, while large numbers can be recognized only by those who understand principles (Li) and they are hard to be handled even by those who know the theory of them. For example, although the science of calendrical numbers became exact and precise in recent years, errors necessarily appear in the long run. (*Complete Works*, Vol. 6, p. 192.)

First of all, the pair-concept of "small numbers" and "large numbers" in the quoted passage reminds us of introductory chapters of Yoshida Mitsuyoshi's (吉田光由) popular book on *soroban*, or abacus, *Jinkóki* (《尘劫记》), first published in 1627. From Hakuseki's autobiography entitled *Oritakushibanoki*, we know that Hakuseki read *Jinkóki*. Chapter One of *Jinkóki* is on "The Naming of Large Numbers" and provides the naming of units of integers in the decimal system, e.g. one, ten, hundred, and so on, i.e. 10^n ($n \geq 0$). Chapter Two is on "The Naming of Small Numbers" and gives the naming of the fractional units, e.g. one tenth, one hundredth, one thousandth, and so on, i.e. 10^{-n} ($n \geq 1$). It is known that the author of *Jinkóki*, Yoshida, referred to Cheng Dawei's *Suanfa Tongzong* and the pair-concept of "small numbers" and "large numbers" are safely considered to have been originated from *Suanfa Tongzong*. But, this use of "small numbers" and "large numbers" is clearly different from Hakuseki's. Rather, Hakuseki's use seems to be similar to "proper (*eigentlich*) representation of numbers" for visible and intuitively manipulatable integral numbers and "symbolic representation of numbers" for invisible integers which can be represented only through some symbols in the German philosopher Edmund Husserl's *Philosophie der Arithmetik*, of which the first edition appeared in 1891. Of course, this is a quite ahistorial deviation.

Then, what did Hakuseki mean by the pair-concept of "small numbers" and "large numbers"? The expression "great arithmetic exists between the heaven and the earth" also appears in autobiographical *Oritakushibanoki*. There Hakuseki comments on this expression in the following: "these things are what human intellects cannot make conjectures and they may argue on whether reasoning (Li) is correct or not."

Some historians have suggested that Hakuseki's argument may be similar to Nakae Tôju's (中江藤树) sentences in *Okina Mondô* (《翁问答》, *Questionings and Answerings of Old Man*), published in 1649. Tóju wrote in this monograph concerning *Ekigaku*, or divination lore as follows: "Nothing between the heaven and the earth cannot escape from the divine principles of *I Ching*, nothing to say of Confucian books. Nothing isn't originated from the principles of *I Ching*." So, for Tóju the Confucian classic *I Ching* was very essential. The same thing can be said for Arai Hakuseki.

It seems to me that Hakuseki's cited passage cannot be understood without some knowledge of Neo-Confucian diagram of the world. According to this diagram which is usually ascribed to Zhou Dunyi (周敦颐, 1017-1073) of the eleventh century, everything begins with "Supreme Pole", which is at the same time understood as "No Pole". The Supreme Pole moves and produces the Yang. When the movement has reached its limit, rest (ensues). Resting, the Supreme Pole produces the Yin. Then the combinations of the Yang and Yin produce the Five elements: water, fire, wood, metal and earth. These elements or rather categories consist of "Qi", pneuma.

"Small numbers" in Hakuseki's passage may be thought to belong to the world of this physical world of "Qi", while "large numbers" are closer to the metaphysical world of "Li". Thus, the mathematical world as a whole including "large numbers" can be only understood by possessors of the knowledge of the divine principles of *I Ching*. We can witness here a kind of Oriental Pythagoreanism. So, Arai Hakuseki may be regarded as the first Japanese Pythagorean in the Edo Period. Hakuseki is known as a prolific author and the historian of Japanese literature named Kato Shuichi compared Hakuseki with Leibniz in a sense justly and in another sense with a certain exaggeration.

The talented disciple of Seki Takakazu named Takebe Katahiro (建部贤弘, 1664-1739) published his first monograph with the title *Kenki Sampō* (《研几算法》), when he was 19 years old. The word "kenki", which literally means "catch an opportunity", came from the most important commentary of *I Ching*. Consequently, we sense an influence of *I Ching* upon the important mathematician.

3. Neo-Confucian rationalists Miura Baien and Kaiho Seiryô

Tokugawa Japan produced some remarkable thinkers who may be considered to have belonged to the Neo-Confucian intellectual tradition in a broad sense with a sceptical and rationalistic orientation. Here we introduce only two thinkers in the 18th century, Miura Baien (三浦梅园, 1723-1789) of Kyushu Island and Kaiho Seiryô (海保青陵, 1755-1817) of Kyoto.

Miura Baien was a physician who wrote a remarkable metaphysical monograph with the title *Gengo*, literally *Black Words* (《玄语》). Baien studied Confucian books seriously and also travelled to Nagasaki, a center of *Rangaku*, or Dutch Learning. In an introductory part of the

Gengo, Beirn states: "From my childhood, everything I touched seems to be questionable." In "Letter Answering to Mr. Taga Bokkei", he also writes: "After I had a mind to doubt things, nothing became agreeable, e. g. visible and auditable things, moving hands and feet, and thinking mind, to say nothing of the going and the coming of the sun and the moon, and the transforming of living things." In an observation of the excellent scholar of Miura Baien Yamada Keiji (山田庆儿), "Baien's natural philosophy was nothing other than a supreme representation of the theoretical elaboration of Yin-Yang philosophy."

Kaiho Seiryô was quite a rationalistic thinker and very remarkable in his critical or sceptical inclination. For him the divinatory side of the *I Ching* was not important at all, and only the rational and theoretical part could be fully appreciated. He may be characterized as a radical sceptic in early modern Japan. In his *Yôrodan* (《养芦谈》, Talking about the Bringing Up of Reeds), he stated: "Doubt things until doubts are entirely solved. To have a doubt is a virtue." He also insisted upon a rationalistic attitude in economic problems. According to him, to disdain merchants using abacus is a serious mistake. Kongzi, or Confucius, the founder of Confucianism never said so. Instead, Seiryô states very clearly: "The heaven is full of arithmetical uses." He is now regarded as a kind of economical rationalist in early modern Japan. He may not be called a Descartes in Japan, but it is certain that he possessed a certain element of Descartes's character.

4. The Neo-Confucianist who converted to a Western Pythagorean: Sakuma Shōzan

Now we proceed to the Bakumatsu, or the late Tokugawa period.

Sakuma Shōzan (佐久间象山), or Zôzan, was born in Matsushiro, near Nagano in 1811 and received a very good education of Neo-Confucian orthodoxy and Wasan, traditional Japanese mathematics. His most favorite classic of Confucian studies was *I Ching*. Soon after the Opium War, he started to think seriously about Japan's maritime defense and to learn how to make cannons in the western way. Further, he studied the Dutch language to learn things Western generally. His general point of view in the style of learning was "Tōyō Dôtoku, Seiryô Geijutsu" (东洋道德, 西洋艺术, Eastern Morals and Western Technology), namely, with the Neo-Confucian ethics being the center, supplemented by European science and technology.

When Perry of the United States Navy arrived in Edo Bay in 1853, he was in charge of the coastal defense of Edo Bay. In 1854, he suggested Yoshida Shōin (吉田松阴) to go to the United States of America with Perry, but Shōin was arrested by the Tokugawa central government, and Shōzan was also placed under restraint for 9 years.

When he was in prison in Edo and penitent in his hometown in 1854, he drafted *Seikenroku* (《省訾录》, *Reflection on My Errors*), which was first published after 4 years of the Meiji Restoration in 1871 by his brother-in-law Katsu Kaishū (胜海舟). In this record, he wrote a remarkable observation on western mathematics.

All learning is cumulative. It is not something that one comes to realize in a morning or an evening. Effective maritime defense is in itself a great field of study. Since no one has yet thoroughly studied its fundamentals, it is not easy to learn rapidly its essential points…

Mathematics is the basis for all learning. In the western world after this science was discovered military tactics advanced greatly, far outstripping that of former times. (*Collected Works*, Vol. 1, p. 9)

The original Chinese or Japanese word for mathematics in the quoted passage is "shōshōjutsu", the art of precise reasoning. This word is supposed to have derived from the contemporary Wasan mathematician Uchida Itsumi (内田五観), who possessed some knowledge of the Dutch. Uchida used the vocabulary "shōshōgaku" which had been used in a monograph on calendrical science published in 1634 for the Dutch word for mathematics "Wiskunde". According to Shōzan, "Mathematics is the basis of all learning" (详证术者万学之基本也), and "now if one wants to learn things military, it is impossible to do so without this science". Mathematics is sine qua non for military science and technology, which Japan was considered to need to defend its nation.

Before Shōzan began to study western science through the Dutch language, the very basis for all learning for Shōzan seems to have been the Neo-Confucian Li, or principles, provided in particular by the classic *I Ching*. After he met the American navy officer, Perry, the very basis for all learning became western mathematics, which was quite essential for the maritime defense. Thus Shōzan seems to have been transformed from oriental Pythagoreanism to Western Pythagoreanism, so to speak.

Towards the end of 1855, Nagasaki Kaigun Denshūjo (长崎海军伝习所, Nagasaki Navy Training Institute) was established by the Tokugawa government, Kaishū and other samurais of the Bakufu began to be trained by Dutch navy officers. On March 21, 1855, Shōzan wrote to Kaishū in Nagasaki. "The art of navigation consists of mathematics. If one isn't familiar with 'wiskunde', one cannot understand it. This is true." Shōzan then quotes the passage of his drafted record *Seikenroku* of 1854, which we have just referred to above including "Mathematics is the basis of all learning". In Shōzan' opinion, "Li" in Neo-Confucianism and "reason" in western science are the one and the same, no difference.

On May 6, 1856, Kaishū wrote to Shōzan on the mathematical instruction at the Nagasaki Training Institute. Among the disciplines, two astronomers of us understood "stuurkunst" or the art of navigation in mathematical sciences. Naturally, since their mathematics is not ultimately different from ours, those who are good at mathematics quickly understand it. People who are not good at mathematics as I feel quite difficult for the disciplines through it. Recently we recognize this thing almost. As we feel difficult about it, we are not happy. What you said about "shōshōjutsu" ("Wiskunde") is quite correct.

To this statement, Shôzan replied to Kaishū in his letter dated on July 10, and wrote a very interesting observation when he compared Japanese mathematics with Western mathematics:

Concerning the method of mathematics, after all, since there is no difference between ours and the Western, it is true that those who are good at our mathematics become quickly familiar with theirs. Nevertheless, problems and their technics in Western countries are all useful, while mathematicians in our country are usually concerned with useless things. Further, as mathematics in our country have been originated in the Chinese method, mathematicians tackle circle problems through squares, based on kōko (勾股, Pythagorean theorem), when discussing the circle principles of the art of tetsujutsu (缀术). To the contrary, western trigonometry treats square problems through circles. Confucian scholars in the Qing Dynasty such as Dai Donyuan, Jiao Xun, and Ruan Yuan insisted that there is no difference between Chinese and western mathematics. But, I cannot agree with them. In my opinion, whereas Chinese mathematics treats circles through squares, western mathematics challenges squares through circles. How do you think of this? I would like to ask you, if you have a high opinion.

"Tetsujutsu", or the art of linking, in the cited passage clearly means the art of induction used, for example, in Takebe Katahiro's masterpiece *Tetsujutsu Sankei* (《缀术算经》) of 1722 which is well-known to have ingeniously calculated π to 41 decimal places by a kind of the method of induction. This discussion shows that Shōzan was familiar with not simply wasan and Chinese mathematics but western mathematics to a considerable degree. In fact, we know from his letter of Kaishū on August 15, 1855, he studies "Wiskunde" since he had a much spare time under restraint.

Shōzan seems to have become the first Pythagorean of Japan in the western sense of the word, because he was an admirer of *I Ching*.

5. From *tenzan* algebra to western arithmetic: Yanagawa Shunsan's Yōsan Yōhō of 1857

Lastly we argue a technical aspect of Wasan which helped much for Wasan practitioners to have shifted rather easily from traditional mathematics to modern western mathematics.

The very first monograph of western mathematics was *Yōsan Yōhō* (《洋算用法》, *Use of Western Arithmetic*) by the Dutch medical doctor in Nagoya Yanagawa Shunsan (柳河春三, 1832-1870). In this small book, the western way of performing the four elementary arithmetical operations and the rule of three in the elementary theory of proportion were explained in detail using the language of Wasan. At the beginning of his book's main text, Yanagawa wrote: "Western arithmetic is more or less similar to our art of Tenzan (点窜, adding and deleting)."

It was Seki Takakazu (关孝和,? -1708) who reformed the Chinese instrumental art named *Tianyuanshu* (天元术, technique of the celestial element) of solving equations by manipulating calculating rods into a kind of symbolic algebra in the written form called bōshō-hō (傍书法, art of side-writing). This specific form of algebra represented both known and unknown quantities by symbols of Chinese ideographs and was later named Tenzan algebra by Matsunaga Yoshisuke, one of Seki's talented followers. In my opinion, Tenzan algebra is comparable to François Viète's symbolic algebra.

Yanagawa may not to have been called a professional mathematician. Even this intellectual amateur in mathematics knew what the art of Tenzan and could translate western arithmetical operations into formulae of Tenzan algebra easily.

Only in 15 years after the publication of Yanagawa's monograph on western arithmetic, the new Japanese central government adopted western mathematics rather than traditional Japanese mathematics for its school system. For such a rather successful transformation of the style of mathematics, we must not forget both philosophical and mathematical foundations of which the origins were Chinese. In this essay I try to show that there existed a kind of scientific Pythagoreanism in Tokugawa Japan. In this sense, it was Chinese intellectual heritage that contributed a lot to the Japanese's rapid transformation to the modern western mathematical tradition.

Algorithm and Principles of Division of Fractions in Chinese Ancient Literature[*]

Sun Xuhua

(Faculty of Education, University of Macau)

Division of fractions is always a difficult topic to discuss in primary school mathematics because its algorithms and principles are not easy to explain in details. This paper is chiefly concerned with a particular historical development in mathematics which must strike anyone who compares Chinese ancient literature on algorithms of division of fraction with those of any modern mathematics textbook. The algorithm of "flip and multiply" of fraction division, which appeared early in the *Suanshushu* (《算数书》), "Writings on reckoning", an oldest Chinese collection of writings on mathematics, in the Western Han Dynasty in 186 BC The algorithm of "reducing to same denominator" of fraction division was introduced in nine chapters, *Jiuzhang Suanshu* (Nine chapters on the mathematical arts, 《九章算术》), which played in China a role similar to that of the Elements of Euclid (fl. c. 300 BC) in Europe. Besides the two algorithms above, the algorithm called "numerator dividing numerator and denominator dividing denominator" was introduced in *Shuli Jingyun* (《数理精蕴》), a 300-year-old classic encyclopedia on mathematics in the early Qing Dynasty which integrated Chinese and western mathematics. The underlying principles embodied in these algorithms were researched. In light of our understanding, the historical development in fraction division might enable us to apprehend the algorithm and its rationale of fraction division in a deeper way.

Does calculation of fraction division mean the procedure of "flip and multiply"? This paper discusses evidence found in the oldest Chinese mathematical literature handed down by the written tradition, to answer this question. Author's goal was to reconstruct the oldest layer of mathematical algorithms of fraction division in the treatise.

1. Algorithm of division of fractions in *Suanshushu*

Suanshushu was discovered in 1983 by Chinese archaeologists in a tomb dating back to the

---

[*] Acknowledge: This paper is the extension version of "Algorithm and Principles of Division of Fractions in Chinese Ancient Literature" presented in the conference "The 12th International Conference on the History of Science in China", Beijing: the Chinese Society for the History of Science and Technology (CSHST). This research was funded by the Education and Youth Bureau, Macau, China (DSEJ Cativo 12 467). The opinions expressed in the article are those of the author.

Han Dynasty at Zhangjiashan in Hubei Province. The book, made from bamboo strips, was identified by scholars to be the oldest book on mathematics. It was written from 202 BC to 186 BC, or more than 300 years earlier than *Jiuzhang Suanshu*.

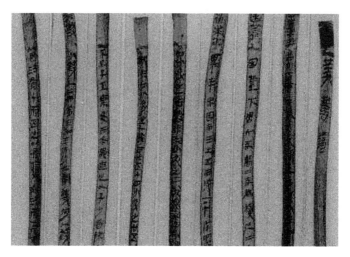

Fig. 1 Some of the bamboo strips on which the *Suanshushu* was written
Counting from the right, the first strip shows the label *Suanshushu*,
"Writings on Reckoning", that described the contents of the original bundle

The problem with the title of "Qicong" (启从) (the Calculation of Width) is that of used the algorithm of fraction division in which the width is to be calculated given the area and length. Whole passage is as follows (Jiangling, 2000)①:

    The calculation the width: given the length of 3/7 step (Bu) and area of 2/4 square step. The result is 1 and 1/6 step. The calculation method: Multiply the numerator of the length by the denominator of the area as the denominator of width, then the numerator of the area by the denominator of length as the numerator of width. When the numbers [on the sides of] the numerator and denominator are equal to one another, then you can replace by 1.

$$Width = \frac{2}{4} \div \frac{3}{7} = \frac{2}{4} \times \frac{7}{3} = \frac{14}{12} = 1\frac{1}{6}$$

The problem directly illustrated the algorithm of "flip and multiply" of fraction division, which broadly used in modern mathematics textbooks. Although it was not written clearly, a principle dealing with fraction division is indicated in the problem with the title of "Jingfen" (径分) chapter, division, in which the fraction is divided by the integer, whole passage is as follows (see Jiangling, 2000)②:

---

    ① 启从。广7分步之3，求田4分步之2，为从（纵）几何？其从（纵）1步6分步之1。求从（纵）术：广分子乘积分母为法，积分子乘广分母为实，实如法一。
    ② 径分，径分以一人命其实，故曰：五人分三、有（又）半、少半，各受卅分廿三。其术曰：下有少半，以一为六，以半为三，以少半为二，并之为廿三，即直（置）人数，因而六之以命其实。有（又）曰，术曰：下有半，因而倍之；下有三分，因而三之；下有四分，因而四之。

Algorithm and Principles of Division of Fractions in Chinese Ancient Literature

"Jingfen", division, refers to the average distribution. For instance, 5 persons shared 3, 1/2 and 1/3. Each one got 23/30. The solution is, if in the lowest [divisor] there is 1/3; take 1 as 6; a half as 3; 1/3 as 2; joining them; 23 to be the numerator; multiply 6 by the number of persons; 30 to be the denominator. The importance of the solution is, in the lowest [place] [divisor] there is 1/2; multiplying 2; there is 1/3; multiplying 3; there is 1/4; multiplying 4.

The detailed procedure is as follows:

$$\left(3 + \frac{1}{2} + \frac{1}{3}\right) \div 5 = (18 + 3 + 2) \div (5 \times 6) = \frac{23}{30}.$$

According to "Jingfen" method mentioned above, emphasis of the solution was the fraction of the divisor or the dividend should be changed into the integer (in the lowest [place] [divisor] there is 1/2; multiplying 2; there is 1/3; multiplying 3; there is 1/4; multiplying 4, namely "deleting denominator", which possibly indicates an underlying principle to solve problems of fraction, the divisor and dividend become integers after being multiplied by the same number and the quotient is not changed, the characteristics of division in modern mathematical theory.

Based on this thinking direction of "deleting denominator", it is easier to guess the omitted procedure in the problem of fraction division in "Qicong" (the Calculation of Width). To delete the denominator, the divisor 3/7 and dividend 2/4 should both multiply with 28, which was totally omitted, the results of which are 12 and 14 independently, shown as $\frac{2}{4} \div \frac{3}{7} = \left(28 \times \frac{2}{4}\right) \div \left(28 \times \frac{7}{3}\right) = (2 \times 7) \div (4 \times 3)$. This thinking direction of "deleting denominator" could provide a clue to understand why the "flip and multiply" algorithm works in a broad sense, namely, divisor and dividend are both multiplied by the denominator of the divisor so it can be removed as follows.

$$number \div \frac{numerator}{deno\min ator} = number \times deno\min tor$$

$$\div \left(\frac{numerator}{deno\min ator} \times deno\min tor\right) = number \times \frac{deno\min ator}{numerator}$$

Here, the direction of "deleting denominator" could provide a possible guidance to guess the underlying rationale of "flip and multiply" algorithm, the divisor and dividend are both multiplied by the same number, and the quotient remains unchanged. This should be considered as a powerful tool to explain the principle of fraction division, rarely be pointed out in modern mathematics curriculum.

One of reasons that the principle was not demonstrated in texts possibly is the medium in ancient China, means of rod numerals, through which the ideas had evolved, was not in a written form①. Another reason is that of ancient Chinese mathematics focused on, not rationale, but its broad application.

---

① The explanations given above by pen and paper, recording the fractions at each step, are more readily understood. It is reasonable to know the ancients did it easier with the counting rods.

Does historical material offer us insight into the explanation? Although the calculating tool is different, we found that the method of "flip division" of today is as valid as those used by the ancients. It is rarely researched in modern teaching the direction of "deleting denominator" could provide a possible explanation for the underlying rationale of "flip and multiply" algorithm.

2. Algorithms of division of fraction in *Jiuzhang Suanshu*

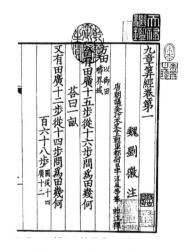

Fig. 2   The first page of *Jiuzhang Suanshu* from the South Song Dynasty

Jiuzhang Suanshu[①], organized fraction knowledge in a systemic manner, including reduction of a fraction, reduction of several fractions to a common denominator, and principles on addition, subtraction, multiplication, and division. These were considered to be the most systematic and soundest theories in the world during the period. It is impressive that *Jiuzhang Suanshu* (See Bai, 1983) presents the algorithm "reducing to same denominator" of fraction division in the problem with the title of "Jingfen" in the chapter of "fangtian" (the Calculation of area). Whole passage is as follows[②]:

"Jingfen", division, refers to the average distribution. For instance, how much will each person obtain by dividing $6\frac{1}{3}$ and $\frac{3}{4}$ yuan by $3\frac{1}{3}$ persons? The result is 2 and 1/8. The calculation method of division: the number of persons should be regarded as the denominator and the amount of money as the numerator. When the numbers [on the sides of] the numerator and denominator are equal to one another, then you can replace by 1. The reduction of fraction to a common denominator is made until

---

① *Jiuzhang Suanshu* written in the first century B. C. is the most important Chinese classical book on mathematics. It integrated mathematical knowledge from the Qin Dynasty to the Western Han Dynasty and manifested the highest level of Chinese mathematics at that time.

② 经分，又有三人，三分人之一，分六钱三分钱之一，四分钱之三。问人得几何？答曰：人得二钱、八分钱之一。经分术曰：以人数为法，钱数为实，实如法而一。有分者通之，重有分者同而通之。

there are no more fractions.

$$\left(6\frac{1}{3}+\frac{3}{4}\right) \div 3\frac{1}{3} = \left(\frac{19}{3}+\frac{3}{4}\right) \div \frac{10}{3} = \frac{85}{12} \div \frac{40}{12} = \frac{85}{40} = 2\frac{1}{8}$$

We note that the Jingfen method above demonstrates the algorithm "reducing to same denominator" of fraction division, in which the fraction unit of divisor and dividend is the same; the quotient is obtained by dividing the two numerators. That is to say, a fraction is divided by another fraction with same denominator to obtain the quotient (same unit of fractions is reduced). For different denominators, it is necessary for it to be transformed into the same denominator.

Liu Hui further explained the principles of algorithm "reducing to same denominator". Namely, for different denominators, it is to be made uniform by the same denominator" so that comparison of its inclusion relation could be conducted if a number is divided into various parts with the same fraction unit. It is in accordance with the addition-subtraction of fraction. This is similar to the principles of integer division; comparison of its inclusion relation could be conducted if a number is divided into various parts with the same unit. Clearly, this principle can be a powerful tool in mathematics teaching to explain the rationale behind the algorithm, which rarely known in the modern curriculum.

Besides, Liu Hui (the commentator of *Jiuzhang Suanshu*) added a new solution, namely, flip multiplication appeared early in the *Suanshushu* (See Bai, 1983)[①]:

Denominator of dividend multiplies the numerator of divisor. Denominator of divisor multiply the numerator of dividend, which is that of flip multiplication by as the denominator and numerator of divisor were reversed to be multiplied by its reciprocal.

$$\left(6\frac{1}{3}+\frac{3}{4}\right) \div 3\frac{1}{3} = \left(\frac{19}{3}+\frac{3}{4}\right) \div \frac{10}{3} = \frac{85}{12} \div \frac{10}{3} = \frac{85}{12} \times \frac{3}{10} = 2\frac{1}{8}$$

Obviously, the algorithm of "flip division" is more convenient to calculate results than that of "reducing to same denominator". It implies one of the possible reasons that the algorithm of "flip and multiply" of fraction division is much more popular than that of "reducing to same denominator" in china and in the world.

3. Algorithms of division of fraction in *Shuli Jingyun*

Shuli Jingyun (*Collected Basic Principles of Mathematics*, 1713-1722) is a 300-year-old classic encyclopedia on mathematics in the early Qing Dynasty which systematically integrated Chinese mathematics and western mathematics, e. g. elements. It is also a book of primary mathematics with abundant content, comprised of two parts and four kinds of mathematics tables. This book was nominated by Kangxi (an emperor in the Qing Dynasty), and was popular during its time. Due to its widespread use, the peak of mathematical research was

---

① 又以法分母乘实，实分母乘法。此谓法实俱有分，故令分母各乘全分内子，故令分母各乘内子，又令分母互乘上下。

formed during the Qian Jia Period (1736-1820), thus providing an important basis for summary of Chinese mathematical knowledge and the spread of western science and technology to China in the late Ming Dynasty.

It is interesting to note that *Shuli Jingyun* ① (See Qing, 1893) presented three algorithms of fractions division in the chapter of fraction by one problem and several application problems with the title of "division" (除法) (the division of fractions) in the unit of fraction. Whole passage is as follows (Fig. 3)②:

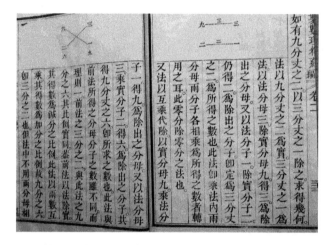

Fig. 3 Content of fraction division in *Shuli Jingyun*

How much is it, if dividing 2/9 zhang (1 zhang = 3 and 1/3 meters) by 1/3 zhang?

2/9 is deemed as the divisor with a dividend of 1/3. The denominator of quotient 3 is derived by dividing denominator 9 by 3, and numerator of quotient 2 is derived by dividing numerator 2 by 1. This is in accordance with the multiplication wherein two numerators and denominators are multiplied with each other independently to obtain the product shown as $\frac{2}{9} \div \frac{1}{3} = \frac{2 \div 1}{9 \div 3} = \frac{2}{3}$.

Another method is flip multiplication which may replace division. For example, the denominator of quotient 9 is the result of the multiplication between denominator 9 and numerator 1. Moreover, the numerator of denominator 6 is obtained from the multiplication between denominator 3 and numerator 2. Finally, 6/9 is obtained. The detailed procedure is indicated as:

---

① The first part includes five rolls; the second part has forty rolls, while four kinds of math tables are contained in eight rolls. The first part is referred to as the *Ligang Mingti*, included *The elements* and *Suanfa Yuanben*, while the second part contains knowledge on areas including arithmetic, algebra, geometry, and trigonometric function on primary mathematics.

② 设如有九分丈之二，以三分丈之一除之，求得几何？
法以九分丈之二为实，三分丈之一为法，以法分母三除，实分母九得三为除出之分母，又以法分子一，除实分子二，仍得二，为除出之分子，即定为三分丈之二，为所得之数也。此法即乘法内两分母两分子各自相乘，为所得之数者转用之耳。此零分除零分之法也。
又法以互乘代除，以实分母九，乘法分母一得九，为除出之分母，又以法分母三乘实分子二，得六为除出之分子，共得九丈之六，即所得之数也。
欲明晰其故，则以两分母九与三，互乘得贰拾七，……是将三分之一变为贰拾七分九，将九分之二变为贰拾七分六，其两分母即等，则其两分子自成比例，故九分之六．以三约之，（岂）非三分之二耶。

Algorithm and Principles of Division of Fractions in Chinese Ancient Literature

$$\frac{2}{9} \div \frac{1}{3} = \frac{2}{9} \times \frac{3}{1} = \frac{2 \times 3}{9 \times 1} = \frac{6}{9} = \frac{2}{3}$$

The reason is: 27 is the multiplication result of the two denominators, namely, 9 and 3. In other words, 1/3 is changed into 9/27, and 2/9 into 6/27, then the two denominators are the same and the two numerators are in proportion. The final result is 6/9, which equals 2/3 after reduction by 3. The procedure is described below.

$$\frac{2}{9} \div \frac{1}{3} = \frac{2 \times 3}{9 \times 3} \div \frac{1 \times 9}{3 \times 9} = \frac{6}{9} = \frac{2}{3}$$

Directly *Shuli Jingyun* illustrated the algorithm of "numerator dividing numerator and denominator dividing denominator, the algorithm of "flip and multiply", and the algorithm of "reducing to same denominator". The first algorithm of "numerator dividing numerator and denominator dividing denominator" is introduced there, which is to say, the numerator of the quotient obtained by dividing between the two numerators, and the denominator of the quotient is the result of the division between the two denominators. This is similar with the method in fraction multiplication where the numerator of product is the multiplication result of two numerators and the denominator of the product is the multiplication result between two denominators. The second algorithm introduced was that of "flip and multiply". It is interesting to note that "reductions of fractions to the same denominator" was also regarded as that the only algorithm illustrating the principle of fraction division, namely, two numerators are in proportion (the same proportionate fraction unit is deleted). In this situation, the quotient is obtained by dividing between two numerators. Apparently, this explanation should be a useful tool for teaching to illustrate its rationale.

Although all three algorithms were introduced in *Shuli Jingyun*, the principles behind three algorithms were neglected because of, not theoretical orientation, but practical orientation in ancient Chinese mathematical tradition. For example, three algorithms were directly illustrated in *Shuli Jingyun* without a general proof of equivalence of three algorithms with the same result in the reductions of fractions to a same denominator, $\frac{b}{a} \div \frac{d}{c} = \frac{bc}{ac} \div \frac{da}{ac} = bc \div ad = \frac{bc}{ad}$, in flip multiplication $\frac{b}{a} \div \frac{d}{c} = \frac{b}{a} \times \frac{c}{d} = \frac{bc}{ad}$, and in the division between the numerator and denominator, $\frac{b}{a} \div \frac{d}{c} = \frac{b \div d}{a \div c} = \frac{\frac{b}{d}}{\frac{a}{c}} = \frac{\frac{b}{d} \times \frac{c}{a}}{\frac{a}{c} \times \frac{c}{a}} = \frac{bc}{ad}$ in several application problems of three algorithms of fractions division.

4. Discussion and conclusion

It deserves to note that the algorithm of "flip and multiply" fundamentally, as valid today using Hindu-Arabic numerals as they were some 2000 years ago using rod numerals, was presented and employed in any modern mathematics textbook now in the world. For example, the six US textbooks (Bolster, Boyer, Butts & Cavanagh, 1996; Abels, Wijers, & Pligge, 2008; Collins, Howard, Drisaa, McClain, Frey, et al., 1999; Education Development Center, 2005; Lappan, Fey, Fizgerald, Friel & Phillips, 2008; Rose; Tourneau, Catherine; Burrows, Ford, 1996) provided "flip-and-multiply" algorithms, but did not explain why the procedure works (Li, 2009). These results are related with the studies repeatedly showed that American pre-service and in-service teachers experienced difficulty in making a conceptual explanation for the underlying principle and meaning of "flip and multiply" (e.g., Ball, 1990; Ma, 1999; Li, 2008; Li & Kulm, 2008)[①]. A typical comment was: "I know this rule perfectly well, but I don't know why." Their understanding consisted of remembering a particular rule, unattached to other ideas on division. Although the four Chinese textbook (Math Textbook Developer Group for Elementary School, 2005, Curriculum and Textbook Research Institute-Elementary Textbook Development Committee [CTRI-ETDC], 2006; Sun, Wang, Lin, Chen & Li, et al., 2007; Xia, Li, Chen, Li & Liang, 2004), and three Japanese textbooks (Kazumatu, Okata, Machita, Aoyanagi & Kiyozawa, et al., 2007; Nakahara, Azemori, Sakitani, Hirabayashi & Ida, et al., 2007; Sugiyama, Itaka, Ito, Aoyanagi & Nakano, et al., 2007) have provided a conceptual explanation that division is equal to the concept of the multiplication divisor's inverse based on concept of reciprocal, the explanation are too abstract to easily teach or learn. These situations make fraction division have become a difficult teaching point for mathematics teachers in primary schools in the world. Looking back to analysis above, the idea of "deleting denominator" in *Suanshushu*, which the fraction is transformed into an integer based on its function through the unification of the denominator, have provided a conceptual guidance for explaining "flip and multiply" practice.

The historical development in fraction division might enable us to understand the algorithms and their rationale of fraction division in a deeper way. In fact, the algorithm of "numerator dividing numerator and denominator dividing denominator" and that of "reductions of fractions to the same denominator" are scarcely known. Therefore, for most of teachers and students, calculation of fraction division means the procedure of "flip and multiply". It is worthy to note that "reducing to same denominator" is an effective strategy to transform fraction division into whole-number division, if necessary by first changing dividend and divisor to have the same denominator. This algorithm makes up some weakness of "flip and multiply", which separates experiences from that of whole-number division and that of fraction computation. It is consistent with that of whole-number division and that of fraction addition and subtraction due to the same procedure of "reducing to same denominator"; It allows easier understanding of the meanings of operations due to the use of inclusion concept same as that of whole-number division; It allows easier recognition of the result size; It allows easy judging of the rationality of the algorithm. It

---

① For example, Ball (1990) reported that few American prospective teachers were able to give a mathematical explanation of underlying principle and meaning.

shows some advantages in making a conceptual "bridge" to the underlying principle of "flip and multiply"[①]. The discussion above shows that historical research can be a powerful tool for mathematics teaching, which gives a deeper understanding of knowledge than that which is provided by textbook knowledge alone. Analysis the oldest layer of mathematical algorithms could suggest us to create an adequate repertoire, which would be helpful to understand this difficult topic in a broader sense.

References

[1] Abels M, Wijers M, Pligge M. Revisiting numbers. In: Wisconsin Center for Educational Research & Freudenthal Institute Ed. Mathematics in context. Chicago: Encyclopedia Britannica, Inc, 2008

[2] Ball D L. Prospective elementary and secondary teachers' understanding of division. Journal for Research in Mathematics Education, 1990, 21: 132-144

[3] Bolster L C, Boyer C, Butts T et al. Exploring Mathematics (Grade 7-9). Glenview, IL: Scott, Foresman, 1996

[4] Collins W, Howard A C, Drisaa L et al. Mathematics: Applications and Connections. Course 1. New York: McGraw-Hill Glencoe, 1999

[5] Education Development Center, the Seeing and Thinking Mathematics Project. MathScape: Course 1. New York: McGraw- Hill Glencoe, 2005

[6] Kazumatu S, Okata I, Machita S et al. Minna to Manabu Shogakkou Sansuu 6nen (gei) (Everybody learning: Elementary arithmetic, sixth grade, 2). Tokyo: Gakkou Tosho Kabushiki Kaisha, 2007

[7] Lappan G, Fey J T, Fizgerald W M et al. Bits and pieces 2: Using fraction operations. In: Michigan State University ed. Connected Mathematics Project. Boston: Person Prentice Hall, 2008

[8] Li Y. What do students need to learn about division of fractions? Mathematics Teaching in the Middle School, 2008, 13: 546-552

[9] Li Y, Kulm G. Knowledge and confidence of pre-service mathematics teachers: The case of fraction division. ZDM—The International Journal on Mathematics Education, this special issue, 2008

[10] Ma L. Knowing and Teaching Elementary Mathematics: Teachers' Understanding of Fundamental Mathematics in China and the United States. Mahwah, New Jersey: Lawrence Erlbaum Associates, Inc, 1999

[11] Nakahara T, Azemori T, Sakitani S et al. Shogakkou Sansuu 6nen (gei) (Elementary Arithmetic, sixth grade, 2). Osaka: Osaka Shoseki Kabushiki Kaisha, 2007

[12] Rose A, Tourneau M, Catherine D et al. New Progress in Mathematics. New York: William H Sadlier, 1996

[13] Sugiyama Y, Itaka S, Ito S et al. Shinhen, Atarashi Sansuu 6 (New arithmetic) 6 (1). Tokyo: Tokyo Shoseki Kabushiki Kaisha, 2007

[14] Sun X H. Curriculum development based on mathematics history: an experiment of novel curriculum design around the division of fractions. In: Numeracy: Historical, Philosophical and Educational Perspectives. Academic conference. Oxford: University of Oxford, 2009

---

① We developed a new curriculum using this strategy and implemented it in twenty-seven classes in six primary schools. Significant positive results in conceptual understanding were obtained (Sun, 2007; 2009).

[15] Sun X H. Spiral Variation (Bianshi) Curriculum Design in Mathematics: Theory and Practice. Field of Mathematics Curriculum. Hong Kong: The Chinese University of Hong Kong, 2007
[16] 小学数学课程教材研究开发中心. 数学. 北京：人民教育出版社, 2005
[17] 孙丽, 王林等. 数学六年级上册. 南京：江苏教育出版社, 2007
[18] 夏明华等. 数学第十一册. 杭州：浙江教育出版社, 2004
[19] 江陵. 江陵张家山汉简（算数书释文）. 文物, 2000, 9: 78-84
[20] 〔清〕乾隆. 御制数理精蕴（清刻本）. 上海：江南制造总局, 1893, 336-356
[21] 白尚恕. 九章算术注释. 北京：科学出版社, 1983

An Exploration of the Original Sources of *Lvlv Zuanyao*

Wang Bing

(The Institute for the History of Natural Science, CAS)

This article is intended to explore the sources of the contents of the work *Lvlv Zuanyao* (《律吕纂要》). The author considers that some contents in this unpublished work originated from the book *Musurgia Universalis* (Rome, 1650) of Athanasius Kircher (1601-1680).

1. The work *Lvlv Zuanyao*

It had been concluded [1] that the western music scores in modern times were mainly constituted by essential factors of stave, clef, note, signature, bar line, accidental and tempo; while the seven essential factors originated from the Middle Ages. As for traditional theory of music, there were completely different systems in Europe and China.

Researches have pointed out that the work *Lvlv Zuanyao* (abbr. to *LLZY* below) was the first work which introduced the European musical theory into China. The work was never published, but its manuscript and transcripts remain up to now. Its author should be the Portuguese Jesuit Thomas Pereira (1645-1708). And it should be completed before 1707, probably in 1680's. Also researches have indicated that work *LLZY* firstly introduced into China the elementary knowledge of musical theory in Europe, which dated from the late period of the Middle Ages to the 17th century. The elementary knowledge was about notation, tone, interval, scale, metre, rhythm, tempo and so on. [2]

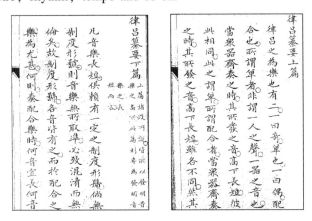

Fig. 1 Photographs of *Lvlv Zuanyao*

2. Athanasius Kircher and his *Musurgia Universalis*

The German Jesuit scholar Athanasius Kircher had high reputation for his great learning and great ability. He studied the humanities and natural sciences when he was young, receiving various disciplines of the 17th century, which contained mathematics, philosophy and theology. Kircher was ordained a priest in 1628. In the beginning of 1630s he left Germany. Afterwards he was appointed some academic positions at Avignon and Vienna, and was engaged in his studies in different fields. He settled down at Rome in 1634.

Fig. 2 Photograph of *Musurgia Universalis* (Rome, 1650)

Kircher's works were very abundant. [①] His studies covered practically many fields both in natural sciences and the humanities. While these studies were carried out using the methods from traditional academicism up to courageous experiments.

As a mathematician and universal scholar, however, Kircher made great contributions towards the dissemination of knowledge. In this aspect he devoted himself to various activities, so that he was in most important position at that time. He widely contacted well-known scientists and important persons of European countries. He collected materials not only about the development of sciences and culture in Europe, but also about the activities of Jesuit missionaries all over the world. Wide social connections made Kircher become a central character

---

① Some 44 books and more than 2000 extant letters and manuscripts.

of the information exchange in the fields of science and culture in his times. [3]

Athanasius Kircher's work *Musurgia Universalis* (abbr. to *MU* below) was published in Roma in 1650. ① Even though it is not very important among his decades works, the contents of this book is still all-inclusive and extremely plentiful. It comprises the knowledge about anatomy and physiology related to sound production and sense of hearing, the knowledge about mathematics and physics related to music and pitch, the knowledge about music and musical instruments in Europe (as well as in the areas of Central Asia and Arab) in ancient and modern times (i. e., to his times), the knowledge about philology and phonology, and so on.

The book *MU* is in two volumes and in ten libers. The first volume has the first seven libers, 690 pages in total; and the second volume has the last three libers, 462 pages in total. There are preface before the text; and the contents, indexes and errata of the book following the text.

3. Music works brought to China by Jesuits

Several works on music were brought to China by Jesuit missionaries. In the book *Catalogue de la Bibliotheque du Pe-Tang* (Peking, 1949) by H. Verhaeren, it was listed 7 kinds of musical books published in Europe during the 16th and 17th centuries, and their amount was 12 copies in total. Among them two were of great importance. One was *Institutioni Harmoniche* (Veneza, 1558) by the Italian musician Gioseffo Zarlino (1517-1590). [4] Another was *Musurgia Universalis* by the German well-known scholar Athanasius Kircher. [5]

Among Jesuits who came to China, the Italian Martin Martini (1614-1661) and the Belgian Ferdinand Verbiest (1623-1688) were associated with Athanasius Kircher. The former had followed Kircher to study mathematics at Roma in his early years. The latter studied theology at Roma in 1652-1653, it was at the celebrated Roman College of the Jesuits that he got acquainted with Kircher and his assistant Gaspar Schottus. [6] Although Ferdinand Verbiest left Roma later, he maintained his contacts with Kircher there, and asking him in the post-scriptum to send over some of his recent works. [7]

Athanasius Kircher's book *MU* — a gigantic compilation of contemporary European musicology — was brought to China by Jesuits in 1650s, and acquired in Beijing. On April 4, 1657, Martin Martini sailed from Lisbon for his second journey to China, travelling together with other Jesuit missionaries. Ferdinand Verbiest and Albert D'orville (1622-1662) were two among them. The latter carried 12 copies of Kircher's *MU* for China. [8] And we know that the Portuguese Jesuit Gabriel de Magalhaens (1610-1677) received a copy from his brother Manoel de Magalhaens. Later it was in the Pei-t'ang Library. [9] At present it is still preserved in the Rare Book Section, Chinese National Library in Beijing.

The association between Athanasius Kircher and the Jesuits who came to China and the

---

① It has a reprint in Hildesheim, Germany, in 1970.

works of Athanasius Kircher brought to China by them, probably helped to bring about the introduction of some contents in his works into China.

4. Some illustrations of the contents in *LLZY* originated from *MU*

In principle, the contents of the work *LLZY* drew some materials from Libers III, IV, V, VI, VII and VIII of the book *MU*. Here the paper gives some brief comparisons between *LLZY*① and *MU*②, based on the sequence of contents in the former.

In *Shangpian* (上篇) of *LLZY* the theory about pitch of Hexachord System is explained.

(1) Basic ideas of stave, and symbols expressed pitch (In I, 1-4)

In *LLZY*, at first, Thomas Pereira described five-line staff of European notation. This form of notation was being perfected in the 17th century. It was completely different from the traditional Chinese notation, such as Gongchipu (工尺谱), Jianzipu (减字谱) and so on. The Sections 1-4 related that recording music should use stave, the (treble, alto and bass) clefs on the left of stave determined tone positions, human voice could be divided into four parts (soprano, tenor, alto and bass), and sharp and flat expressed variety of the tones higher and lower. These could be found corresponding contents in *MU*.

(2) Musical tones, pitch of tones in Hexachord, and interval and harmonic intervals (In I, 5-8)

The establishment of the Hexachord System was attributed to the Italian musical theoretician Guido d'Arezzo (ca. 995-1050). He chose ut, re, mi, fa, sol, la as names of six tones of the hexachord, which derived from the first syllable of the first six sentences of the Latin hymn of St. John. [10]

In the work *LLZY*, the six tones were called as "musical tones". Their names were transliterated as wu (乌), le (勒), ming (鸣), fa (乏), shuo (朔), la (拉) (Fig. 3).

In Liber III of the book *MU*, the names of the six tones was described in Chap. VIII (A, 114) (Fig. 4, Fig. 5); and conjunct and disjunct of tones was described in Chap. V (A, 130-132).

In the late Middle Ages, in Europe, the interval construction in the hexachord scale, which consists of the six tones mentioned above, was determined 1-1-1/2-1-1. That means only one semitone appears in the middle, i. e., between mi and fa. While whole tones appear in other steps between two successive tones. In the book *MU*, a lot of concepts, such as whole tone and semitone, major tone and minor tone, major semitone and minor semitone, minimum

---

① For the contents in *LLZY*, in the following of this article, I and II represent its *Shangpian* (上篇) and *Xiapian* (下篇) separately, and the following number represents the Section(s) of Jie (节). Besides, the author of this article chooses clear copies of paragraphs and figures from different versions of *LLZY*.

② For the contents in *MU*, in the following of this article, A and B represent its first and second volume separately, and the following number represents the page(s) in the volume.

補立形號以曉示樂工
然非嫻於律呂精於作樂之人斷不能窺測
樂之音皆見疊出使之和美絕倫也剛樂形號
如下 □ 者象剛樂强毅之性而方其形也柔樂形
號如下 6 者象柔樂流動之性而圓其形也
樂音說
樂音者樂之種種高下粗細之音節也奏樂時無論
起於何音或自下上讀或自上下讀其自然連貫音
止有八音第八音者隔八相生決定與首音合又九
音之合於第二音十音之合於第三音亦如之凡樂
之樂調皆括於八音故音有八音雖有八然音之名
止於六首曰烏次曰勒三曰鳴四曰乏五曰朔六曰
拉讀此六音時自下而上讀烏字起上至勒鳴乏朔拉
云
拉朔乏鳴勒烏
此六音內自烏上起至勒爲一全音勒上起者同烏
亦爲全音鳴上起者同勒亦爲全音讀至乏音則不
然矣乏之起於鳴音乃爲一半自乏至朔上起亦爲
一全音拉上起者同朔亦爲全音故六音內謂之爲
半音
作樂用爲勒二音慮代以勒鳴二音或乏朔二音
或朔拉二音俱無不可用鳴乏二音時以別二音代
之則不可即用別二音時亦不可代以鳴乏二音共
可代用與不可代用者何以鳴乏二音之相爲上下

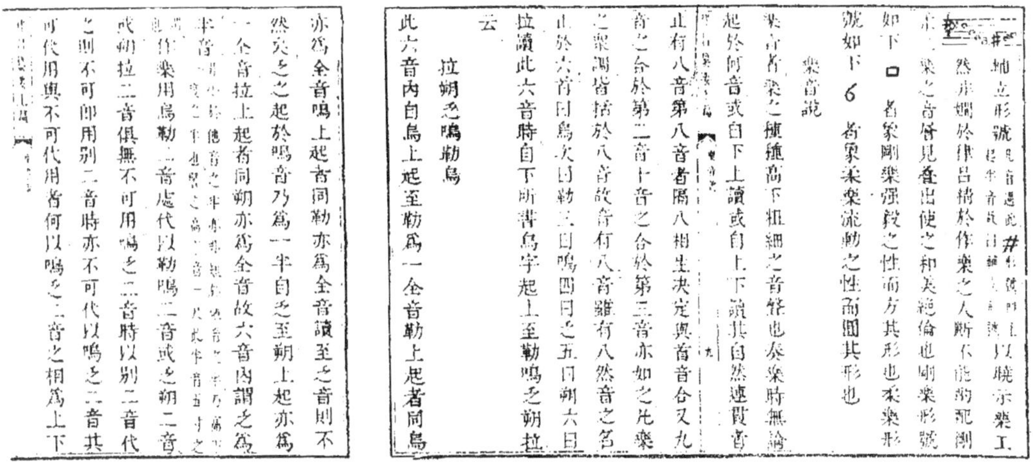

Fig. 3 "On musical tones" (*LLZY*-I)

Guido voces notarũ musicaliũ inuenit. Porrò Guido Aretinus eximius sui temporis Musicus, vt Musicam cultiorem, & vocibus aptiorem redderet, vocũ quædã signa inuenit, quas partim figuris (quas nunc Notas vocant) partim syllabarũ appellatione expressit; ex quibus sex sunt in vsu. *vt, re, mi, fa, sol, la.* ex hymni ecclesiastici capitalibus syllabis desumptæ, vt sequitur.

Vt queant laxis.
Re sonare fibris.
Mi ra gestorum.
Fa muli tuorum.
Sol ue polluti.
La bij reatum.
O Pater alme.

Quas

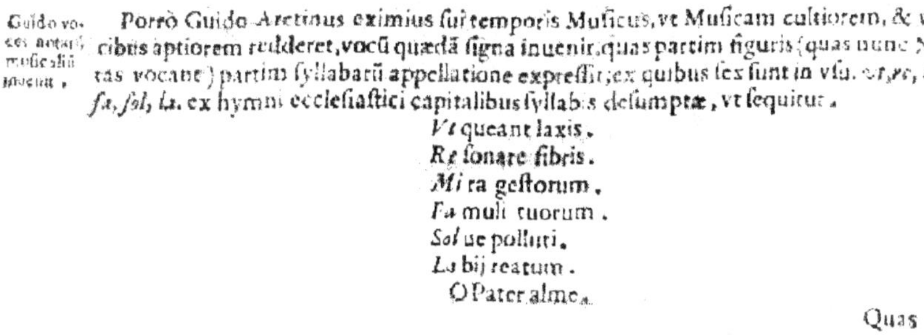

Fig. 4 Musical tones (*MU*-A-114)

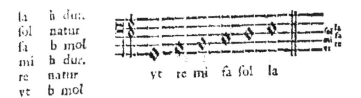

Fig. 5 Musical tones (*MU*-A-127)

comma, were explained in Chap. V (A, 95ff.) and Chap. VI (A, 101ff.) of Liber III. The schemata about interval were showed in Chap. IX (A, 125) of Liber III and in Chap. I (A, 557[①]) of Part II of Liber VII. (Fig. 6, Fig. 7) And in fact, the problems about interval and harmonic interval were explained in detail at many places in Liber III and Liber IV of *MU*.

(3) *Yuemingxu* (乐名序), the change of the tone name caused by semitone, and transposition of the hexachord (In I, 9-13)

① In this book it mistakes page 557 for page 617.

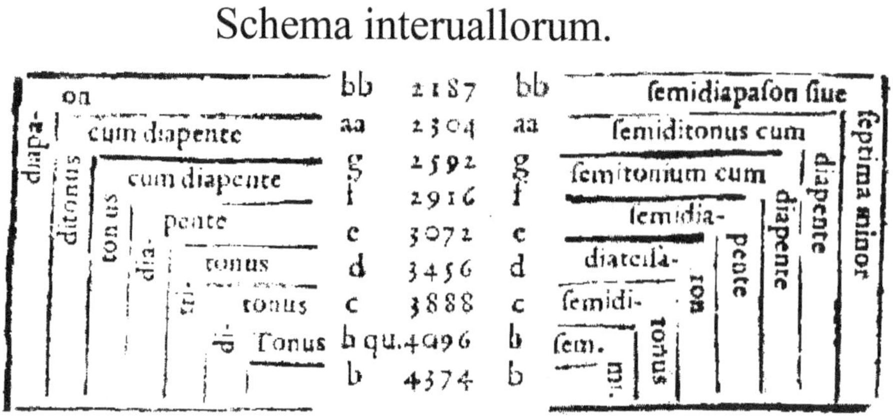

Schema interuallorum.

Fig. 6 Interval (*MU*-A-125)

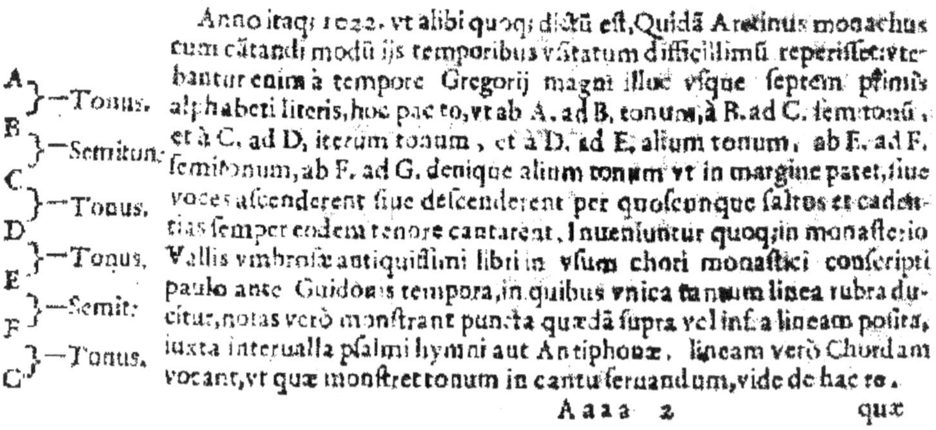

Fig. 7 Interval (*MU*-A-557)

In the *Shangpian* of *LLZY*, the transposition of the hexachord was explained (Fig. 8). This problem arises from the semitone exiting in the middle of the hexachord, that is, it must be semitone between the third and fourth tones. In result, three kinds of hexachords generates: "hexachordum naturale" starting from C, "hexachordum durum" starting from G (and with sharp #) and "hexachordum molle" starting from F (and with flat b).[11] In great stave, a series of overlapping hexachords form the whole compass. There are seven hexachords from G to EE (e^2) in total: G A B c d e, c d e f g a, f g a^b b c^1 d^1, g a$^\#$ b c^1 d^1 e^1, c^1 d^1 e^1 f^1 g^1 a^1, f^1 g^1 a$^1{}^b$ b^1 c^2 d^2, g^1 a$^1{}^\#$ b^1 c^2 d^2 e^2 (Fig. 9). It is evidently that the progression of music would be over the scope of a hexachord, so that the "modulation" (i.e., scale transposition) is needed for linking two or more than two hexachords. The two hexachords linked together would share one tone at least, but the syllable of the tone would be changed when the transposition takes place from one scale to another scale. This is exactly the situation of "changing the name" of the tone expounded in *LLZY*.

An Exploration of the Original Sources of *Lvlv Zuanyao*

Fig. 8 "Yuemingxu" (*LLZY*-I)

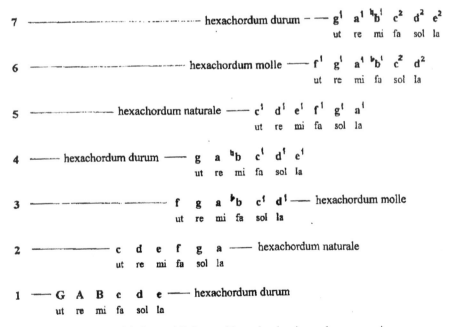

Fig. 9 Modulation and link up of hexachords, in modern expression

The modulation and the link up of hexachords were illustrated in A, 116 of the original book *MU* (Fig. 10). It could be seen clearly from middle and right parts of the illustration that seven hexachords overlapping each other form the whole compass from G (r) to e^2 (Ee). Among the tones, from G to f, one could read the tones with same pitch horizontally so that

· 89 ·

seven *Yumingxu* called in the work *LLZY* would be gotten immediately. They were sol — re — ut, la — mi — le, fa — mi, sol — fa — ut, la — sol — re, la — mi and fa — ut.

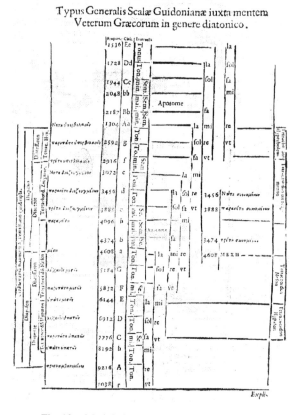

Fig. 10　Modulation and link up of hexachords
(*MU*-A-116)

These contents originated from Chap. VIII (A, 115ff.) of Liber III of *MU*. The so-called "Guidonian Hand" was illustrated in A, 115 of the original book (Fig. 11). It was invented by Guido d'Arezzo in order to easy to teach and memory the hexachord scale. This method took finger tips and joints of left hand to symbolize tones, which included all tones in the compass from G to e^2 in natural septemchord scale. The figure of the "Guidonian Hand" had a great vogue in Europe during the Renaissance and in a period afterwards.[12] The Section *Zhangzhong Yuemingxushuo* (掌中乐名序说) in *Shangpian* of *LLZY* introduced precisely the "Guidonian Hand" (Fig. 12). It should be emphasized here that this is the unique introduction in China about the knowledge of the "Guidonian Hand", because the contents of this Section was not compiled into the book *Lvlv Zhengyi Xubian* (《律吕正义·续编》) later.

In *Xiapian* (下篇) of *LLZY* the signs and theory about value of Mensural Notation are explained.

An Exploration of the Original Sources of *Lvlv Zuanyao*

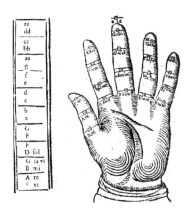

Fig. 11 "Guidonian Hand" (*MU*-A-115)

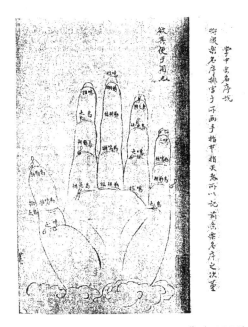

Fig. 12 "*On Zhangzhong Yuemingxu*" (*LLZY*-I)

(4) Notes, note values and their applications (In II, 1-5)

In the *Xiapian* of *LLZY*, at first, Thomas Pereira described that "signs of the eight forms" (八形号) were used to represent the values of tones. Their names were as follows: "the longest", "long", "short", "half of short", "little", "half of little", "fast" and "fastest", while their forms could be seen in Fig. 13. Among them, the tone represented by "half of short" corresponded to a whole note; and "little", "half of little", "fast" and "fastest" corresponded to a half note, quarter note, eight note and sixteenth note separately. In fact, the eight kinds of notes were those of "mono notes" in the mensural notation which had tended to regularization during the 16th century. They were as follows: Maxiam (abbr. Mx), Longa (L), Brevis (B), Semibrevis (S), Minima (M), Semiminima (Sm), Fusa

(F) and Semifusa (Sf) (Fig. 14).

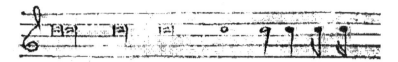

Fig. 13 Notes and rests (*LLZY*-II)

Fig. 14 Notes in mensural notation (*MU*-A-217)

Fig. 15 Types of pauses (*MU*-B-53)

These contents originated from the relation in Chap. IV (A, 216ff.) of Liber V of *MU*. If we compare the note forms both in *MU* and in *LLZY*, we know that the latter used round head in representing "half of short" and other small notes, while the former still used square head of the regular mensural notation in representing semi-brevis and other small notes (See the illustrations in A, 217 and B, 52). This is main difference in representing notes between the two books. Meanwhile in the original book by Kircher, the writing method of Minima and other small notes in stave was illustrated in A, 217, that is, the regulation of writing note stem. It was also stated and illustrated in *LLZY*.

The concepts about dotted note and rest (Fig. 13) were also introduced in *LLZY*. The similar forms (Fig. 15) could be found in Chap. II (B, 53) of Part III of Liber VIII in *MU*.

(5) Rhythm and tempo of music (In II, 6-10)

The work *LLZY* introduced to China the western concepts about rhythm and tempo of music. It explained that there were two kinds of division between notes in notation, that is, dividing them into two equal parts (平分度) or into three equal parts (三分度). In result the perfect time and imperfect time (common metre and divisional metre) produced. It related three kinds of metres in the case of two equal division, which were perfect metre "O", common metre "C" and divisional metre "₵", as well as two kinds of metres in the case of three equal division, which were common metre " ₵₃ " and divisional metre " ₵₃ " (Fig. 16).

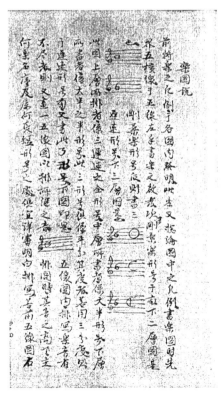

Fig. 16 On musical sheet (*LLZY*-II)

The mensural notation was generated and developed in Europe between the 13th and 17th centuries. Based on this notation, between notes, two kinds of division, "perfect" and "imperfect", were appeared. The "perfect" meant to divide them into three equal parts; and the "imperfect" meant to divide them into two equal parts. So, when two notes were in succession, four kinds of their combination produced. As a result, four kinds of metre generated, marking as the signs "⊙", "⊆", "O" and "⊂" separately. In fact, they indicated separately the metre of nine eighth notes (quaver), the metre of six eighth notes, the metre of three quarter notes (crotchet) and the metre of two quarter notes. The contents could be seen in Chap. X (A, 676) of Part III of Liber VII in *MU* (Fig. 17).

Fig. 17 "Tempo of music" (*MU*-A-676)

(6) Main points and methods of writing and reading music (In II, 11-13)

The last three sections of the work *LLZY* explained main points and methods of writing and reading music score. When one was writing music, he had to draw stave at first, then to indicate clef, sharp or flat, time signature and etc. at the beginning of the stave successively, afterwards to write notes on the stave. When one was reading music, he had to pay attention to clef, time signature, sharp or flat, *Yuemingxu*, the change of the tone name caused by semitone, and etc. successively. And at last the work gave examples of ascending and descending, conjunct and disjunct of the six tones in hexachord, to explain briefly the pitch and rhythm which notes indicated in the music score.

The contents of these sections hardly exact corresponded to that in the book *MU*. It could be considered as the summarization made by Thomas Pereira, the author of the work *LLZY*.

5. Concluding remarks

As indicated above, the contents of *MU* was all-inclusive and extremely plentiful. The abundant musical knowledge in this book was related to the knowledge in many fields from ancient to Kircher's times. It is obvious that in fact very small part of its contents were explained in the work *LLZY*.

On the other hand, it could be undoubtedly considered that the sources of the contents of the work *LLZY* did not limited to the book *MU*. However, Thomas Pereira probably took it as main reference when he compiled his work.

We could also take into account their orders of arranging the contents. As to the book *MU*, the order was: the general ideas of sound, voice and music—mathematical and physical knowledge related to music—theoretical problems about music and musical instruments—rhythm and metre of music—problems of generalized harmony. As to the work *LLZY*, the order was: elementary ideas and expressions of music and musical tones—theory of the pitches of Hexachord System—theory of the durations and their expressions. Although there were great differences of complication and simplicity between them, their orders of arranging the contents showed similar to a certain extent.

It is sure that now we could not deduce Thomas Pereira's arrangement of chapters and sections and his decision of choosing the contents, when he compiled the work *LLZY*. In any case, there were evident differences of detail and sketch among the sections of this work.

Besides, it could be aware that some contents in the work *LLZY* were changed to some extent, contrasting those in the book *MU*. For example, the heads of notes were expressed in round shape, instead of in square shape in regular mensural notation. It indicated that in Europe not only the Middle Ages was flourishing evolution period of the stave, but also the 17th century was repaid progress period of the knowledge of music. In any case, the European musical notation was recorded in the Chinese manuscript *LLZY* for the first time.

Acknowledgements

This research was carried out in the Centre for Asian Studies of the Secção Autónoma de Ciências Sociais, Juridicas e Politicas at the University of Aveiro, and supported by the Orient Foundation, Portugal, to which many thanks are due.

The author is grateful to all persons for their help, especially to Prof. Manuel Serrano Pinto for his concern and to Ms. Shao Xiaoling for her collecting materials and other aspects.

References and Notes

[1] 王光祈 (1891-1936). 东西乐制之研究. 上海: 中华书局, 1926. 228-232

[2] 王冰.《律吕纂要》之研究. 故宫博物院院刊, 2002, (4): 68-81; 徐日昇和西方音乐知识在中国的传播. 文化杂志, 2003, (47): 71-90

[3] Cf. Athanasius Kircher's biography wrote by Hans Kangro. See Dictionary of Scientific Biography. VII. New York: Charles Scribner's sons, 1973. 374-378

[4] Verhaeren H. Catalogue de la Bibliotheque du Pe-Tang. Peking. 1949. nr. 3542／1142

[5] Verhaeren H. ibid. nrs. 1921／743, 1922／744, 1923／745

[6] Cf. Golvers N. The ASTRONOMIA EUROPAEA of Ferdinand Verbiest, S. J. (Dilligen, 1687) — Text, Translation, Notes and Commentaries. Steyler Verlag. 1993. Introduction, 17

[7] Cf. Golvers N. ibid. Notes and Commentaries, (3): 135; (8): 136

[8] Golvers N. ibid. Notes and Commentaries, (8): 311, 136. See Fletcher J. Athanasius Kircher and the distribution of his books. The Library. 5th ser. 1969, (23): 108-117

[9] Verhaeren H. ibid. nr. 1922／744

[10] Cf. The International Cyclopedia of Music and Musicians. 11th ed., New York: Dood, Mead & Company, 1985. 883; Baker's Dictionary of Music. 383

[11] Cf. The International Cyclopedia of Music and Musicians. 883

[12] Cf. The International Cyclopedia of Music and Musicians. 883; 缪天瑞. 音乐百科词典. 北京: 人民音乐出版社, 1998. 227

On Delisle's Correspondence to and from China Through the Archives of the Paris Observatory

Suzanne DÉBARBAT

(SYRTE, Observatoire de Paris, CNRS et UPMC)

Delisle Joseph-Nicolas (1688-1768) was the ninth child of Claude Delisle and Charlotte Millet de La Croyère; he had an older brother Guillaume (1675-1726) known under Delisle l'aîné. They had a younger brother Louis (? -1741), known under the name of his mother de La Croyère not to be mixed with Joseph-Nicolas known as Delisle le cadet or le jeune. Their father was a geographer and all of them made carrier in this field. Guillaume established his fame on all the nine maps he made, being given, in 1718, the title of *Premier Géographe du Roi* at the time Louis XV was a boy being born in 1710. Joseph-Nicolas was also known as a geographer but he was mostly considered as an astronomer. Louis also an astronomer went to Russia with Joseph-Nicolas but he is better known as a traveller, going through Siberia to Aratcha, in Kamchatka were he died; he is buried there and it is said that his grave is still maintained in good state. All the three have been members of the French Académie Royale des Sciences created, in 1666, by Louis XIV and his Minister Colbert.

1. Short biography of J. -N. Delisle

After his studies at the Collège Mazarin known also as the Collège des Quatre-Nations, from the four provinces added to the French Kingdom in 1661, and from 1806 the seat of the Institut de France, composed of five academies and including the Académie Royale des Sciences. Delisle took interest in mathematics. Astronomy was in this field and he came frequently to the Observatoire de Paris, where Cassini (1625-1712) was installed from 1671. Not far from the Observatoire was a dome in the center of a palace built for Marie de Médicis, wife of Henri IV, around 1612 later offered to Louis XIV. The dome of this Palais du Luxembourg, from the name of a previous palace built there by a man of this name, was offered to Delisle and he installed an observatory where he could observe a lunar eclipse on 1712 January 23; but in 1715 he had to move to another place, coming back in Luxembourg a few years later. Meanwhile Delisle was nominated as an élève astronome in 1714 being adjoint in 1716, associé in 1719, associé vétéran in 1741 and pensionnaire vétéran in 1761, having during almost half a century been acquainted to the Académie; he published several memoirs in the

Mémoires of this academy which led him to a position at the Collège Royal nowadays Collège de France.

When Peter the Great made his Grand Tour in occidental Europ, being in 1717 at the Observatoire de Paris visiting Cassini (1677-1756), son of the first one, and his colleagues, he had in mind to have an observatory installed in the town Petersburg. After his death, his wife Catherine succeeded him and she maintained the invitation and Delisle went for a few years in Russia. Leaving Paris in 1725, he arrived at Petersburg on March 5/February 22 1726.

There, besides an observatory installed in the dome of the building of the Academy for Sciences created by the same Catherine in 1725, he formed many students. This observatory is nowadays included in a museum, the academy having been moved to Moscow in 1934. Instead of a few years as he was invited for, Delisle was absent from France during more than twenty years. Back in Paris, in 1747, regaining his position at the Collège Royal and very soon having a new observatory on the tower of the Hôtel de Clugny, nowadays *Musée du Moyen-Âge*. Having sold all his manuscripts to the French Government he was given the title of *Astronome de la Marine*. Leaving the Collège Royal, he went to the close Abbaye de Sainte-Geneviève; Delisle died on 1768 September.

Delisle was then 80 years old; he had accumulated a great amount of manuscripts and among them many letters he exchanged with people, more or less all around the world. They contain numerous astronomical data and/or informations, Delisle having in mind to publish a *Traité complet d'astronomie exposée historiquement et démontrée par les observations*. This treatise never came to existence, but most of his correspondence is preserved in Paris, most of his books and manuscripts being in the *Dépôt de la Marine* up to 1795. Nowadays one part of Delisle's correspondence is in the Archives of the Observatoire [ref. B1 1-8] for volumes I et II, VIII-XII, supplément aux portefeuilles XIV et XV; the other part, Volumes III-VII, XIII-XVI, which were still in the *Dépôt*, have been moved in the *Archives Nationales*.

2. Delisle's correspondence between Paris and Beijing

The paper concerns the part of Delisle's letters exchanged between Paris and Beijing (at his time Pékin or Péking), before and after his sejourn in Russia. For this study have also been used the manuscript E1 13. 146 he mentionned as *Table des dates et des noms de ceux qui m'ont ecrit depuis 1709* showing that he began to collect letters when he was only twenty-one. Delisle also adds in this manuscript *Pour la liste des Personnes nommées dans les recueils de mes lettres pour leurs observations le tems de leur mort etc*, and *Histoire et details de ma Correspondance avec les Jesuites missionaires de la Chine...*

The list of such letters shows that, during the first period, the correspondence was mostly exchanges with European relatives. As an example Delisle had received a letter on 1724 August 16 he had to forward to Father Koegler in China, but he had written that he could not send it before the 28th of December 1730. A letter received before his departure, was sent six years later from Russia.

During Delisle's sojourn in this country some letters were sent, around 1731 and 1732, to several Jesuits in Péking asking them to provide astronomical data. Such were Fathers such as Gaubil, Kögler, Slavisck... The delays were more or less the same depending of the opportunity to have someone to bring the letters. In Gaubil's case one can find a letter sent in 1743, received in 1748 by Delisle. As he received it on February 19, Delisle answered the same year on November 15. From the same Gaubil he received a letter, dated 1749 September 19, the following year (1750) on July 25 and on August 5 a letter from 1749 September 8. At that time he was a new time in Paris from 1747.

3. Father Gaubil, Delisle's main correspondent

Father Antoine Gaubil was born, on 1689 July 14, in Gaillac a town from the south of France. He became superieur de la résidence française de Pékin, known as an interpret, an historian and a geographer, from 1742 to 1748. Gaubil had an important rôle in astronomy in the observatory of the French mission; this observatory was built in 1754 before his death on 1759 July 24.

Delisle began, as mentionned above, to exchange letters with Gaubil during the years he was in Russia. Indeed he addressed his request for observations to several missionaries in Péking in 1731 and 1733 and by December 31, on that year, he sent letters explicitly to Gaubil, Kögler and Slavisck having, meanwhile, received from everyone answers to his previous request. The same process followed up to his departure from Russia. Concerning his return to Paris, on 1748 November 29, Gaubil wrote to Delisle, who received the letter on 1749 August 12, saying that he thought he was in Petersburg while he learned that he was not more there. The following day 1748 November 30, Gaubil wrote several letters, all of them to Delisle, but concerning different subjects. We learn there that the letters were sent by russian caravans from Péking to Petersburg and that sometimes, the letters are lost. In parallel Gaubil sends astronomical data, beginning with the coordinates of what he names *notre maison françoise*; as an example he deduced its latitude, 39°55′ after more than fifty years of such observations.

Some other letters can be very long; an example is given on 1752 October 22, on four pages beginning with *Je viens de recevoir votre lettre du 25 décembre 1751* meaning that, in October 1752, he was answering Delisle's letter from December 1751. For them Delisle has ordered a quadrant in view of new latitude determinations. He also gave informations about ancient eclipses, discussing their local conditions. This letter was received by Delisle in 1753 on August 13. The same day, he had received a note signed Gaubil with observations beginning with the eclipse of the third galilean satellite of the Jupiter; they were followed by some other ones during January 1752 about similar eclipses followed by occultation of these objects by the moon on February 22. The last sentence says that Gaubil has not more astronomical observations to send at the given date 1752 November 2.

On 1753 August 11, Delisle had received, most probably through a different way, a letter dated 1752 August 28. This very long letter, provides Delisle, not only astronomical informations, but also historical points concerning ancient astronomy in China. Two months after having received this letter, on November 11, Delisle answered several letters of Gaubil sent during the second part of 1752. From the copy of his letter, it is seen that between August 11 and October 7, many letters arrived, having been sent between August 12 and November 12, one year earlier. Some letters were duplicate of others, showing that they were sent through different ways.

In his answer, Delisle gives informations from his side, wrote memoirs to make better known how astronomy was developed during the old times, how to proceed to better used available observations... The end of the letter concerns the clock he promised to send to Péking, adding that—fortunately—he found one which, being achieved, will be sent to replace old ones. The very last sentence is promising to send it the following year—the date is November 1753—having found a solution through Father La Tour.

Most of the letters written by Gaubil contains astronomical observations and also informations concerning the astronomical equipment such as quadrants, some being equipped with micrometers, with two fixed refractors to observe in the south as well as in the north, requesting oculars... What is obvious that the letters needed at least one year, sometimes two, to catch Delisle.

4. Delisle's other correspondents

Among other correspondents of Delisle, when being returned to Paris, Father August von Hallerstein, who was born in 1703, August 27, in nowadays Slovenia; he made his studies in Vienna before being sent to China in 1736. He arrived in Péking in 1739 and, in 1746, he became président du Tribunnal des Mathématiques. Hallerstein died in Péking on 1774 October 29.

As an example of their exchange of letters, one from Hallerstein, dated 1750 November 2, was received by Delisle on 1751 September 23. On December 25, the same year, Delisle answered that three months ago he had received this letter, granting Gaubil who had spoken to Hallerstein about possible observations, for which a certain Sanchez had provided and sent a transit instrument. Being able to compare the observations of a solar eclipse, both by Hallerstein and by Gaubil, and seeing a small difference in the timings, Delisle advises Hallerstein about the way the observer has to proceed to have his clock checked with the meridian transit of the Sun. Delisle will also take care of the place from where the observations are made: the observatory of the College or the Imperial Observatory which had also the name of Public Observatory, the difference being of 14 seconds.

Answering questions of Hallerstein, about La Caille (1713-1762) and Le Monnier (1715-1799), and their publications of stars positions, Delisle added that La Caille is at the Cape of

Good Hope from about one year [La Caille left Paris in 1750] and that he will observe stars which cannot be seen from Europe. Delisle provided informations about some new celestial maps with constellations which will be published in England for the northern hemisphere while, after La Caille's return, the same maps would be made for the southern part of the sky. He announced to Hallerstein that, been able to have a good clock from the parisian horologer Julien Le Roy, it will be sent to Hallerstein. Delisle added that a similar clock was made for the Paris Observatory and for La Caille who brought it with him. With the transit instrument and the clock Delisle had no doubted that Hallerstein would be able to provide good observations.

Besides the observations to come, Delisle asked Hallerstein about what he calls Calendar but which appears to be, among other things, the predictions of astronomical phenomena for the coming year, being published each year. He explained his request, willing to be informed about Chinese astronomy in view of the Treatise he had in preparation. This explains why Delisle have collected so many papers, observations, documents... but we know, nowadays that—having spent so many days in putting in order his valuable collection—he died before publishing his project.

In a very long post-scriptum, Delisle recommended Hallerstein, the transit of Mercury over the Sun which will occur on May 5 of the year 1753. And in 1753, on August 13, Delisle received a letter from Hallerstein dated October 20, 1752, answering the letter dated December 25, 1751. To this letter, Delisle answered on November 11, 1753. From the answer it is known that, in October 1752, Hallerstein, asked by the Emperor to go to Macao to welcome the ambassador from Portugal, has received there the clock announced by Delisle. Hallerstein will send data from his predecessor, Father Koegler, he had put in good order and Delisle was waiting for them plus, perhaps, some from Father Verbiest (1623-1688). He informed Hallerstein that a book he wrote about astronomy and geography will be given to him, to Gaubil to which he sent three; the third one will be for Father Gogails. Delisle also hoped that Hallerstein was back in Péking for the Mercury transit for which he provides the numerical values for the parisian observations, which were made simultaneously.

As an example of other correspondents of Delisle, Father Chanseaume (1711-1756) whose letters were mostly sent to Gaubil, from Macao where he is from 1750, to Péking and, there, added to Gaubil's correspondence. As another example, Chanseaume could not send any observations to Gaubil during one year. This example is not, except the signature, from Chanseaume.

Other Fathers were also sending data to Gaubil-Hallerstein from whom they were directed to Delisle. Such was a note about a comet, included in a letter dated 1748 november 9, from Gaubil to Dortous de Mairan (1678-1771), under the form Dans la maison des jesuites français à Pékin, observation d'une Comète en 1748. But indeed, from the note it is seen that the comet was also observed à la tour des mathematiques, by Chinese astronomers, on 1748 April 26, with their zodiacal armillary sphere. What is now named the old Beijing observatory was, at that time, the mathematical tower and the instrument is still there.

5. Delisle and Chinese ideograms

At the end of Delisle's correspondence is a portfolio including *Memoires pour les Lettres à Ecrire*, documents for further letters, and *Lettres à ecrire Commencées* meaning that some of them have not been brought to their end. One is dated April 1758, while Delisle had still ten years to live before his death. Among this set of such letters, some are in a group which title is *Lettres de Monsieur de L' isle au Reverend Père Berthier à l' occasion de la Dispute litteraire Sur l' origine des Chinois*, entre M^{rs} De Guignes [(1721-1800), professor for Oriental languages] et Deshauterayes [(1724-1795)].

The Father Berthier, for whom these letters were written, was Guillaume-François born, in 1704 in Issoudun, in the center of France. He was a Jesuit who died in 1782. The letter begins explaining that De Guignes considers that an egyptian colony went, more than one thousand years before our era, to Persia, from Persia to India, from there to China and later to Japan. Among the objections to these assent the fact that Chinese characters were in use quite a long time earlier... Against, it was also said that, if such was the case, why some letters—such as B. D. R. and Z. —are not known in Chinese. By the end, it is clear that all this rather long text, corresponds to an attempt to be able to read Egyptian characters through an eventual similarity with Chinese ideograms under the form donc les caractères chinois ne peuvent être d'aucun secours pour l'intelligence des Inscriptions Hieroglyphiques des Egyptiens.

Was this sort of memorandum by Delisle? Indeed what is clear that it is not from his writing. But what is clearly seen that he made, in some parts, small corrections. The most important one concerns some positions. The writer has written *Nommé commissaire pour l' examen...* while Delisle introduced *de la part du Collège Royal l' un des...* meaning that this collège, now Collège de France, had appointed him to be one of the experts to examine the various points of the discussion. This must be explained through the fact that Delisle was among the professors in the Collège Royal and that, through his large exchange of letters with a great number of Jesuits in China, he had become a sort of specialist for this country. Nowadays we know that it is only from 1822, with the French Champollion (1790-1832), that people can read Egyptian characters. It must be added that the Father Gaubil was not favourable to De Guignes.

Besides this very curious letter, they are more serious not achieved letters, some are to Father Patouillet, to Father Bertrand in Lyon, to Father Ximenes in Firenze (with the date 1755!), to Wargentin in Sweeden (no date but about observations from 1753!) ... From these remarks it is clear that in these times, 18th Century, people were not hurrying as nowadays when, not answering e-mail within one hour, a new message is sent!

6. Delisle's correspondence in the Paris observatory collections

Delisle's correspondence was put in rather right order by Delisle himself dividing it into letters received and letters sent. He also wrote a two pages document to which he gave the title *Histoire et details de ma Correspondance avec les Jesuites missionnaires de la Chine et des Indes Orientales* meaning that he gave a special treatment to this subject.

The last letter Delisle from Gaubil dated 1758 November 20, was received in 1759, October 14; Father Gaubil was dead from the previous July. Indeed the letter, as it is often the case, contains two documents. The first part gives general informations and news from the other Fathers, such as Portugueses, some French, Hallerstein, Gogails (1701-1771), from Germany arrived in China in 1738, Kögler (1680-1746) from Bavaria, arrived in Péking in 1717, suggestions to send letters through Spain if there is no boat from France. The second part is related to astronomical observations, eighteen positions of locations in latitude and, in longitude, under the form of the difference between the places and Péking. Having left France in 1721 at the time (Kangxi 1654-1722) was emperor, Gaubil arrived in Beijing in 1723, under Yong-tcheng, suceeded by K' ien-long and thus he stayed thirty-six years in this town.

In the book *Correspondance de Pékin (1722-1759) de Gaubil*, Renée Simon (published in 1970, one thousand pages) numbered 342 entries, some of them including two or more when the date is the same. From her alphabetic list of names, it is seen that Delisle was second with the number of quotations, almost one thousand. There is no question that the first, for such quotations in Gaubil's letters was Father Parrenin, about one thousand six hundreds ! Under the first name Dominique (1665-1741) he arrived in China in 1698 and was official interpret for the Chinese and Tatar languages.

Among European astronomers, Cassini the third (1714-1784) from Paris Observatory is found with about seven hundreds and fifty quotations; in the future it could be of interest to look at them from astronomical point of view. Among other astronomers from other countries are noticeable Flamsteed (1646-1719) and Halley (1656-1742) from Greenwich Observatory, Ch. Kirch (1694-1740) from Berlin Observatory in Germany.

In 1795, after being in the *Dépôt de la Marine* after the exchange Delisle made, around 1750, to get some money for the rest of his life, and the title of *astronome de la Marine*, his collection was divided and the astronomical part was sent to the Paris Observatory. This explains why many researchers, from all over the world, are interested by Delisle's archives. Two publications have been established by Isaia Iannaconne for the Chinese part of this important documentation. Delisle's correspondence, the object of this paper, is included in a list of ten entries; but each one contains many documents. More information are given on the site of the library, which is, in French, *Bibliothèque*. To get it, first call the Paris Observatory at obspm.fr, then bibliothèque and alidade.

Notes

The main references correspond to Delisle's manuscripts preserved in the Archives of the *Observatoire de Paris* given below:

[1] B1 1-8, Correspondance

[2] C2 14-15. 113-114. Observations astronomiques faites à Paris, 1712-1725

[3] E1 13-146. Noms et dates des Lettres qui composent sa Correspondance

Some details concerning these references can be obtained from the site of the *Observatoire de Paris* at < obspm fr > , Bibliothèque, Patrimoine, Alidade, *Inventaires disponibles*, *Bibliothèque de l' Observatoire de Paris*, *Archives*, Inventaire général... Bigourdan (1895) and Complément à l' inventaire Bigourdan

The Transmission of Western Astrolabe in Late Medieval China

Fung Kam-Wing

(The University of Hong Kong)

1. Greek innovative views on projections

Ancient Greeks were familiar with the mechanics of the projective transformation of one surface (or plane) onto another. Thales of Miletus (624 BC? -547 BC) and Anaximander (610 BC? - 546 BC) had applied "gnomonic projection" in cartographic drawings.[1-3] In his influential treatise *De Architectura* (*Ten Books on Architecture*; probably written around 25 BC), the renowned Roman architect Vitruvius (died after 27 AD) mentioned three types of projections that were widely used in his time: the horizontal and the vertical projections of buildings (i. e. *ichnography and orthography*) and the perspective images shown on sceneries in the theatres (*scenography*).[4,5]

Greek sundial maker Diodorus (first century BC), in his work *Analemma*, described the method of orthogonally projecting the celestial sphere onto a plane as "*analemma*". Ptolemy of Alexandria (85? -165) also wrote a treatise entitled *Peri Analemmatos* but offered a different interpretation of projection. The cartographical writings of Eratosthenes (276 BC-196 BC) and Marinus of Tyre (second century AD) contained depictions of the cylindrical projection of the inhabited part of the Earth onto a plane.[6] In Ptolemy's *Geographike Hyphegesis* (written around 150 AD), the methods of conic projection and pseudo-conic equivalent projection were employed to put the earth onto a plane.[7-9]

2. Ptolemy's *Planisphaerium* and stereographic projection

Astrolabe (Greek: ἀστρολάβον astrolabon), literally means "star-taker".[10-13] An astrolabe imitates the motion of the heavens. The astrolabe was first designed to represent the positions of the sun, the moon and the fixed stars with respect to the local horizon in two dimensions rather than three by means of stereographic projection.[14,15] (Fig. 1) It may safely be said that without stereographic projection the astrolabe was inconceivable. It is, however, difficult to determine the origin and development of the method of stereographic projection. The *Planisphaerium* by Ptolemy (85? -165) is the only Hellenistic work on stereographic projection

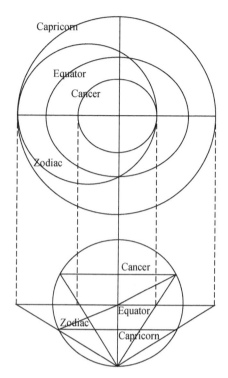

Fig. 1　Stereographic projection

that has come down to the present day.[16-18] In Ptolemy's *Planisphaerium*, the theory of stereographic projection of a sphere onto a plane was explicitly indicated as a projection of a sphere from one of its points either onto a plane tangential to the sphere at the point opposite to the chosen one, or onto a plane parallel to it.[19-22]

3. Hipparchus of Nicaea (190 BC-125 BC) and the discovery of stereographic projection

Historians of mathematics argued that the discovery of stereographic projection could be traced back to its affinity with mathematics. Recent researches claimed, however, that the discovery of stereographic projection was closely connected with astronomical observations.[23] The theory of stereographic projection, developed by Hipparchus of Nicaea ca. 150 BC, holds the key to solving problems of spherical astronomy such as the determination of the rising times (in terms of right and oblique ascensions) for the zodiacal signs. Synesius of Cyrene (373-414), who certainly had access to the works of Hipparchus through his teacher Hypatia, daughter of Theon of Alexandria (335? -395?), described how Hipparchus had determined the positions of stars on a spherical surface by means of a two-dimensional instrument based on stereographic projection.[24] Hence, historians of instrument-making generally ascribed the invention of the astrolabe to Hipparchus.[25]

4. Roman architect Vitruvius (died after 27 AD) and the anaphoric clock

Roman architect Vitruvius described a horological mechanism called "the anaphoric water clock" (literally meaning the clock based on rising times during the season when the sun is not shining) in his work *De Architectura*. In the Book IX of *De Architectura*, Vitruvius wrote the following:

There are also made winter clocks of another kind, which are called Anaphorica, and they make them in the following fashion. An *analemma* is described, and the hours are marked with bronze rods, beginning from a centre on the clock face. On this circles are described which limit the spaces of the months. Behind these rods there is a drum, on which the firmament and zodiac are drawn and figured: the drawing being figured with the twelve celestial signs. Proceeding from the centre the spaces are greater and less. On the back part in the middle of the drum is fixed a revolving axle. On the axle a pliable brass chain is coiled. On one end hangs a cork or drum raised by the water; on the other, a counterpoise of sand equal in weight to the cork. [26-28] This clock, which featured a large bronze disk engraved with a flat star map (or *planisphere*) and marked with constellation figures, was a stereographic projection of the celestial sphere from its south pole on the plane of the celestial equator. And it was made to simulate the diurnal motion of the heavens. Rotation was achieved by fixing the centre of the disk (corresponding to the north celestial pole) to the end of a wooden roller that turned on a horizontal axis. A cord or light chain winding round the roller had one end attached to the float of a water clepsydra and the other to a carefully counterpoised bag of sand. [29-31] A fragment of bronze dial (engraved figures representing the constellations Triangulum, Andromeda, Perseus and Auriga) of an anaphoric clock (radius 42cm, thickness 0.3cm, weight 5.5kg; dating around 1st -2nd century AD) was found in Salzburg in 1897 and preserved in the Museum Carolino Augusteum of Salzburg, Austria. (Fig. 2a, 2b) Two bronze fragments of bronze calendar-plate (diameter 37cm) were also found at a 2nd century Roman site at Grand, in north-eastern France and were preserved in the Musée des Antquités Nationales, St. Germainen-Laye. [32,33] (Fig. 3)

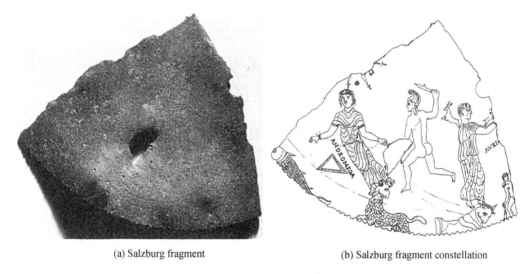

(a) Salzburg fragment   (b) Salzburg fragment constellation

Fig. 2 Hellenistic anaphoric clock

5. Roman portable sundial and stereographic projection

In recent decades, a Roman portable sundial which is engraved with the stereographic projection of the tropics of Cancer and Capricorn, the equator, and the unequal-hour lines, and which is preserved in Kunsthistorisches Museum at Vienna, reveals the early use of stereographic projection in the time of Ptolemy. [34-37] The complete horological instrument

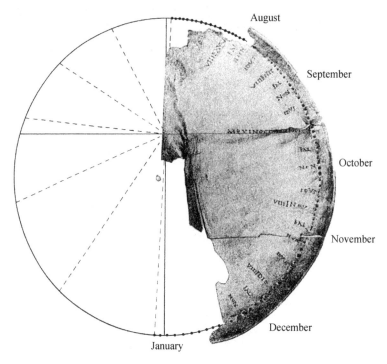

Fig. 3　Hellenistic anaphoric clock grand fragment

consists of a small circular box 39 mm in diameter and approximately 14 mm thick. The top and bottom of the box are made from the medallion of Roman Emperor Antoninus Pius (86-161; r. 138-161); the stereographic projection is on the inside of the lid. (Fig. 4) The front top of the box bears the portrait of Antoninus Pius. The box contains four plates, each engraved on both

Fig. 4　Roman portable sundial stereographic projection

sides with sundial scales of measurement and each scale for use in different single latitude. Fixed to the base of the box is a vertical pin which passes through a hole in the plates, thus securing them in position (Fig. 5, 6). The four features of this instrument bear remarkable resemblance to those of a planispheric astrolabe:

Fig. 5 Roman portable sundial vertical pin

Fig. 6 Roman portable sundial different latitudes

(i) The portable box, containing a set of plates;
(ii) Each plate being drawn for a different latitude;
(iii) The plates being secured by a pin;
(iv) The use of stereographic projection.

In the late 4th century, Theon of Alexandria (335? -395?) compiled a treatise on the

Hellenistic Roman style astrolabe, but it was no longer extant. As it was cited in early Roman texts, the essential characteristics of Theon's astrolabe which was similar to surviving examples of the 9th or 10th century, can be identified as follows:

(1) A stereographically projected map of the heavens in which the projection of the stars and ecliptic is engraved on the skeletal *rete* which rotates over local co-ordinate plate that usually carries hour lines.

(2) A set of plates, each projected for a different latitude.

(3) A sighting device.

(4) A suspension apparatus.

6. Monophysitist Bishop Severus Sēbōkht (575? -666/667) and Hellenistic-Roman style astrolabe

In succeeding centuries, treatises on astrolabe began to poliferate amongst Roman and Syrian scholars such as Ammonius (*fl.* 390), John Philoponus (490-570) and monophysitist Severus Sēbōkht, Bishop of Nisibis.[38] Severus Sēbōkht, a famous Nestorian priest, was committed to research in astronomy. His key writings, mostly based on Greek sources, included *On the Phases of the Moon and Lunar Eclipses*, *Treatise on the Constellations and the Figures of the Zodiac*, *Treatise on the Astrolabe* and commentary on Aristotle's (384BC-322BC) *De Interpretatione*.[39-43] In Severus Sēbōkht's *Treatise on the Astrolabe*, the astrolabe was basically modeled on the Hellenistic-Roman style.

7. Islamic astrolabe and *Qiblah*, the sacred direction of Mecca

Sēbōkht's *Treatise on the Astrolabe* was frequently translated into Arabic in the mid-to late 8th century. A Treatise on the astrolabe, *Kitāb al- 'amal bi' l-asturlāb al-musattah* (Book of the Use of the Plane Astrolabe), was ascribed by the famous Islamic bibliographer Ibn an-Nadīm, compiler of the *Kitāb al-Fihrist* (The Catalogue) in 988, to al-Fazārī (? -777), who was generally regarded as the first Muslim to make an Islamic astrolabe.[44] In early 9th century Baghdad, the Abbasid astronomers 'Ali b. 'Isa (*fl.* c. 830) and Muhammad ibn Mūsā al-Khwārizmī (780? -850?) wrote treatises on the use of astrolabe respectively and the acclaimed astronomer al-Farghānī (? -861?) compiled a set of tables displaying the radii and centre distances of both altitude and azimuth circles for each degree of the terrestrial latitude.[45-48] (Fig. 7) In the year around 970 AD, the celebrated Persian astronomer 'Abd al-Rahmān ibn 'Umar ibn Muhammad al-Sūfī (903 or 904-986) composed at least two treatises concerning astrolabe and its operations entitled *Kitāb al- 'amal bi' l-asturlāb* (On the Construction and Use of the Astrolabe) (Fig. 8), whose *Kitāb Suwar al-kawākib al-Thābita* (Book of Images of the Fixed Stars) constituted an important link between the astronomy of the fixed stars as known to the Greeks, and that of modern times.[49-51]

In the *Qur'ān*, Muslims are enjoined to face the *Qiblah*, the sacred direction of Mecca during their daily five prayer-times, namely, *Maghrib* (from sunset to evening), '*ishā*' (from evening to dawn), *Fajr* (from dawn to sunrise), *zuhr* (noon time when the sun is at its highest point), '*asr* (is calculated when the shadow of any object becomes equal to its length; sunset). [52-54] Telling the time and establishing the *Qiblah* are therefore important matters, and a variety of methods has been used to calculate both. One convenient means of telling the time that accommodated the unequal hours is using Islamic-style planispheric astrolabe. In addition to measuring celestial bodies and their movement, Islamic astrolabists even further developed their astrolabes to measure inaccessible earthly locations such as the heights of mountains and depths of wells. [55,56] (Fig. 9) Hence, the development of Islamic astronomy and the making of astrolabes owed much to its significance to religious worship.

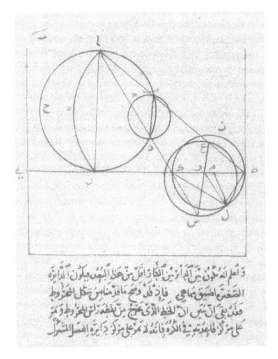
Fig. 7 Diagram in al-Farghani's proof of fundamental theorem of stereographic projection

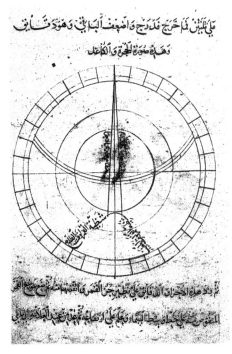
Fig. 8 Al-Sufi (903 or 904-986) The Use of the Astrolabe

8. Foreign astronomers in the Tang court and Persian astronomer Li Su (李素) (?-796)

Emperors of the early Tang appeared to have been enthusiastic patrons of astronomy and related learning. It was not rare for astronomers of non-Chinese descent to serve in the Tang court. Notable examples were the Indian astronomer Qutan (瞿昙) family and the Persian astronomer Li (李) family.

In the early days of the Tang Dynasty, among those who served in the capacity of

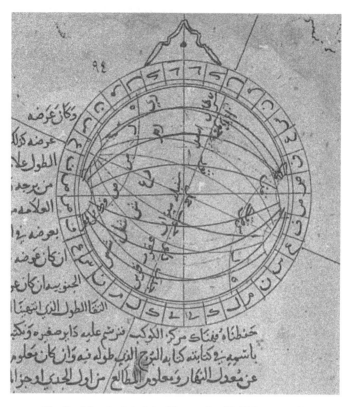

Fig. 9 Islamic Astrolabe Manuscript (9-10th century)

directorship of astronomical bureau were the three clans of Indian calendrical experts, Jiaye (迦叶) (Kāsyapa) or Jiaye bo (迦叶波), Qutan (Gautāma) and Jumoluo (俱摩罗, 拘摩罗) (Kumāra) or Jiumoluo (鸠摩罗).[57] During the last quarter of the seventh century, i. e. from the second year of Linde (麟德) reign-period (665) to the second year of Shengong (神功) reign period (698), the Hindu astronomer Qutan Luo (瞿昙罗) (fl. 7th century) held the post of directorship of astronomical bureau for more than thirty years.[58-60] He produced two calendar systems known as the *Jingweili* (《经纬历》) and the *Guangzhaili* (《光宅历》) in the years 665 and 698 respectively. Then at the turn of the next century Qutan Xida (瞿昙悉达) (Gautāma Siddhārtha, ? -724?), the Indian celebrated astronomer and compiler of *Kaiyuan Zhanjing* (《开元占经》) (Kaiyuan reign-period Treatise on Astronomy, 718-724), also succeeded to the post of directorship in the sixth year of Kaiyuan (开元) reign period (714).[61] Far from being a translation, however, *Kaiyuan Zhanjing* indeed constitutes an integral part of the great collection of Chinese astronomical texts dating from fourth century BC. The section on the *Jiuzhili* (《九执历》) (Indian *Navagraha* calendrical system; literally means Nine planets) contains elements of Indian calendrical science. It also seems that an early form of Indian trigonometry and methods of mathematical computation have been introduced in this treatise. Towards the first year of Yongtai (永泰) reign period (765), Qutan Zhuan (瞿昙

譔）（712-776）, the fourth son of Qutan Xida, was promoted to the director of astronomical bureau[62] (Fig. 10). In the time of Emperor Dezong（德宗）(r. 779-805）, Qutan Yan（瞿昙晏）(fl. the last quarter of 8th century) was Chief Officer of the Winter Agency for the astronomical bureau (Dongguan zhen, 冬官正).

Fig. 10 Indian astronomer in China Qutan Zhuan（瞿昙譔）(776)

The stone epitaphs (engraved with Chinese inscriptions dated 817 and 823 respectively) buried in the tombs of Persian Nestorian astronomer Li Su and his Turkish wife Beishi（卑失）, unearthed in 1980 in Xi'an（西安）, are extremely valuable historical records. (Fig. 11, 12) It tells us that Li Su studied astronomy and astrology under his father's direction in the period of 756-770 in Guangzhou（广州）where a large community of Muslims had been established since High Tang period. In around 770, Li Su was recruited as a Hanlin Imperial Attendant（翰林待诏）in the Tang imperial court and because of his expertise in foreign astronomy and astrology (very likely of Islamic tradition), he was also appointed as a director of astronomical bureau.[63,64] Li Su might also have access to the newly translated *Duli Yvsi Jing*（《都利聿斯经》）or *Yvsi Simen Jing*（《聿斯四门经》）by another Persian Nestorian Li Miqian（李弥乾）. In fact, recent scholar speculated that both *Duliyvsi jing* and *Yvsi Simen Jing* were translated from Ptolemy's *Tetrabiblos* (*Four-part Book of Astrology*).[65] If this bibliographical link was really the case, Li Su could have acquired his knowledge of stereographical projection and related instrument such as astrolabe from Ptolemy's work.

Fig. 11　Islamic astronomer in China Li Su（李素，819）

Fig. 12　Islamic astronomer in China（卑失氏）Li Su'wife（823）

9. Arabic astronomer Ma Yize and Islamic astronomy

The late Professor Lo Hsiang-lin（罗香林）(1906-1978) recounted how the militarist-emperor Song Taizu（宋太祖）(r. 960-976) summoned an Islamic astronomer Ma Yize（马依泽）(921-1005) who excelled in astrology to his throne in the second year of Jianlong（建隆）reign period (961) and who was appointed as astronomical official in the Bureau of Astronomy.[66,67] A careful examination of the *Huaining Mashi Zongpu*（《怀宁马氏宗谱》）(*Genealogy of the Ma Clan of Huaining*, 1875), *Hebei Qingxian Mashi Menpu*（《河北青县马氏门谱》）(*Genealogy of the Ma Clan of Qingxian, Hebei Province*, 1486) and *Dacetang Maguazhou*（《大测堂马挂轴》）(*Hanging scroll of the Dace Hall of the Ma Clan*, 1920?-1930?) and *Juzhentang Mashi Zongpu*（《聚真堂马氏宗谱》）(*Genealogy of the Ma Clan of Juzhen tang*, 1928) reveals that Ma Yize, an ancestor of the clan, arrived from *Lumuguo*（《鲁穆国》）which was not recorded in the *Songshi*（《宋史》）. (Fig. 13) This *Lumuguo* is

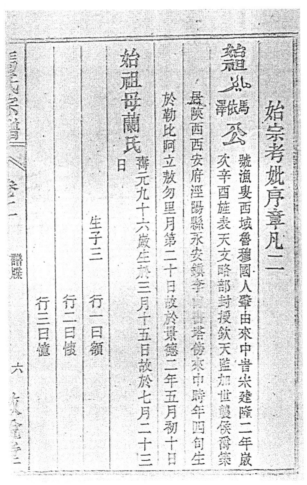

Fig. 13　Ma Yize genealogical document

generally thought to be the *Lumiguo* (《鲁迷国》) in the *Mingshi* (《明史》) and *Mingshilu* (《明实录》) (Veritable Records of the Ming Dynasty), which presented eleven tributes to the Ming court. For the location of the *Lumiguo*, my recent research on the World Map in Giulio Aleni's (艾儒略, 1582-1649) *Zhifang Waiji* (《职方外纪》) (Account of countries not listed in the Records Office) discovered that *Lumiguo* was situated 120 miles southeast of Mecca and stretched for hundred miles to the north-eastern part of Medieval Yemen.[68,69] According to the *Huaining Mashi Zongpu*, Ma Yize collaborated with the eminent astronomer and mathematician Wang Chune (王处讷) (913?-982?), the then Deputy Director of the Astronomical Bureau, in compiling the new *Yingtianli* (《应天历》) (Calendar of Corresponding Heavens). (Fig. 14) Ma focussed on astronomical computation and finished all the calculation in 964. He then let Wang Chune present the *Ying Tian* Calendar to the throne. Emperor Taizu wrote an imperial preface for the new calendar. Although parts of the *Yingtian* Calendar are preserved in the *Lvlizhi* (《律历志》) (Treatise on Pitch-pipes and Calendars) of the *Songshi*, it is difficult to determine whether this calendar incorporated Islamic calendrical elements or not. The late

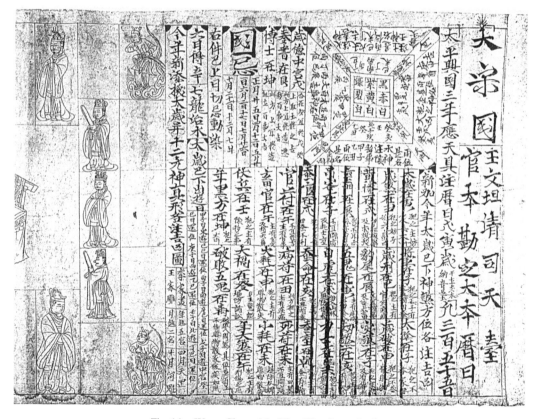

Fig. 14 Wang Chune Ma Yize Yingtian calendar

Professor Wang Yingwei (王应伟, 1877-1964) has noted that firstly the method of "determining the true (weekly) entry days of the mean new moon, crescents and full moon in accordance with Seven Luminaries" (tui ding shuo xian wang ri chen qi zhi,推定朔、弦、望日辰七直) in the *Yingtian* Calendar implies the utilization of the concept of 7-day week. Secondly the computation of the calendrical epoch (i. e. the beginning point of time when the calendrical cycle was aligned) of the *Yingtian* Calendar tallied with three significant Islamic calendrical characteristics, namely, the beginning of the sexagenary cycle, the associated day of Venus (Arabic: ādīnah, literary means Friday, Year One of the Muslim era beginning on Friday 16 July AD 622) and the dawn of the first day of astronomical month at winter solstice.[70] Wang Yinglin's (王应麟, 1223-1296) *Yuhai* (《玉海》) (Sea of Precious Jade) and Xu Song's (徐松, 1809) *Song Huiyao Jigao* (《宋会要辑稿》) (Drafts for the History of the Administrative Statues of the Song Dynasty) contains two entries about the *Yingtian* Calendar:

Lijing Yijuan (《历经一卷》) (The Calendrical Canon)

Suancao Yijuan (《算草一卷》) (The Mathematical Draft)

Wugeng Zhongxing Licheng Yijuan (《五更中星立成一卷》) (The Table of licheng of Meridian Transits at the Fifth watch of the night)

Chenhunfen Licheng Yijuan (《晨昏分立成一卷》) (The Table of licheng of Dusk and Daybreak)

Zhouye Richuru Licheng Yijuan (《昼夜日出入立成一卷》) (The Table of licheng of the Sun rising or setting at Day and Night)

Guiying Licheng Yijuan (《晷影立成一卷》) (The Table of licheng of the Shadows)[71,72] These Licheng (立成) tables are "quickly computed tables for astronomical calculation" and are similar to Islamic astronomical tables (Arabic: zījes, astronomical handbook with tables and explanatory text).[73-76] Most Islamic *zījes* typically contain tables of various parameters such as solar, lunar and planetary mean motions, lunar and planetary latitudes, solar and lunar eclipses, determination of the sacred direction of Mecca and mathematical astrology.[77,78] As Islamic astronomer Ma Yize came from Medieval Yemen, it may be safely assumed that he might know how to use the Islamic astrolabe to measure celestial bodies and their movement and he might have also been influenced by his predecessors or contemporaries such as Muhammad ibn Mūsā al-Khwārizmī (780?- 850?), Abū Ma'shar Ja'far ben Muhammad ben 'Umar al-Balkhī (Latin: Albumasar, 787-886), Abū'Abdallāh al-Battānī (Latin: Albategni, 858-929) and Yemeni mathematician-geographer Al-Hamdānī (900?-946?/950?) and might have facilitated the introduction of their works to China[79-88] (Fig. 15). Al-Battānī, a celebrated author of *zījes* such as *Kitāb al-Zīj* (Al-Battani sive Albatenii Opus astronomicum, 880) and *al-Zīj al-sāb ī* (The Sabian Tables), carried out observational activities in Raqqa in North Syria.[89-91] Al-Hamdānī compiled a Zīj and a treatise on mathematical astrology entitled *Kitāb Sarā'ir al-hikma*. It is noteworthy that Al-Hamdānī's geographical work *Sifat jazī rat al-'Arab* has given an account of the "Easterners" (ahl al-mashriq), the Indians and those who followed them in measuring base meridians west from the eastern boundary of China.[92-95] Because of his dedication and outstanding contributions in compiling the *Ying Tian* Calendar, Ma

Fig. 15　Islamic Astrolabe 1079

Yize was promoted to the directorship in the astronomical bureau and was conferred a degree of nobility by Emperor Taizu in the eighth month of the fourth year of Qiande (乾德) reign period (966).[96] Ma Yize's sons, Ma E (马额) and Ma Huai (马怀), continued to serve as astronomical officials in the Bureau of Astronomy and inherited their father's title in the years 997 and 1001 respectively.

References

[1] Aujac G. The Foundations of Theoretical Cartography in Archaic and Classical Greece. In: Harley J B, David Woodward eds. The History of Cartography V. 1: Cartography in Prehistoric, Ancient, and Medieval Europe and the Mediterranean. Chicago: University of Chicago Press, 1987. 130-147, esp 134

[2] Snyder J P. Flattening the Earth: Two Thousand Years of Map Projections. Chicago: University of Chicago Press, 1993. 18-19

[3] Dilke O A W. Greek and Roman Maps. Baltimore: Johns Hopkins University Press, 1998. 21-26

[4] Buchner E. Antike Reiseuhren. Chiron, 1971, (Band1): 457-482

[5] Field J V. The Invention of Infinity: Mathematics and Art in the Renaissance. Oxford: Oxford University Press, 1997. 117-142

[6] Pédech P. La géographie des Grècs. Paris: Presses universitaires de France, 1976. 89-95

[7] Neugebauer O (1899-1990). Mathematical Methods in Ancient Astronomy. Bulletin of the American Mathematical Society, 1948, 54 (11): 1913-1041, esp 1037-1039

[8] Neugebauer O. Ptolemy's Geography. Book VII (Chapters 6 and 7). Isis, 1959, 50 (159): 22-29

[9] Berggren J L, Jones A. Ptolemy's Geography: An Annotated Translation of the Theoretical Chapters. Princeton, N J: Princeton University Press, 2000. 31-41

[10] 同 [7]

[11] Neugebauer O. The Exact Sciences in Antiquity. New York: Dover Publications, Inc, 1969. 159-161

[12] Fung K W. Christopher Clavius (1538-1612) and Li Zhizao (李之藻, 1565-1630). In: Celina A Mendoza C A L, Nicolaïdis E, Vandersmissen J eds. The Spread of the Scientific Revolution in the European Periphery, Latin America and East Asia (Proceedings of the XXth International Congress of History of Science, Liège, 20-26 July 1997, Vol V) [De Diversis Artibus, Collection de Travaux de L' Académie Internationale D' Histoire des Sciences, Tome 45 (NS8)]. Belgium: Brepolis Publishers, 2000. 147-158

[13] Fung K W. Mapping the Universe: Two Planispheric Astrolabes in the Early Qing Court. In: Chan A K L, Clancey G K, Loy H-C eds. Historical Perspectives on East Asian Science, Technology and Medicine. Singapore: Singapore University Press, 2002. 448-462

[14] Kunitzsch P. Celestial Maps and Illustrations in Arabic-Islamic Astronomy. In: Forêt P, Kaplony A eds. The Journey of Maps and Images on the Silk Road. Leiden: Brill, 2008. 175-180

[15] King D A. Islamic Astronomical Instrumemts and Some Examples of Transmission to Europe. In: Calvo E, Comes M, Puig P et al. eds. A Shared Legacy: Islamic Science East and West: Homage to Professor J. M. Millàs Vallicrosa. Barcelona: Universitat de Barcelona, 2008. 321-349

[16] Drecker J. Das Planisphaerium des Claudius Ptolemaeus. Isis, 1927, 9: 255-278

[17] Kunitzsch P. The Second Arabic Manuscript of Ptolemy's Planisphaerium. Zeitschrift für Geschichte der arabisch-islamischen Wissenschaften, 1994, (Band 9): 83-89

[18] Sidoli N, Berggren J L. The Arabic Version of Ptolemy's Planisphere or Flattening the Surface of the Sphere: Text, Translation, Commentary. SCIAMVS, 2007, 8: 37-139

[19] 同 [7]

[20] 同 [8]

[21] Berggren J L, Alexander Jones. Ptolemy's Geography: An Annotated Translation of the Theoretical Chapters. Princeton: Princeton University Press, 2001. 31-41

[22] Rosenfeld B A, Youschkevitch A P. Geometry. In: Rashed R ed. Encyclopedia of the History of Arabic Science Volume 2: Mathematics and The Physical Sciences. London: Routledge, 1996. 447-493, esp 476

[23] Neugebauer O. A History of Ancient Mathematical Astronomy (Part 2. Egypt, Early Greek Astronomy, Astronomy during the Roman Imperial Period and Late Antiquity). Berlin: Springer-Verlag, 1975. 868

[24] 同 [23]. 869

[25] 同 [7]. 1036-1037

[26] Granger F ed. Vitruvius: On Architecture. Cambridge, Massachusetts: Harvard University Press, 1998. 260-265

[27] Rowland I D, Howe T N eds. Vitruvius: Ten Books on Architecture. Cambridge: Cambridge University Press, 1999. 177-178

[28] Hill D R (1922-1994). Arabic Water-Clocks. Aleppo, Syria: Institute for the History of Arabic Science, University of Aleppo, 1981. 2-5

[29] 同 [23]. 869-870

[30] Price D J de S. Precision Instruments: To 1500. In: Charles Singer ed. A History of Technology. Vol3. Oxford: The Clarendon Press, 1957. 601-603

[31] King H C. Geared to the Stars: The Evolution of Planetariums, Orreries, and Astronomical Clocks. Toronto: University of Toronto, 1978. 10-12

[32] Drachman A C. The Plane Astrolabe and the Anaphoric Clock. Centaurus, 1954, 3 (1): 183-189

[33] 同 [23]. 870

[34] Price D J de S. Portable Sundials in Antiquity, including an Account of a New Example from Aphrodisias. Centaurus, 1969, 14 (1): 242-266

[35] Buchner E. Römische Medaillions als Sonnenuhren. Chiron, 1976, (Band6): 329-348

[36] Turner A J. The Time Museum Volume 1: Time Measuring Instruments Part 1: Astrolabes, Astrolabe Related Instruments. Rockford, Illinois: Time Museum, 1985. 10-12

[37] Fung K W. 西方星盘传入中国小考. 华学. Vol 9 &10. 学艺兼修·汉学大师——饶宗颐教授九十华诞颂寿文. Part II. 上海：上海古籍出版社, 2008. 703-716

[38] Johannes P. Traité de l'astrolabe: De usu astrolabii eiusque constructione libellus. Paris: Société Internat. de l'Astrolabe, 1981. 15-171

[39] Nau F-N (1864-1931). Le traité sur l'astrolabe de Sévère Sabokt, écrit au VIIe siècle d'après des sources grecques et pub. pour la première fois. Journal Asiatique, 1899, 13 (9): 56-101, 238-303

[40] Nau F-N. Notes d'Astronomie Syrienne. Journal Asiatique, 1910, 16 (10): 209- 228, esp 219-224

[41] Nau F-N. La cosmographie au VIIe siècle chez les Syriens. Revue de l'orient Chrétien, 1910, 18 (2): 225-254

[42] Peters F E. Aristotle and the Arabs: The Aristotelian Tradition in Islam. New York: New York University Press, 1968. 48

[43] De Lacy Evans O'Leary. Arabic Thought and Its Place in History. London: Routledge & Kegan Paul Ltd, 1957. 47-48

[44] Pingree D. The Fragments of the Works of al-Fazārī. Journal of Near Eastern Studies, 1970, 29: 103-123, esp 103-104

[45] Tagi-Zade A K, Vakhabov S A. Astrolyabii Srednevekovogo Vostoka (Astrolabes of the Medieval Orient). Istoriko-astronomicheskie issledovaniya (Studies in the History of Astronomy), 1975, 12: 169-204

[46] Saliba G. A Sixteenth-Century Drawing of an Astrolabe Made by Khafif Ghulam 'Ali b. 'Isa (c. 850 AD). Nuncius, Annali di Storia della Scienza, 1991, 6: 109-119

[47] Charette F, Schmidl P G. Al-Khwārizmī and Practical Astronomy in Ninth-Century Baghdad. The Earliest Extant Corpus of Texts in Arabic on the Astrolabe and Other Portable Instruments. SCIAMVS, 2004, 5: 101-198

[48] Lorch R. Al-Farghānī on the Astrolabe (Arabic text edited with translation and commentary). Stuttgart: Steiner, 2005. 1-19

[49] Sezgin F. Editor's Introduction. In: Sezgin F ed. Two Books on the Use of the Astrolabe, by 'Abd al-Rahmān al-Sūfī, Abū' l-Husain 'Abd al-Rahmān ibn' Umar ibn Muhammad (d. 986AD), 1986. i-ii

[50] Publications of the Institute for the History of Arabic-Islamic Science, Series C, Vol23, Frankfurt am Main, 1986

[51] Facsimile edition of MSS 3509 Ahmet III Collection, Topkapi Sarayi Library, and 2642 Ayasofya Collection, Süleymaniye Library, Istanbul

[52] Haleem M S A. The Qur'an: A New Translation. Book 2, Section 144. Oxford, New York: Oxford University Press, 2004. 28

[53] Berggren J L. A Comparison of Four Analemmas for Determining the Azimuth of the Qibla. Journal for the History of Arabic Science, 1980, 4: 69-80

[54] Rius M. Finding the Sacred Direction: Medieval Books on the Qibla. In: Rubiño-Martín J A, Belmonte J A, Prada F et al. eds. Cosmology Across Cultures: Proceedings of a Workshop held at Parque de las Ciencias, Granada, Spain, 8-12 September, 2008. San Francisco, Calif: Astronomical Society of the Pacific, 2009. 177-182

[55] Mayer L A (1895-1959). Islamic Astrolabists and Their Works. Genève: Albert Kundig, 1956. 13-21

[56] Meyer W. Instrumente zur Bestimmung der Gebetszeiten im Islam. In: Istanbul Technical University ed. Proceedings of International Congress on the Histiry of Turkish-Islamic Science and Technology, 14-18 September 1981. Istanbul: Istanbul Technical University, 1981. 9-32

[57] Needham J, Wang L. Science and Civilization in China. Vol 3: Mathematics and the Sciences of the Heavens and the Earth. Cambridge: Cambridge University Press, 1959. 201-202

[58] 阮元. 畴人传. 卷13. 长沙: 商务印书馆, 1939. 159

[59] Kiyosi Y (薮内清). Researches on the Chiu-chih li — Indian Astronomy under the Tang Dynasty. Acta Asiatica (Tokyo), 1979, (36): 7-48

[60] 陈久金. 中国古代天文学家. 北京: 中国科学技术出版社, 2008. 216-225

[61] 同[58]. 160-161

[62] Tansen S. Gautama Zhuan: An Indian Astronomer at the Tang Court. China Report, 1995, 31 (2): 197-208

[63] 荣新江. 一个入仕唐朝的波斯景教家族. 载: 叶奕良编. 伊朗学在中国论文集. Vol 2. 北京: 北京大学出版社, 1998. 82-90

[64] 赖瑞和. 唐代的翰林待诏和司天台——关于《李素墓志》和《卑失氏墓志》的再考察. 唐研究, 2003, 9: 315-342

[65] Michio Y. A Note on Ptolemy in China. In: Akira H ed. Documents et archives provenant de l'Asie Centrale: Actes du Colloque Franco-Japonais, Kyoto (Kyoto international conference hall et univ. Ryukoku, 4-8 octobre 1988). Kyoto: Association Franco-Japonaise des Études Orientales, 1990. 217-220

[66] Hsiang-lin L. A Study of Chinese Genealogies: An Inaugural Lecture from the Chair of Chinese. Supplement to the Gazette, University of Hong Kong, 1966, 16 (1): 1-7

[67] 陈久金. 回回天文学史研究. 南宁: 广西科学技术出版社, 1996. 52-66

[68] Fung K W. 罗香林教授对宋初入华西域历法家马依泽的研究 (Professor Lo Hsiang-Lin's Researches on Islamic Calendrical Astronomer Ma Yize [921-1005]). In: Ma C J (马楚坚) ed. 罗香林教授与香港史学——罗香林逝世二十周年纪念论文集 (Professor Lo Hsiang-Lin and Hong Kong Historical Studies: Selected Essays in memory of the Late Professor Lo Hsiang-Lin). Hong Kong: Organizing Committee for the Conference in memory of the Late Professor Lo Hsiang-Lin, October, 2006. 87-99

[69] Fung K W. From al-Farghānī (? -861?), al-Battani (858? -929) to Christopher Clavius (1537-1612) and Matteo Ricci (1552-1610): A Documentary Journey. A paper presented in Colloquium on Information Science: HKUST Library Series No. 6: Celebrating Special Collections: Scholarship and Beauty (organized by University Library, Hong Kong University of Science and Technology), June 6, 2002. 16

[70] 王应伟. 中国古历通解. 沈阳: 辽宁教育出版社, 1998. 534

[71] 王应麟. 玉海 (photo dupl. j. 10). 上海: 上海书店, 1988. 25a-26b

[72] 徐松. 宋会要辑稿 (photo dupl.) (An evidence of the Ying Tian Calendar of the year 978 was preserved in good condition in Dunhuang Manuscripts). 北京: 中华书局, 1957. 2130

[73] Kennedy E S. A Survey of Islamic Astronomical Tables. Transactions of the American Philosophical Society, 1956, 46 (2): 123-177

[74] Xu S. The Chinese-Uighur Calendar as Described in the Islamic Sources. Isis, 1964, 55: 435- 443

[75] Kennedy E S. Studies in the Islamic Exact Sciences. Beirut: American University of Beirut, 1983. 652-660

[76] Hu T Z. The "Quickly Completed" (Licheng) Astronomical Tables in Chinese Calendars after the Eighth Century. In: Hashimoto Keizo, Catherine Jami eds. East Asian Science: Tradition and Beyond. Osaka: Kansai University Press, 1995. 403-409

[77] King D A. Islamic Mathemathical Astronomy. 2nd revised edition. "II: On the Astronomical Tables of the Islamic Middle Ages". London: Variorum, 1993. 37-56

[78] Saliba G. A History of Arabic Astronomy : Planetary Theories during the Golden Age of Islam (New York: New York University Press, 1994), I. 2. " Astrology / Astronomy, Islamic ". 66-81

[79] Sāʻid al-Andalusī. Science in the Medieval World: "Book of the Categories of Nations". Semaʻan I Salem, Alok Kumar trans. Chapter 12 "Science in the Arab Orient". Austin, Texas: University of Texas Press, 1991. 46-57

[80] King D A. Al-Khwārizmī and New Trends in Mathematical Astronomy in the Ninth Century. New York: The Hagop Kevorkian Center for Near Eastern Studies, New York University, 1983. 7-32

[81] King D A. "Astronomy " and David Pingree, "Astrology". In: Young M J L, Latham J D, Serjeant R B eds. Religion, Learning and Science in the ʻAbbāsid Period. Cambridge: Cambridge University Press, 1990. 274-300

[82] Charette F, Schmidl P G. Al-Khwārizmī and Practical Astronomy in Ninth-Century Baghdad. The Earliest Extant Corpus of Texts in Arabic on the Astrolabe and Other Portable Instruments. SCIAMVS, 2004, 5: 101-198

[83] Stanton C M. Higher Learning in Islam: The Classical Period, AD 700-1300. Savage, Maryland: Rowman & Littlefield Publishers, 1990. 127-135

[84] North J. The Norton History of Astronomy and Cosmology. Chapter 8 "Eastern Islam". New York: W W Norton & Company, 1995. 177-192

[85] Lemay R. Abu Maʻshar and Latin Aristotelianism in the Twelfth Century: The Recovery of Aristotle's Natural Philosophy Through Arabic Astrology. Beirut: American University of Beirut, 1962. i-xl

[86] Kunitzsch P. Abū Maʻsar Johannes Hispalensis und Alkameluz. In: Kunitzsch P. The Arabs and the Stars. Northampton: Variorum Reprints, 1989. XVII, 103-125

[87] Saliba G. The Role of the Astrologer in Medieval Islamic Society. In: Annick Regourd, Pierre Lory eds. Bulletin d'Études Orientales, Tome XLIV Année 1992 (Sciences Occultes et Islam) [Damas, 1993]. 45-67

[88] Yamamoto K. Charles Burnett edited and translated. Abū Maʻsar on Historical Astrology. Vol1. Leiden: Brill, 2000. 115-117

[89] On al-Battānī, see Carlo Alfonso Nallino (1872-?), al-Zīj al-sābī (Paris, 1893) and "Al-Battānī", In Gibb H A R, Kramers J H eds. The Encyclopaedia of Islam. Volume 1. Leiden: E J Brill, 1960. 1104-1105

[90] Hartner W. Al-Battānī. In: Charles C Gillispie ed. Dictionary of Scientific Biography. Volume 1. New

York: Charles Scribner's Sons, 1970. 507-516

[91] Ragep F J. Al-Battānī, Cosmology, and The Early History of Trepidation in Islam. In: Josep Casulleras, Tulio Samsó eds. From Baghdad to Barcelona: Studies in the Islamic Exact Sciences in Honour of Prof. Juan Vernet. Barcelona: Instituto "Millás Vallicrosa" de Historia de la Ciencia Arabe, 1996. 195-252, 267-298

[92] Al-Hamdānī. Al-Hamdānī's Geographie der Arabischen Halbinsel. Leiden: E J Brill, 1968. 27, 45

[93] King D A. Mathematical Astronomy in Medieval Yemen. Malibu, California: Undena Publications, 1983. 19-20

[94] Morelon R. Eastern Arabic astronomy between the eighth and the eleventh centuries. In: Rashed R ed. Encyclopedia of the History of Arabic Science. Vol1. London: Routledge, 1996. 20-57, esp 46-48

[95] Kennedy E S. Mathematical Geography. In: Rashed R ed. Encyclopedia of the History of Arabic Science. Vol1. London: Routledge, 1996. 185-201, esp 188-189

[96] 怀宁马氏宗谱. 卷1, 卷2. 25b-26a

Theories of Solar Motion in *Chongzhen Lishu*, *Yuzhi Lixiang Kaocheng* and *Lixiang Kaocheng Houbian*

Lu Dalong

(The Institute for the History of Natural Science, CAS)

At the end of the Ming Dynasty (1368-1644) and in the early Qing Dynasty (1616-1911), Four Chinese Calendrical Books, of which the titles are *Chongzhen Lishu* (revised and compiled as *Xiyang Xinfa Lishu* in 1645 and as *Xinfa Suanshu* in *Qinding Siku Quanshu* in 1784), *Kangxi Yongnian Lifa* in 1668, *Yuzhi Lixiang Kaocheng* in 1724 and *Yuzhi Lixiang Kaocheng Houbian* in 1742, are of great significance to the introduction of western astronomy into China.

Based upon textual interpretation of the theories in *Xinfa Suanshu* and *Yuzhi Lixiang Kaocheng*, the following conclusions have been reached. The characters of the leap years in the former three calendrical books are consistent. The geometrical models for calculating the Sun's equation of center were an equidistant eccentric orbit in *Xinfa Suanshu* and an oblique-epicyclic orbit in *Yuzhi Lixiang Kaocheng*. The secular variation of the obliquity of the ecliptic was distinctly expressed in *Xinfa Suanshu*, of which the retarding rate is 45″.454545 per 100 years, being slightly smaller than the numerical value 46″, and the obliquity of the ecliptic was 23°31′30″ in *Xinfa Suanshu* and 23°29′30″ in *Yuzhi Lixiang Kaocheng*.

In the second year of the Chongzhen Emperor (1629), astronomical solar eclipse was not correctly predicted by the Qintianjian (Royal Observatory), and the Ministry of Rites of the Ming Dynasty presented a memorial about repairing the calendar, which was approved by the Emperor. Xu Guangqi (1562-1633), Li Zhizao (1565-1630), Li Tianjing (1579-1659) and the Jesuits Nicolas Longobardi (Long Huamin, 1559-1654), Jean Terrenz (Deng Yuhan, 1576-1630), Jacques Rho (Luo Yagu, 1593-1638), Johann Adam Schall von Bell (Tang Ruowang, 1591-1666) and some Chinese astronomers had put forward time after time ideas to reform the calendar and compile new calendrical books. At the end of the seventh year of Chongzheng Emperor (1634), the Books, which were called *Chongzhen Lishu*, *Chongzhen reign-period Treatise on Calendrical Science*, which form the first Jesuit astronomical encyclopaedia, were classified five times with a total of 46 entries in 137 volumes. They were been presented to the Emperor for his consideration and judgment. Today, this original version of the Books is no longer complete, but is scattered both at home and abroad.[1]

In the second year of the Shunzhi Emperor (1645) of the Qing Dynasty, Johann Adam Schall von Bell presented a revised version of *Chongzhen Lishu* with a new title: *Xiyang Xinfa*

Theories of Solar Motion in *Chongzhen Lishu*, *Yuzhi Lixiang Kaocheng* and *Lixiang Kaocheng Houbian*

Lishu, *Treatise on Calendrical Science According to the Western Method*, of 32 entries in 103 volumes to the Emperor. Then the official almanacs of Chinese tradition, based on the calendrical books, were put into use. In 1784, *Xiyang Xinfa Lishu* was renamed as *Xinfa Suanshu*, *Mathematical Treatise According to the new Method*, in order to respect the Qinglong Emperor's style name *Hongli*, and compiled into *Qinding Siku Quanshu*, *Complete Books in Four Treasuries Royally Determined*. All the above-mentioned treatises select the year of 1628 as the epoch of the Calendar and give the astronomical parameters for 200 years from 1628 to 1827[2], though some of the main parameters for the solar motion were revised (Fig. 1).

In 1669, *Kangxi Yongnian Lifa*, *Eternal Calendrical Method for the Kangxi Emperor*, was compiled by Ferdinand Verbiest (Nan Huairen, 1623-1688). It gives the astronomical parameters from 1828 to 3827 in 32 volumes, 4 volumes each for the Sun, the Moon, Five Planets (Saturn, Jupiter, Mars, Venus and Mercury) and Eclipses.[3]

In 1683, *Jiaoshi Lishu*, *The Calendrical Book of Eclipses*[4], was compiled by Ferdinand Verbiest. It was of historic significance in the development of Chinese Calendars in the early Qing Dynasty, and has not been meticulously investigated until now, and is not even mentioned in volume 3 of *Science and Civilisation in China* (Cambridge University Press, 1959) and the volume of astronomy in *The History of Science and Technology in China* (Beijing, Science Press, 2003).[5,6] *The Calendrical Book of Eclipses* consisted of *Huangdao Jiushidu Biao*, *Tables of Ninety Degrees of the Ecliptic*, and *Taiyang Gaodu Biao*, *Tables of Solar Latitude*,

(a) [1]:582

歷元後二百恆年表

紀年	度十分	十秒	十微	最高衝度	十分	十秒	宿	紀日
戊辰	五三	三五	三九	五	五九	五九	井	己卯
己巳	三九	一六	一七	六	○○	四四	鬼	甲申
庚午	二四	五六	五五	六	○一	二九	柳	己丑
辛未	一○	三七	三三	六	○二	一四	星	甲午
壬申	五五	二六	三一	六	○二	五九	翼	庚子
癸酉	四一	○七	○九	六	○三	四四	軫	乙巳
甲戌	二六	四七	四七	六	○四	二九	角	庚戌

欽定四庫全書

新法算書 卷二十五

紀年	度十分	十秒	十微	最高衝度	十分	十秒	宿	紀日
乙亥	一二	二八	二五	六	○五	一四	亢	乙卯
丙子	五七	一七	二三	六	○五	五九	房	辛酉
丁丑	四二	五八	○一	六	○六	四四	心	丙寅
戊寅	二八	三八	三九	六	○七	二九	尾	辛未
己卯	一四	一九	一七	六	○八	一四	箕	丙子
庚辰	五九	○八	一五	六	○八	五九	斗	壬午
辛巳	四四	四八	五三	六	○九	四四	女	丁亥
壬午	三○	二九	三一	六	一○	二九	虛	壬辰

(b) [2],788:402

Fig. 1 The main parameters for the solar motion in (*Chongzhen Lishu* and *Xinfa Suanshu*)

though both of them have distinctive page numbers. *Tables of Ninety Degrees of the Ecliptic*, specifically called *Shengjing Jiushidu Biao*, *Tables of Ninety Degrees for Shengjing* in *Qingshigao*, *the Miscellany on the History of the Qing Dynasty* was ordered by imperial edict *to be eternally followed* (*Yongyun Zunshou*). It has two parts, *Huangdao Jiushidu Biao Tushuo*, *the Explanation through Diagrams*, in three leaves, and the relative tables in six leaves. *Tables of Solar Latitude* are on 11 leaves without explanation. As examined, *the Explanation through Diagrams* only revealed the three main calculating steps for the compilation of *Tables of Ninety Degrees of the Ecliptic* (Libiaofa zhiyao yi you san), but the eight steps are absolutely necessary. The obliquity of the Ecliptic of 23°32′ was applied in *Tables of Ninety Degrees of the Ecliptic* and otherwise the obliquity of the Ecliptic of 23° 30′ (Er shi san du ban) was permutated in *Tables of Solar Latitude.* Furthermore, the methods of calculation and permutation in *Tables of Solar Latitude*, of which the error is 4′, and *Tables of Solar Latitude* in *Yuzhi Lixiang Kaocheng*, *Through Investigation of Calendrical Astronomy Imperially Composed*, in which the obliquity of the Ecliptic is 23°29′30″, are rather different. *Yuzhi Lixiang Kaocheng* was issued in the second year of the Yongzheng Emperor (1724) in 3 parts of 42 volumes and is usually called *Jiazi Yuanli.* It selected the year 1684 as the epoch of the Calendar and gave astronomical parameters for 300 years from 1684 to 1983. [7]

In the seventh year the Qianlong Emperor (1742), *Yuzhi Lixiang Kaocheng Houbian*, *Supplement to Through Investigation of Calendrical Astronomy Imperially Composed*, was issued in 10 volumes and is usually called *Guimao Yuanli*. It selected the year 1723 as the epoch of the calendar and gave astronomical parameters for 300 years from 1723 to 2022. [8]

Therefore, from 1645 to 1742, the four calendrical books, *Xiyang Xinfa Lishu*, *Kangxi Yongnian Lifa*, *Yuzhi Lixiang Kaocheng* and *Yuzhi Lixiang Kaocheng Houbian*, which are abbreviated *XYXFLS*, *KXYNLF*, *LXKC* and *LXKCHB*, were successively put into use. All the almanacs during the Qing Dynasty were imperially given a general name, *Shixianli*, which included *Jiazi Yuanli* and *Guimao Yuanli*.

In 1959, Dr. Joseph Needham pointed out:

> The historical vicissitudes of the Chinese calendar will, indeed, remain a matter of some difficulty, as the definitive monograph on this subject has not yet been written, either in Chinese or a western language. Fortunately, however, it is not of primary scientific importance. The various shifts to which the calendar experts were put by their inaccurate knowledge of precession, planetary cycles, etc., need not delay us too much. What seem really interesting in Chinese astronomy are such questions as the ancient and medieval cosmic theories, the mapping of the heavens and the coordinates used, the understanding of the great circles of the celestial sphere, the use of circumpolar stars as indicators of the meridian passages of invisible equatorial constellations, the study of eclipses, the gradual development of astronomical instruments (which by the +13th century had attained a level much higher than that of Europe), and the through recording of observations of important celestial phenomena. [5]

The characters of the leap year, the equation of the center of the Sun and the secular variation of the obliquity of the ecliptic discussed below could be regarded as a definitive monograph on the calendrical science in the Qing Dynasty and give a preparatory resolution of the fifth "question", the study of eclipses.

1. The Characters of the leap years of calendrical treatises in *XFSS*, *KXYNLF*, *LXKC* and *LXKCHB*

Based upon *Liyuan Hou Erbai Hengnian Biao* of *Richan Biao Juan Yi*, volume 1 of *the Solar Tables* in *Chongzhen Lishu*, abbreviated *CZLS*, and that of *Richan Biao Juan Ershiwu*, volume 25 in *Xinfa Suanshu*, which is abbreviated as *XFSS*, *Jiaoshi Juan Yi*, *Er*, *San* and *Si*, volumes 1, 2, 3, 4, in *KXYNLF*, *Taiyang Niangeng Biao* of *Richao Biao* in *LXKC* and *Taiyang Niangeng Biao* of *Richao Biao* in *LXKCHB*, the leap years, which are related to the tropical year, in the four calendrical Books are outlined as Table 1 and Table 2.

Tab. 1 The leap years in *XFSS*, 1628-1827, in *KXYNLF*, 1828-, and in *LXKC*, 1684-1983

	1644*	1677	1710	1743	1776	1809	1842	1875	1908	1941	1974
	48	81	14	47	80	13	46	79	12	45	78
	52	1685	18	51	84	17	50	83	16	49	82[3)]
	56	89	22	55	88	21	54	87	20	53	86
	60	93	26	59	92	25[1)]	58	91	24	57	90
1631	64	1697	30	63	1796	29[2)]	62	1895	1929*	1962*	1995*
35	68	1701	34	67	1801*	1834*	1867*	1900*	33	66	1999
1639	1673*	1706*	1739*	1772*	05	38	71	04	37	70	2003

The leap years with * have 4 year intervals, and the others are of 3 years

1) For *XYXFLS* 200 years from 1628 to 1827
2) For *KXYNLF* 2000 years (1828-3827), based on the 4 volumes for ecliptic tables
3) For *LXKC* 300 years from 1684 to 1983

Tab. 2 The leap years in *LXKCHB*, 1723-2022 [7]

	1743	1776	1809	1842	1875	1908	1941	1974	2007
		47	80	13	46	79	12	45	78	11
		51	84	17	50	83	16	49	82	15
		55	88	21	54	87	20	53	86	2019
1726	59	92	25	58	91	24	57	90	
30	63	1796	29	62	95	28	61	94	
34	67	1800	33	67	1899	32	65	1998	
1739*	1772*	1805*	1838*	1871*	1904*	1937*	1970*	2003*	

Therefore, the leap years, being intercalated in *XYXFLS* and successively in *KXYNLF* have been duplicated in *LXKC*. *LXKCHB* is based on the 33-year pattern of leap years (there is a rather exact accord between days and years over this interval, with eight days being intercalated per 33 years). 1900 in *KXYNLF* and *LXKC*, and 1800 in *LXKCHB*, were selected as the leap year. So the four calendars are uniquely Chinese creations. [7,8]

2. Models for calculating the annual equation of the mean motion of the Sun in *XFSS*, *LXKC* and *LXKCHB*

The theory of solar motion in *XFSS* was derived from the model of an eccentric circle. The distance between its center and another center is 3584, where the mean distance between the Earth and the Moon was taken to be 100 000, the eccentricity of the orbit was assumed to be 0.017 92.

The two illustrative examples were only given in *Suan Jiaojian Biao Shuo*, *Explanation on the calculation of the annual equation of the mean motion of the Sun*, in *Richan Biao Juan Er* of *XFSS*, but were not included in the orginal version of *CZLS*.

As noted in Fig. 2, Yinshu *M* is measured from the perigee of the Sun assumed to be 30°. Then,

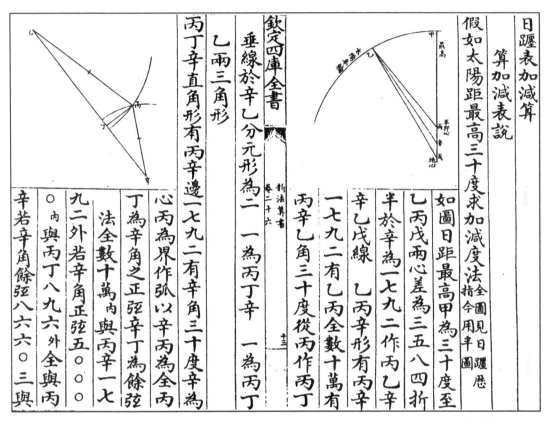

Fig. 2 The illustrative model for the calculation of the equation of the center in *XFSS*[2], 788:437

$\alpha = \arcsin (0.01792 \sin M) = 30'48''.1573932$;

$\beta = \text{arctg} [0.01792 \sin M \div (\cos\alpha - 2e \cos M)] = 0°.5298216083 = 31'47''.35779$;

$\alpha + \beta = 30'48''.1573932 + 31'47''.35779 = 1°02'35''.5151832$.

The corresponding values are $30'46''$, $31'44''$, $1°02'30''$ in the text[2], 788:437-438 and $1°02'33''$ in the Table of *XFSS*[2], 788:431. So, the formula for calculating the equation of the mean motion of the Sun in *XFSS* is given as follows:

the equation of the center

$= \arcsin (e \sin M) + \text{arctg} \{e \sin M \div [\cos (\arcsin (e \sin M)) - 2e \cos M]\}$.

The relative values calculated by the formula, in the Tables of *XFSS* and in *Tychonis Brahe Astronomiae Instauratae Progymnasmata* (1602) are juxtaposed in Table 3.

Tab. 3 The relative values of the equation of the center by the formula, tables of *XFSS* and Tycho (1602)

M	Values by the formula				Tables of XFSS[1]	Tycho's Tables[2]
	α	β	α + β	In °, ′ and ″		
0	0	0	0	0°	0°	0°
15	0.2657409146	0.275270404	0.5410113186	0°32′27″.6407472	0°32′28″	0°30′48″
30	0.5133770537	0.5298216083	1.043198662	1°02′35″.5151832	1°02′33″	0°59′44″
45	0.7260345073	0.744912064	1.470946571	1°28′15″.4076556	1°28′14″	1°24′56″
60	0.8892189389	0.905443817	1.794662756	1°47′40″.7859216	1°47′40″	1°44′47″
75	0.99180457	1.001090286	1.992894856	1°59′34″.4214816	1°59′32″	1°57′52″
87	1.025387991	1.027314844	2.052702835	2°03′09″.730206	2°03′08″	2°02′45″
88	1.026169766	1.027454836	2.053624602	2°03′13″.0485672	2°03′11″	2°02′55″
89	1.026638926	1.027281451	2.053920377	2°03′14″.1133572	**2°03′12″**	2°03′03″
90	1.026795329	1.026795329	2.053590658	2°03′12″.9263688	2°03′10″	2°03′09″
91	1.026638926	1.025997205	2.052636131	2°03′09″.4900716	2°03′06″	3°03′13″
92	1.026169766	1.024887907	2.051057673	2°03′03″.8076228	2°03′00″	**2°03′15″**
93	1.025387991	1.023468353	2.048856344	2°02′55″.8828384	2°02′52″	2°03′13″
105	0.99180457	0.9826895156	1.974494086	1°58′28″.1787096	1°58′28″	2°00′04″
120	0.8892189389	0.8735652558	1.762784195	1°45′46″.023102	1°45′46″	1°48′36″
135	0.7260345073	0.7080900452	1.434124553	1°26′02″.8483908	1°26′02″	1°29′23″
150	0.5133770537	0.4979225533	1.011299607	1°00′40″.6785852	1°00′40″	1°03′33″
165	0.2657409146	0.25684914	0.5225900546	0°31′21″.3241968	0°31′20″	0°33′01″
180	0	0	0	0°	0°	0°
210	-0.513377053	-0.4979225533	-1.011299606	-1°00′40″.6785816	-1°00′40″	-1°03′33″
240	-0.8892189389	-0.8735652558	-1.762784195	-1°45′46″.023102	-1°45′46″	-1°48′36″
270	-1.026795329	-1.026795329	-2.053590658	-2°03′12″.9263688	-2°03′10″	-2°03′09″
300	-0.8892189389	-0.905443817	-1.794662756	-1°47′40″.7859216	-1°47′40″	-1°44′47″
330	-0.5133770537	-0.5298216083	-1.043198662	-1°02′35″.5151832	-1°02′33″	-0°59′44″
360	0	0	0	0°	0°	0°

1) [2], 788: 431-436

2) *Tychonis Brahe Astronomiae Instauratae Progymnasmata*, Bragae Bohemiae, M. DC. II., 60-61

In *JiaoShi LiZhi Juan Er* of *CZLS*[1]:241-242, two equivalent formulas for calculating the equation of the mean motion of the Sun were implied, i. e.,

the equation of the center

$$= \arcsin(z e \sin M \div \sqrt{1 + 4e^2 + 2e\cos M}), \text{ or}$$

$$= 180° - \frac{M}{2} + \text{arctg}\left[(1-2e)\,\text{tg}\,\frac{M}{2} \div (1+2e)\right].$$

The theory of the solar motion in *LXKC* was derived from the model of an eccentric circle, as illustrated in the model of oblique-epicyclic circles (Benlun-Junlun). The diameters of the epicycle and oblique circle are assumed to be 268 812 and 89 604, respectively, when the mean distance between the Earth and the Moon was taken to be 10 000 000. The equation of the mean annual motion of the Sun depends on the eccentricity of the Earth's orbit around the Sun, which is 179 208 of such parts, where the Earth's mean distance from the Sun would be 10 000 000. The maximum value for of the Equation of the Sun's center is 2°03′11″.

the equation of the center = arctg $[0.0358416 \sin M \div (1 - 0.0179208 \cos M)]$.

The theory of solar motion in *LXKCHB* was derived from "the simplified elliptic", not solved in the form of Kepler's equation, and the eccentricity was taken as 0.0169. [9] Then,

the equation of the center = $2 \arcsin (e \sin M \div \sqrt{1 + e^2 - 2e \cos M})$
$\pm [\text{arctg} (10000000 \times \text{tg}M \div 9998571.85) - M]$.

The models in *XFSS*, *LXKC* and *LXKCHB* are an eccentrical circle, oblique-epicyclic circles and "the simplified elliptic". The calculated value for the maximum of the equation of the center of the Sun in *XFSS* is 2°03′14″, and the value given in the table is 2°03′12″, while the calculating and given values are 2°03′11″ in *LXKC* and 1°56′13″ in *LXKCHB*.

3. Secular variations in the precession and obliquity in *XFSS*

In the history of observations showing that the precession of the equinoxes and solstices is not uniform, chapter 2, Book Ⅲ of *On the Revolutions*, Nicholas Copernicus pointed out:

> To 1° there will have to be assigned, as will be seen, not 100 years at all, but 66 years. Moreover, in the 741 years from Ptolemy [to Al-Battani], only 65 years are to be assigned to 1°. Finally, if the remaining period of 645 years is compared with the difference of 9°11′ of my observation, 1° will receive 71 years. Hence in those 400 years before Ptolemy, clearly the precession of the equinoxes was slower than from Ptolemy to Al-Battani, when it was also quicker than from Al-Battani to our times. [10]:122

And Eduard Rosen annotated in his translation and commentary version of *On the Revolutions*, this wholly imaginary nonuniformity in the precession of the equinoxes was discarded by Tycho Brahe. In his *Astronomiae instauratae mechanica* (Wandsbek, 1598; *Opera omnia*, Copenhagen, 1913-1929, V, 113, lines 9-17) the great Danish astronomer remarked:

> I have also noticed that in the longitudes [of the stars] the intricacy of the nonuniformity is not as great as Copernicus believed. For, what he imagined in this regard insinuated itself through defects in the observations, both ancient and recent. Hence the precession of the equinox in these times is also not as slow as he indicated. For at present the fixed stars traverse 1°, not in 100 years in accordance with his computation, but in only 71 1/2 [years]. In the past they always uniformly completed very nearly this [motion], if the observations of [our] predecessors are properly delimited, with only a trivial irregularity, arising accidentally from another source. [10]:384

In *XFSS*, the precession of the equinoxes is given as 51″ per year, i.e., 1° is assigned to 69 years 91 days 73 Ke (1 Ke equals 15 minutes), which is different from that of Copernicus and Tycho Brahe. [11]

The secular variation of the obliquity of the ecliptic once was an issue relative to the precession of the equinox. Nicholas Copernicus pointed out:

> Likewise in the motion of the obliquity a difference is discovered. For, Aristarchus of Samos found the obliquity of the ecliptic and equator to be 23°51′20″, the same as Ptolemy; Al-Battani, 23°36′; Al-Zarkali the Spaniard, 190 years after him, 23°34′; and in the same way 230 years later, Profatius the Jew, about 2′ less. But in our time it is found not greater than 23°28 1/2′. Hence it is also clear that from Aristarchus to Ptolemy, the motion was a minimum, but from Ptolemy to Al-Battani a Maximum. [10]:122

The corresponding values for the obliquity of Profatius (c. 1236-1305, Ibn Tibbon, Jacob ben Machir) 23°32′, of George von Peurbach (1423-1461) 23°28′, of N. Copernicus 23°28′24″ and of T. Brahe 23°31′30″ were included in *XFSS*.[12] Furthermore, the secular variation of the obliquity is implied in *Taiyang Pingxing Yongbiao* (not included in *CZLS*) of *XFSS*, of which the retarding rate is 45″.454545 per 100 years, i. e., 1° in 7920 years (132 Jiazi years), being slightly smaller than the numerical value 46″.[13]

References and Notes

[1] Chongzhen Lishu, Chongzhen reign-period Treatise on Calendrical Science, 2 volumes, Edited by Xu Guangqi, Compiled by Pan Nai. Shanghai: Shanghai Ancient Books Press, 2009

[2] Xiyang Xinfa Lishu, was renamed as Xinfa Suanshu, Mathematical Treatise accoding to the new Method, in Qinding Siku Quanshu, Complete Books in Four Treasuries Royally Determined. Vol 788. Taibei: The Commercial Press, 1983

[3] In 1669, Kangxi Yongnian Lifa, Eternal Calendrical Method for the Kangxi Emperor, compiled by Ferdinand Verbiest, gives astronomical parameters from 1828 to 3827 in 32 volumes, and 9 volumes of KXYNLF are kept in the National Library of China

[4] The Calendrical Book of Eclipses, compiled by Ferdinand Verbiest in 1683

[5] Needham J. Science and Civilisation in China. Vol 3. Cambridge: Cambridge University Press, 1959. 173

[6] 陈美东. 中国科学技术史·天文学卷. 北京: 科学出版社, 2003

[7] 鲁大龙. "癸卯元历"闰年的特点. 中国科技史料, 1998, 19 (3): 87-94, 90

[8] Kollerstrom N. How Newton Inspired China's Calendar? Astrnomy and Geophysics. The Journal of the Royal Astronomical Society, 2000, 41 (5): 21-22

[9] 鲁大龙.《历象考成后编》的"均数"和"推日躔法". 中国科技史料, 2003, 24 (3): 244-254, 247

[10] Copernicus N. On the Revolution, Edited by Jerzy Dobrzycki, Translation and Commentary by Edward Rosen. The Macmillan Press Ltd, 1978

[11] 关于"岁差",《新法算书·日躔历指》卷首曰:"岁差者,日与恒星比论乃得之也;未论恒星,未可论岁差也。"[2],788: 367

《崇祯历书·恒星历指一卷》"恒星历叙目"对"岁差"的论述是:

黄赤二道,位置不等,其各两极不等,二经二纬,纵横不等,交互不等,故令星行不等,其差亦不等。有名为有差,而绝不可谓差者,黄道之经度是也。恒星依黄道东行,如载籍相传尧时冬至日躔约在虚七度,今躔箕四度,四千年间,而日退行若干度者,即星之进行若干度也。古历谓之岁差,各立年率。郭守敬以为六十六年有奇而差一度,今者斟酌异同,辨析微眇,定为每岁东行一分四十三秒七十三微二十六纤,六十九年一百九十一日七十三刻而行一度,凡二万五千二百〇二年九十一日二十五刻而行天一周，终古恒然也。此立名为差，而实有定法，不可谓差者也。
薄树人．《崇祯历书·恒星历指一卷》"恒星历叙目"．中国科学技术典籍通汇·天文卷（第八分册）．郑州：河南教育出版社，1997．1371-1375，1372．《新法算书·恒星历指卷一》未收录上述的"恒星历叙目"。

[12]

《日躔历指》西历名家关于"黄道与赤道之距度（黄赤交角）"的测定

	"古今各测"术文[1)]	Times	Astronomers	Obliquity
1	周显王二十五年丁丑，迄崇祯元年戊辰为一千九百七十二年，西古史亚理大各	BC 344	（萨摩斯的）阿里斯塔克（Aristarchus，约公元前310～约前230）[3)]	
2	秦二世三年甲午，迄崇祯元年戊辰为一千八百四十七年，西史陑（通"厄"）腊多	BC 207	埃拉托色尼（约公元前276～前194，Eratosthenes）	
3	汉景帝中元元年壬辰，迄崇祯元年戊辰为一千七百七十七年，西史意罢阁	BC 149	喜帕恰斯（Hipparchus，约公元前190～前125）	23°51′20″
4	汉光武建武十七年辛丑，迄崇祯元年为一千四百八十八年，西史多勒某，其书为历家之宗。已上四家测定黄赤相距为二十三度五十一分二十〇秒，于中分为二十三度八十五分	AD 40[2)]	托勒密（约90～168，Claudius Ptolemaeus）[13]	
5	唐僖宗广明元年庚子，迄崇祯元年为七百四十八年，西史亚耳罢德测定二十三度三十五分，于中分为二十三度五十八分三十三秒	AD 880	阿尔—巴塔尼（Al Battani，约858～929）	23°35′
6	宋神宗熙宁三年庚戌，迄崇祯元年为五百五十八年，西史亚杂刻测定二十三度三十四分，于中分为二十三度五十六分六十七秒	AD 1070	阿尔—卡尔扎里（约1029～1087，Al Zarkali）	23°34′
7	宋高宗绍兴十年庚申，迄崇祯元年为四百八十八年，西史亚尔满测定二十三度三十三分，于中分为二十三度五十五分	AD 1140	Al-Mamun[4)]	23°33′
8	元成宗大德四年庚子，迄崇祯元年为三百二十八年，西史波禄法测定二十三度三十二分，于中分为二十三度五十三分三十三秒	AD 1300	普罗法提阿斯（约1236～1305，Profatius；Ibn Tibbon，Jacob ben Machir）	23°32′
9	天顺四年庚辰，迄崇祯元年为一百六十八年，西史襃尔罢测定二十三度二十八分，于大统历为二十三度四十六分六十七秒	AD 1460	皮欧巴赫（George von Peurbach，1423～1461）	23°28′

· 133 ·

续表

	"古今各测"术文[1]	Times	Astronomers	Obliquity
10	正德十年乙亥，迄崇祯元年为一百一十三年，西史歌白尼测定二十三度二十八分二十四秒，于大统历为二十三度四十八分一十二秒	AD 1515	哥白尼（N. Copernicus, 1473~1543）	23°28′24″
11	万历二十四年丙申，迄崇祯元年为三十二年，西史第谷造铜铁测器十具，甚大甚准，又算地半径差及清蒙差，岁岁测侯，定为二十三度三十一分三十○秒，西土今宗用之，于大统历为二十三度五十二分三十○秒。第谷覃精四十年，察古史测法，知从来未觉有清蒙之气及地之半径两差，又旧用仪器，体制小、分度粗，窥筒孔大，所得余分不过四分度或六分度之几而已，且古来测北极出地之法未真未确，故相传旧测，俱不足依赖以定太阳躔度	AD 1596	第谷（T. Brahe, 1546~1610）	23°31′30″

1)［2］,788：372-373；本表参考：严敦杰. 明清之际西方传入我国之历算记录. 梅荣照主编. 明清数学史论文集. 南京：江苏教育出版社, 1990. 114-181
2) 拟应改正为"汉顺帝永和五年庚辰"（A. D. 140）
3) 如果这里的数据取自哥白尼的《天体运行论》，哥白尼在此"所犯的历史性错误"，使"亚里大各"指"阿里斯塔科"（或《天体运行论》中译本的"阿里斯塔尔恰斯"），而不是正确的"阿里斯泰拉斯"（Aristyllus）[10]:385
4)［哈里发］Al-Mamun［时代的天文学家］测定23°33′于830年. 时间不一致，汉译时间当伪"（严敦杰, 133）. [13]

太阳平行永表[2],788:408-409

甲子数	距冬至		最高冲			宿数	纪日
	度	分	宫	度	分		
一	○	二二	一○	一七	一二	一四	四○
二	○	二二	一○	一七	五七	○五	五五
三		二三		一八	四二	二四	一○
四		二三		一九	二七	一五	二五
五		二四		二○	一二	○六	四○
六五	○	五一	○	五	一二	二六	四○
天启四年							
六六	○	五二	○	五	五七	一七	五五
百三一	一	二二	一	二四	四二	二○	一○
百三二	一	二二	一	二五	二七	一一	二五

Further References

潘鼐. 西洋新法历书提要. 载：薄树人. 中国科学技术典籍通汇·天文卷（第八分册）. 郑州：河南教育出版社, 1993. 643—650；潘鼐.《崇祯历书》的成书前后.《中国天文学史文集》编辑组. 中国天文学史文集（第六集）. 北京：科学出版社, 1994. 1-29

江晓原. 第谷. 载：席泽宗. 世界著名科学家传记（天文学家I）. 北京：科学出版社, 1990. 8-34；江晓原, 钮卫星. 天文西学东渐集. 上海：上海书店出版社, 2001. 259-413

陈美东. 古历新探. 沈阳：辽宁教育出版社, 1995

付邦红, 石云里.《崇祯历书》和《历象考成后编》中所述的蒙气差修正问题. 中国科技史料, 2001, 22（3）. 260-268；石云里.《历象考成后编》中的中心差求法及其日月理论的总体精度——纪念薄树人先生逝世五周年. 中国科技史料, 2003, 24（2）. 132-146

祝平一. 西历东渐：晚明中西历法的初步接触与历法形式的冲突. 载：《法国汉学》丛书编辑委员. 法国汉学（第六辑, 科学史专号）. 北京：中华书局, 2002. 336-344

宁晓玉.《新法算书》中的月亮运动理论. 自然科学史研究, 2007, 26（3）. 352-362

大桥由纪夫.《历象考成》中的太阳运动论. 内蒙古师范大学学报（自然科学中文版）, 2007, 36（6）：662-665

哥白尼. 天体运行论. 叶式辉译, 易照华校. 武汉：武汉出版社, 1992

邓可卉. 希腊数理天文学溯源——托勒玫《至大论》比较研究. 山东：山东教育出版社, 2009

Hashimoto K. Hsü Kuang-ch'i and Astronomical Reform: The Process of the Chinese Acceptance of Western Astronomy, 1629-1635. Kyoto: Kansai University, 1988

Shi Y L. Reforming Astronomy and Compiling Inperial Science in the Post-Kangxi Era: The Social Dimension of the Yuzhi lixiang kaocheng houbian, in East Asian Science, Technology and Medicine, 2008, 28: 36-81

The Vacancy of Error Ideas about the Calculation of the Chinese Traditional Calendar

Wang Yumin

(Beijing Ancient Observatory, Beijing Planetarium, Beijing 100005, China)

In the light of the theory of modern error and approximate calculation, this article organizes and analyzes the vacancy of error ideas about the Chinese traditional calendar. On the basis of the marked features of the calendar using fractional, mysterious number and the grand origin, the article firstly disserts that the Chinese traditional calendar is an exploration activity of grasping the "perfect" celestial body motion. And secondly, the article concludes that the Chinese traditional calendar always lacked the unified understanding of observational accuracy, model precision and calculating accuracy. Thirdly, according to the error analysis of the past dynasties' main astronomical instruments, we find that the method of measuring the length of winter solstice shadow to define tropical year is a backward technology, comparing with measuring the summer solstice in early stage. Lastly, the article analyzes the deviation of calculating accuracy of some calendars through some ancient calculating cases, especially the *Shoushi Calendar*.

中国古代历法推算中的误差思想空缺

王玉民

(北京天文馆古观象台)

研究中国古代历法时，我们时常为古人计算动辄精确到小数点后六七位，或使用巨大分子分母构成的精确数而惊叹，按说，用如此精密算法修成的历法，其可靠性也应该同样令人惊叹，然而，纵观各朝各代，一部历法经常使用几十年就因不合天而被废弃，这与历法推算中的表现精度完全不成比例。

在科学研究中，对精度的追求是一项系统活动，具体在对天体运行规律的把握中，需要观测精度、模型精度、计算精度的高度统一才能达到理想状态。古代历法家由于缺少误差思想的指导，在对观测精度不高、对天体运行的规律掌握尚有限的情况下，单纯追求计算的高精度，多是徒劳无功的，因为计算结果比计算本身更重要，没有意义的数字位数参与运算是一种资源的浪费。

The Vacancy of Error Ideas about the Calculation of the Chinese Traditional Calendar

1. 中历很长的历史时期内都在追求对"完美"天行的把握

古代各朝频频改历，小部分是政治因素，大部分是验天不合或不精造成的。

在"天不变，道亦不变"思想的指导下，古人存在一种能够对天行"终极把握"的理想，认为天体的运行周期是可以通约的（至少在中古前是这样），战国、秦汉时代的"四分历"，认为一回归年为 $365\frac{1}{4}$ 天，闰周为 19 年 7 闰，于是现代人根据这两个数值替古人推出他们的一朔望月为 $29\frac{499}{940}$ 天，其实那时的历法家不可能测量出这么精确的朔望月平均数值，古人只是通过当时尽可能精密的观测认为阴阳合历有 19 年 7 闰的关系而已，这是古人对天体运行通约关系的一种良好设想。按此前提，实际上回归年与朔望月长度是不能各自独立确定的。

在追求对天行终极把握理想的推动下，各朝历法都设法使其精度在前人的基础上有所提高，这种提高是符合科学精神的，但也由此形成了一种观念，认为历法就应该是一种不断改变的东西。甚至清代，历法精度大幅提高时，历法家王锡阐仍说："历之道主革，故无百年不改之历"[1]:592，把改变历法当成了常态。可是，由于天体运行的复杂性，运行周期找到通约的理想是无法实现的。

于是古人试图寻找另外的途径作为补充，乐律是有明显周期的，既然日长、年长都有周期，所以他们认为天体运行也会像乐律一样遵守简明的数理关系，因此从很早开始，历法就和乐律联系在了一起，称"律历"，历法的某些参数需要符合黄钟、律吕之数；另外，根据古代根深蒂固的数字神秘观念，古人还认为，某些数字存在万能的功用，包含着万物之理，所以人们经常让历法的某些参数与一些神秘数字相合（如乾象、大衍之数等）。综合这些因素，古人认为一部完美的历法应该是既合天，又满足乐律和神秘数字，三者兼得。

随着观测精度的提高，人们发现简单的通约关系（如"19 年 7 闰"）并不能合天，但多数人仍相信天体运行周期是有通约关系的，只是更复杂而已。于是古人使用了一些更为"精密"的参数，如西汉《太初历》取朔望月为 $29\frac{43}{81}$ 日，东汉刘洪《乾象历》定回归年长为 $365\frac{145}{589}$ 日等。这些数据包含实测成分，但也包含有靠律、神秘数制造虚幻精度的成分，来自对完全精密理想的一种向往。

上元积年是古人追求对完美天行把握的最典型体现。上元是古代历法家寻找的一个理想推算起点，他们相信，按照他们观测获知的日、月、五星的运行周期，叠加回推，最后总能找到一个理想时刻：这一刻是一个甲子日的夜半，又是日月合朔的时刻，又恰是太阳过冬至点，同时在冬至点发生罕见的"五星联珠"。正式推算上元积年的工作从西汉的刘歆开始，他在《三统历》中推到 143,127 年前，才找到这个理想时刻。[2]:77 后来随着交点月、近点月等周期的发现，这些参数也常被加入到上元推算中去，随着各种参数的精密化，理想推算起点变得越来越难找，只好继续向远古推去，求出的上元积年数字，到唐代的《大衍历》达到 96,661,740 年[2]:623，金代的《重修大明历》更达到

了383,768,657年。[2]:1266

古人这么孜孜以求庞大的上元积年，除了基于天行完美的假设外，还有这样一种观念：想为历法推算、天体位置和天象的推算找到一个通用程序，真正达到孟子说的"千岁之日至，可坐而致也"[3]的境界。可是古代通常的演算方法是"筹算"，用摆放和移动算筹（小竹棍）来排列算式，非常费时费力，这样，太大的计算量就会成为历法家沉重的负担。

在对中国古代历法的研究中，推算上元积年的方法因为涉及解一次同余式，是数学史上的一大贡献，被大加赞赏，但因其推算的思想不甚"科学"而经常被忽略或淡化。实际上，推算上元积年从方法和思想上都是古代历法最核心的内容。陈遵妫曾说："一部中国历法史，几乎可以说是上元的演算史。"[4]上元积年的确定需要同时考虑许多长短不齐的周期，尤其是这些周期达到小数点后许多位的精度时，回推上去想找到一个共同的"起跑线"，具有非常高的难度，精确的位置恐怕永远也找不到，今人研究发现，古人设定上元时并不完全是客观运作的。

曲安京经过研究证实，古代围绕着上元积年的计算，存在着一个天文常数系统，推导上元积年时只用60干支周期、朔望月和回归年长度进行推导。如果布列的同余式组无解，或求出的上元积年超过1亿年，说明取的朔望月和回归年长度"不合用"，需要"设计"新的朔望月和回归年长度，重新计算，若所得结果小于1亿年，说明该上元符合要求，于是定出上元和上元积年。至于恒星年、近点月、交点月、五行星会合周期等数值，按这亿年数量级的尺度作出调整，总是能找到一个既满足上元起点又非常符合今日位置的长度周期的。[5]且不说取朔望月和回归年长度也靠改换数字、闭门造车来迁就理想上元，就说恒星年、近点月、交点月、五行星会合周期等数据在上亿年的漫漫历史长河中一分配，虽然得到一个与当时观测长度相近的值，但有没有误差，误差究竟是多少，根本不去考虑，结果这些参数仅符合了当时和较近的过去，行用日久必然会不合天。

上元积年本来是一种良好设想，但后来人们过分沉溺于这种设想，它就成了一种对理想的病态偏执和对精度的虚妄追求了，在这种追求过程中，历法家不惜削足适履，将那些本该完全靠实测求出的天体运行周期安排成"导出常数"，这是古人不懂得误差出现的必然性和使用近似计算的必要性造成的，追求完美的结果是极度不完美——历法几十年一换。

由于后世历法家的质疑，到中、近古，经唐宋时代的曹士蒍、杨忠辅等作了尝试性的改革以后，元代郭守敬在创制《授时历》中彻底废除了上元积年。

王锡阐在评价一部历法的优劣时，曾提出这样的原则："合则审其偶合与确合，违则求其理违与数违，不敢苟焉以自欺而已。"[1]:607上元积年的凑数字，就是求偶合，用这种方法产生的历法，也只能是偶合而不是确合，结果造成理不违而数违。

2. 中历一直对观测精度、模型精度、计算精度的不统一认识不足

在追求把握完美天行观念的指导下，古代历法家对误差观念十分模糊，特别是唐代以前，只用绝对精确的数字——整数和分数来表示天体运行的各种周期及其关系，计算时也尽可能追求绝对的精度。实际上，计算只是历法编制的一个方面，从科学的角度讲，只有

做到观测精度、模型精度、计算精度的统一，才能编制出一部好的历法来。

按现代误差理论，在科学研究中，定量的误差主要有3种：

（1）观测误差。就天文观测来说，仪器、方法、人员的因素，都会产生误差。仪器的安装准确程度、调准程度、刻度精密程度、瞄准精确程度、读数精确程度对误差都有重要影响，我们只能将其控制在一定范围内而不可能将其彻底消除。

（2）模型误差。就天体运行模型来说，古人对日、月、五星的运动描述时总是进行抽象简化，往往没发现或忽略掉次要因素，如对行星的南北向运动、早期对天体的不均匀运动等，这样得到的数学模型只是近似描述。

（3）截断和舍入误差。在作天体位置推算的过程中，用数值方法求的总是近似解，比如用有限过程逼近无限过程只能取有限项，计算过程的四舍五入等。这些误差积累起来也是很惊人的，这是计算方法造成的误差，又称方法误差。

我国古代，对天体位置、运行以及相应时间的测量精度如何呢？随着测量仪器的出现、完善，以及测量方法的不断改进，测量的精度也在随着历史的演变逐渐提高，但其精度与同时代的"计算精度"相比，总是非常低的。

最早出现的天文仪器是圭表和浑仪。圭表由直立的表（生成日影）和水平放置、朝向正北的圭组成，圭上标刻度，正午时，表的影子恰好完全落在圭上，通过圭上的刻度可以读出影子的长度。开始人们是寻找正午影子最短的一天，这天即为夏至，两次夏至的间隔即为一年。[7]

冬至是正午影子最长的一天，测量冬至表影同样可以确定年长。后代人们改为专测冬至表影长来确定回归年。这就需要把圭的长度大大增加，至少在汉代，圭表已经定型为"圭"长于"表"的制式（《三辅黄图》："长安灵台……有铜表，高八尺，长一丈三尺。"[8]）。对这种改变的原因，典籍中没有记载，按现代误差理论，用同一方法测量长的样本时，比测量短的样本会获得更高的精度，因为样本越长，读数的相对误差就越小。古人可能也是这么想的，这是古人改进测量精度的良好愿望，被后代认可，所以沿用下来，直到清末。

但实际上，这种改进不但没有提高，反而降低了测量精度。日影长是一种特殊的样本，具体地说，一个物体的影子越长，其影端由于太阳圆面造成的半影带就变得越宽，模糊的区域会按比例增加，所以与测量短影相比，测量长影并不会改善测量精度。而对于圭表来说，由于冬至投影的倾斜角大大减小，会进一步加剧影子的弥散，使半影带在圭面上变得更宽，测量精度不但不会维持在短影的水平，相反会降得更低。例如在北京，太阳冬至投射角约为 26.5°，夏至为 73.5°，取二者余弦的倒数之比，可知圭面上冬至半影的弥散度是夏至半影弥散度的 3.2 倍。后世长圭的圭表除了用于冬至测影外，当然还可以测量各节气的日影长度，这是非常必要的，但就利用圭表专门测量冬至影长求回归年这一点来说，这种技术"改进"实际上是一种倒退。

由于这种测冬至影长的传统，直到元代以前，圭表测影的精度一直很低，直到郭守敬发明景符后，利用小孔成像原理获得清晰的影端，测影长的精确度才明显提高。在1985～1986年，北京古观象台利用复原的郭守敬八尺铜表，加上景符进行不间断的仿古模拟观测，测得夏至前后影长误差值为 0.1～0.2 厘米，冬至前后误差为 0.4～0.7 厘米。[9]这也证明了我们上文的分析：使用景符，冬至日半影模糊带成比例增加的毛病被克服了，但不同投射角造成的弥散度问题依然存在。

据郭盛炽等人利用元代高表加景符的模拟观测，发现其误差平均也在0.47厘米左右[10]，与北京古观象台的模拟测影精度相仿，看来使用景符可以获得一个较高、较稳定的绝对精度，但不知古人意识到了没有，即使采用景符，测量冬至影长的误差仍远远高于测夏至时的误差。

在有可定量测量的圭表以前，年长的数据精度非常低，《尚书·尧典》中一年的长度为"期三百有六旬有六日"[11]，这不仅是举成数，也是测的不准，日下余分是未定之数[6]，等于只准确到"天"。圭表的观测目的是求出冬至时刻，进而求出回归年长，古人掌握了用多年测量求平均数的办法来改进年长数值，以求达到较高精度。"四分历"的回归年长为$365\frac{1}{4}$日，精度相对提高了。东汉刘洪的《乾象历》，取年长为$365\frac{145}{589}$日，更接近了我们现在知道的回归年长度。以后各朝一直到元明，回归年长的精度逐渐在提高，但比起同时算术的发展水平和计算精度，仍然是很低的。至于测冬至的时刻，精度则一直很低，南朝宋何承天测冬至的误差在50刻左右，随后祖冲之采用比例归算的方法，测冬至的误差仍有20刻，宋代误差为10刻，元代郭守敬发明高表、景符，才使冬至测定的误差降到1刻。[12]:61-64

浑仪是按照浑天说宇宙体系，将地平、子午、赤道、黄道等圈环嵌套组成的仪器，观测者通过照准天体，可利用环圈上的刻度来测量天体的坐标位置。在读数上，清以前浑仪的精度一直处在一个很低的水平。唐和唐以前的仪器都以"度"为最小单位，观测时仅可精确到半度[13]，北宋期间才出现了度以下的单位"少"（1/4度）、"半"（1/2度）、"太"（3/4度）[14]:80，元代郭守敬的简仪，刻度刻到1/10度，可以估量到1/20度。[14]:86由此可见，大多数朝代，还是停留在"一度少弱、一度少强"的水平上，据研究，宋及宋以前的仪器观测精度在某种意义上甚至不如目视直接估测的精度。[15]

古代已经有人意识到了仪器的测量误差问题，明代礼部尚书范谦说："历家以周天三百六十五度四分度之一，纪七政之行，又析度为百分，分为百秒，可谓密矣。然浑象之体，径仅数尺，布周天度，每度不及指中，安所置分秒哉？至于臬表之树，不过数尺，刻漏之筹，不越数寸，以天之高且广也，而以尺寸之物求之，欲其纤微不爽，不亦难乎？"[2]:1468表明了当时人对仪器测量精度的怀疑。可惜没有人把它与天体运行模型，以及计算程序的精度联系起来考虑。

至于漏刻测时，误差也很可观。汉代漏刻至少有1/3刻的误差，到元代《授时历》才提高到1/20刻。[12]:205-210时间的测量是"牵一发而动全身"的事，其误差对观测数据、计算结果都会有重要甚至层层放大的影响。现代误差理论证明，起始数据冬至时间的测定误差如果是一刻，那么无论用什么精密方法进行的历法推算，其预报结果的平均精度也不会比一刻更高。

而且，今人研究有时得出的古人观测精度也是虚假的。如唐梁令瓒在公元732年测的黄赤交角为24度，只精确到"度"，但被今人换算成现代度数，就成了23°39′57″，仿佛精确到了秒，然后与现代推算的当时黄赤交角理论值23°36′20″比，只差3′30″，于是虚假的精度就出来了。[12]:99 其实"24度"这值说明其精度不会好于1/2古度。

《后汉书·律历志》引《石氏星经》的话,提到黄道轨"去极百一十五度",这也不过是个精确"度"的值,但经今人手里点化:一个象限弧为 $365\frac{1}{4}$ 度 $\div 4 = 91\frac{5}{16}$ 度,115度 $-91\frac{5}{16}$ 度 $=23\frac{11}{16}$ 度,化成小数,就成了 23.6875 度[16],"去极百一十五度"成了精确到小数点后4位的数。实际上古代是到唐中叶才有十进小数的,古人很可能以为他们那"去极百一十五度"是完全精确的,不知什么是"精确度",更谈不上精确到"小数点后4位"。

古人的历法活动被后人赞为"先以密测","继以数推"[12]:30。模型精度是与观测精度紧密关联的,没有精密的观测结果,就没有建立精确模型的基础,也发现不了旧有模型的偏差。中国描述天体运行的模型是"代数体系",根据一定间隔的天文观测,将观测数据列表,在了解了天体的运行周期和盈缩后,辅之以内插等代数方法来预测天体的位置,多是来自观测数据的经验外推。较低的后续观测精度至多也只是改良一下过去的方法而已,难以发现旧模型的问题,因此这样的"继以数推"就大为可疑了。虽然有一次、二次内插法等创造,但总的来说古人更重视推算的准确性而非天文模型的合理性,反而严重影响了精度。

隋代刘焯首次把太阳的不均匀运动引入历法推算,这在历法史上是一个创举,值得高度肯定,因为这样做会使历法的精度大大提高,也为深入了解天体的运动打下基础。可是,刘焯设立的太阳不均匀运动的模型却很荒唐:冬至时太阳运行最快,然后减慢,又加快,到春分点前的开区间达到最快,过了春分点又突然改成最慢,然后又波动到了秋分……(图1)即便考虑到古人不懂得动力学上加速必须是一个渐进过程这一原理,我们也不能听任这样的模型而一言不发,按日常经验来说,即使太阳真是由羲和驾车驱动的,怕也经不起这样大起大落的跳跃式"无级变速"。这种模型不像是观测所得,而像是主观臆测,即使是观测所得,也说明观测精度太低,观测者把观测的"噪音"、误差都当成真实的太阳运动了。且不说它与今天观测到的太阳不均匀运动符合多少,只说这样一个丑陋的模型,刘焯还在用他发明的插值法精确地计算,结果只能是南辕北辙,从实用的角度来说,这种计算还不如不算。

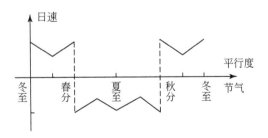

图1 刘焯的太阳不均匀运动模型[17]

历代天文观测、天体运行模型的精度在以很有限的速度提高。与此相比,古人在历法计算上一开始就追求高精度,而且越古,这种追求高精度的欲望就越强。唐及唐以前,天文历算一直使用分数,今天的历法史家经常把这些分数展成小数,然后把古人追求的精度随心所欲地定位在小数点后4位或7位上,实际上今人这样做还是低估了古人的"志向",古人对天体推算的目标是"绝对精确",所以才使用分数,因为分数是实现这一目的的最好手段——用今天的术语说,分数的有效数字是无穷多位,有真正理想的精确度。

所以说,古人用分数计算历法并不是一种单纯的求简心理(希望有个简单的分数,比十进小数省力),而是对绝对精确和完美的追求,这实际是陷入了一个不懂误差的"误区"。因为计算只是历法制定过程中的一个手段,它不是目的,计算的精确度应与所给数据的精确度相当,而计算得出数据的精确度应与结果的实际需要或实际能观测到的精确度相当,否则精确度再高也无效。

保存到现在最早的真正传世历法(不光历谱,也包括"七政"的推算),是明朝钦天监大统历科编的《大明嘉靖十年岁次辛卯七政躔度》,其中所有的表格中,推算天体位置是宿度只精确到"度",时间则精确到"刻"——这是真正可用的推算结果,至于计算过程中那么高的精度到此全部抹平了,又回到观测时的"一度少弱、一度少强"水平。明代尚且如此,再古的历法就更可想而知了。[18]

可拿今日的科学活动与古代比较:现代科学技术在误差和近似理论的指导下,测量、模型和计算三者的精度可以在很高的水平上合理匹配。现在,卫星导航的时间准确度要求在±10纳秒之内,而原子钟完全可以做到这一点;对天体轨道的确定需要天文观测的精度达到10^{-3}角秒,现代望远镜也完全可以做到这一点。当然,现代的巨型电子计算机的运算可以达到极高的精确度,比如可以把圆周率计算到小数点后超过2.7万亿位(法国科学家2010年结果),但天文学家不可能用这样的圆周率去进行天体轨道的计算,因为就现有的观测、模型精确度来说,圆周率超出小数点后十几位的计算,对精度的提高已经没有任何帮助了。

到了清代,历法家们对历法误差问题开始有了比较清晰的认识。比较有代表性的是王锡阐,他曾对历法误差问题做过较为系统的分析,他认为导致历法误差的主要原因,是制定历法的基本指导思想的失误,"创法之人,不能深推理数,而附和于蓍卦、钟律以为奇,增损于积年、日法以为定。"[1]:592

王锡阐提出误差分三方面:数差、法差和理差,其中数差相当于测量误差,法差相当于计算误差,理差相当于模型误差。不过,王锡阐只强调了测量误差,称"器精于制而不善于用"、"一器而使两人测之,所见必殊"、"一人而用两器测之,所见必殊"、"所测之时,瞬息必有迟早也"等,对理差只泛泛说"数非理也,而因理生数,即因数可以悟理"[1]:615,593,而对法差基本没提。事实上,法差,也即推算中的误差是古代历法推算中的很关键因素,也一直是被古今学者在相当程度上忽略的因素。当代学者在研究古代历法时,曾对历代冬至太阳所在宿度的测算作了详尽的误差分析,总结出了九种可能的误差,但主要考虑的是观测误差和模型误差,几乎没有考虑计算方法误差。[12]:90-92

3. 古代不掌握近似计算的规律,中历的高精度仅流于表面和一厢情愿

近、现代学者们对古代历法的研究,对古人观测、模型、参数制定的精度作了很多有

益的探讨，证明历代以来，历法的精度越来越高（其综合成果见文献[12]，第180~184页），但很少有人从原理、理论上分析历法计算过程的精度、误差问题。本节主要以《授时历》为案例，以古代其他历法的部分事例为补充，通过历法中的某些具体计算过程来对此问题作一初步分析。

对误差问题的理性思考来源于在计算过程中对计算结果和原始数据比较的深入认识，按说中国古代历法的"代数体系"本应及时发展出对误差问题的要求，但实际上，直到明末西法传入以前，古代历算家一直没有这方面的明确认识，包括元代郭守敬的《授时历》也是如此。日本学者中山茂曾对此现象说过一句意味深长的话：《授时历》是数值至上主义。[19]其意思很明了：授时历过分依赖数值方法，不关心天体的真实运动模型，也忽略了可能的观测误差，只沉浸在数值的精度中。其实不光《授时历》如此，以前的历法基本都如此。

在实用计算中，不掌握近似计算的规律会严重加大计算量，在手工计算的时代该问题尤其突出。对天体位置的推算，人们总是希望精度越高越好，但在对时刻、天体位置度数的初始精度都很低，模型也远不能精确反映天体运行状况的条件下，不懂近似计算就不止是"事倍功半"了，按笔者模拟推算的经验，不考虑近似计算时，计算量会多十倍甚至几十倍，而精确度毫无提高。古代的计算工具是算筹，速度很慢，想必很多时间都耗费在无意义的计算中了。

举例来看，北宋《仪天历》中的漏刻计算冬至后损益差：

$$G_2 = \left(\frac{126875}{20126374}\right)n^2 - \left(\frac{6250000}{20126374} \times 522009\right)n^4$$

n为太阳行度，测值仅精确到$\frac{1}{4}$度，但计算所给值精确到$\frac{1}{10000}$，从93.7412开始取。将$n=93.7412$代入，得$G_2=91.5844$。[12]:195

其实只要保留两位小数$n=93.74$，就可达到测得精度91.6，也就是说，没有必要用精确到4位的小数作数据，n取两位小数，就可得到精确至0.1度的结果，而计算量只有原来的几十分之一。

观测误差和模型误差在一系列的计算程序中可能会被传播、放大。如果数据的初始误差小、模型合理，某一步运算对后一步的影响较小，随着计算的继续进行，可使结果偏离真实位置的时刻来得较晚或很晚，较长久计算结果仍是合天的。反之，如果初始数据误差本来就突出、模型偏差较明显，再加上计算程序中对误差过大的步骤浑然不觉（如不合理的舍入等），一些步骤会对后一步的影响加大，会导致误差逐步积累，计算结果就远离了应有的答案。

对此，曲安京曾一针见血地指出："古代历算误差太高，很多殚思极虑、苦心孤诣的东西，都被淹没在误差中，使历法几十年一换。"[21]

《授时历》是中国古代历法的顶峰，下面我们就通过对《授时历》有关内容的分析，看一看在这顶峰时期，历法家对误差思想和近似计算认识到了什么程度。

《授时历》是中国古代最好的历法，清代梅文鼎曾作过这样的赞叹："《授时历》不用积年，一凭实测，故自元迄明，承用三四百年无大差，以视汉、晋、唐、宋之屡差屡改，不啻霄壤。故曰：《授时》集诸家之大成，盖自西历以前，未有精于《授时历》者也。"[22]:615《授时历》元朝用了 88 年，明代沿用，改名《大统历》，元、明共用 382 年，是历代历法中使用时间最长的一部。

但是，即使是在这样一部优秀、自觉摒弃了完美主义上元积年观念的历法中，对误差思想和近似计算也是相当忽略的。试举几例来看：

载于《元史·历志》的《授时历》以 365.2425 日为回归年，每百年减少 0.0001 日，以 365.2575 度为周天（相当于恒星年），以上的精度都是万分之一日。

而朔望月（朔策）是 29.530593 日，精度提高成百万分之一日，交点月（交终）27.212224 日，精度又降到十万分之一日，近点月（转终）27.5546 日，又降回万分之一日，精度各不相同。这样，在计算过程中，实际上精度高于万分之一日的部分根本无效，一般情况下仅起到加大计算量的作用。

对回归年的分割更能说明问题，《授时历》中一年的节气长是这样确定的：气策 = $\frac{岁实}{24}$ = 15.2184375 日，精度居然高到千万分之一日，这是在回归年（岁实）= 365.2425000 日的假设下得出的，而那时对回归年的测定不可能会到这样高的精度，所以气策这样的高精度只有"数值至上"的意义而没有任何实际意义。

还有，在对太阳不均匀运动的校正计算中，取"缩初盈末限"春正至夏至为 93.712025 日，冬至至春正为 88.909225 日，也都比"基本常数"岁实的精度毫无必要地高了 100 倍。

在步日躔具体计算中有这样的算式：

每日定行度 = 四正后每日行度 ± 日差

计算时，原有的基数（行积度）取小数点后 4 位，"每日行度"取小数点后 6 位，"日差"则取小数后 7 位，精度不一，前后差了 1000 倍，相加时后者的高精度完全被前面的低精度吞没。[22]:656,660

计算月行疾迟时有同样的问题：

初限损益分 = 定差 0.1111 - 平差 0.000281 - 立差 0.00000325 = 0.11081575

三项的精度也差了 1000 倍。[22]:673

对于月平行度（每天运行的度数），《授时历》取的是 13.36875 度，比周天度 365.2575 的精度也高了 10 倍。在"步交食"的推算中，月平行度取 13.36875 度，而交终度取 363.7934 度，正交度又取 357.64 度，日食阳历限取 6 度，精度愈来愈低，如果一个算式以上几个值同时出现，其结果只能准确到"度"，可谓极度的不完美了。[22]:663

这些事例都说明，在中国传统历法中，即使最顶尖水平的历法家，对误差和近似计算观念也是混乱的，古代一以贯之的"绝对精确"观念仍然占主导地位。

最后，我们再看一看著名的"弧矢割圆术"问题。

弧矢割圆术是中国特有的一种球面三角理论，是继先人的相似勾股术、沈括会圆术后郭守敬的一项杰出创造。它在中国传统历算上的贡献和影响，史界早有定论，本文不再赘述，只分析一下弧矢割圆术使用者在误差思想上的一些空缺。

郭守敬在推演弧矢割圆术时，采用的一周天度数为 365.2575 度，周天半径为 60.875

度（弧矢割圆术中，线长度也用"度"），后者是怎么来的呢？原来，他取圆周率 π 为 3，这样 365.2575÷3 = 121.7525，原文为"径 121 度七十五分少"，将"少"略去，近似为 121.75，取其半就成了 60.875 度。[2]:1479

在南北朝时，祖冲之已经把圆周率 π 的值计算到了小数点后 7 位，虽然这个结果几乎被后人忘记，但人们在进行有关圆周与直径关系的运算时还是经常使用比 3 更精确的 3.14 的。郭守敬用了最粗疏的古率 3，这实在让我们坐立不安，因为用 3 或用 3.1416，此一项误差就达 4.7%。求周天半径时，求出的 121.7525 已经偏差很大了，接着又把 0.0025 略去，只保留两位，除以 2 时又得 60.875，保留 3 位——对精度诸如此类的种种设定，实在是过于随心所欲了。

在随后的演算中，郭守敬又取黄赤大距为 24 度，与周天半径为 60.875 度相比，精度又陡降为"度"，按此极不准确的周天半径和精度极低的黄赤大距推算（图 2），求出了 OK 即黄赤大股 q = 56.0268 度，DK 即黄赤大勾 p = 23.8070 度。[2]:629

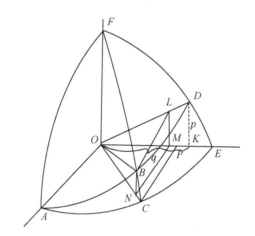

图 2 弧矢割圆术的求黄赤道差和内外度图[22]:629

而郭守敬实测的黄赤大距本为 23.90 度，如利用此值，同时求周天半径时 π 取 3.1416，经计算可得 q = 53.2880 度，p = 23.2328 度，郭守敬的值与之相比，误差太大。再用他求出的值算黄道各度下赤道度，亦即用勾、股，从弧度求弦、矢，会导致更大的误差。

在和白道有关的常数中，郭守敬又取黄白交角为 6 度，精度也降到了"度"。当白道降交点与春分点重合时，得出"半交"（最低点）在黄道外 6 度，在赤道内 18 度——郭守敬甚至不顾实测的数，无意中又把前人"通约"的理想搬到了黄赤、黄白交角上。[2]:1346

有学者认为，郭守敬弧矢割圆术中的古率 3 是"会圆术"这个统一的单位系里独特的比率，而不是圆周率，因为无论是弧矢割圆术，还是更古的会圆术，都是近似公式，但其最后结果与用现代三角学方法求出的相近，在特殊点甚至与现代值完全相等。[23] 那么，是否那些近似公式包含了 π 的修正值？近似公式抵消了 3 为"圆周率"的部分误差？但据笔者使用部分值验算，使用古率 3 是次要的，主要是会圆术、弧矢割圆术本身是近似公式，如果用现代三角法算，即使圆周率用 3，其结果的误差仍小于授时历的方法；相反，弧矢割圆术用更精确的π值时，得到的值精度更高，可见古率3并没有起到修正作用。

对于很多运算，误差是会积累的，所以这么优秀的一部历法，用了300余年后，在明末清初终于变得不可收拾，最后让位于西法。细分析《授时历》中推算过程的缺憾，都不是"瑕不掩瑜"所能开脱的，本来是孜孜以求精密的天文历法计算上居然率意使用3为圆周率，不顾实测的数据把天体运行轨道交角定成"通约"式的理想状态，在同一算式中不同项的精度居然相差达1000倍，都说明历法制定者在对算法的认识上有着严重偏差。

本文仅对古代历法在误差思想方面的空缺作一初步而粗陋的分析，难免挂一漏万。以往学者对历法的研究，多集中于重建历法的推算结构，而对其运算过程的误差，特别是误差思想重视不够，甚至在复原计算时也常不考虑误差和近似结果，无意中在继续重演古人的虚幻精度。重建古人的工作是必要的，但在重建的同时，利用现代科学理论为指导，高屋建瓴地了解古人的成就，则更为必要。就误差分析来说，如果我们在重建古代一部历法的演算结构时，同时对之作比较详尽的误差分析，想必我们会得到许多新的认识和发现。

参 考 文 献

[1] 薄树人. 中国科学技术典籍通汇·天文卷. 第6册. 郑州：河南教育出版社，1993
[2] 薄树人. 中国科学技术典籍通汇·天文卷. 第3册. 郑州：河南教育出版社，1993
[3] 孟子. 四部丛刊初编. 经部. 卷10册. 卷8. 上海：商务印书馆（缩印宋刊本），1936.68
[4] 陈遵妫. 中国天文学史. 第3册. 上海：上海人民出版社，1984.1391
[5] 曲安京. 中国数理天文学. 北京：科学出版社，2008.56
[6] 高平子. 高平子天文历学论著选. 台北：中央研究院数学研究所，1987.90
[7] 伊世同. 元代圭表复原探索. 自然科学史研究，1984，3（2）：128
[8] 何清谷. 三辅黄图校释. 北京：中华书局，2005.279
[9] 崔石竹，李东生. 仿古测影探索. 自然科学史研究，1987，6（4）：335
[10] 郭盛炽. 元代高表测景数据之精度. 自然科学史研究，1992，11（2）：151
[11] 尚书. 四部丛刊初编. 经部. 卷1册. 卷1. 上海：商务印书馆（缩印宋刊本），1936.7
[12] 陈美东. 古历新探. 沈阳：辽宁教育出版社，1995
[13] 潘鼐. 中国恒星观测史. 上海：学林出版社，1989.140
[14] 张柏春. 明清测天仪器之欧化. 沈阳：辽宁教育出版社，2000
[15] 王玉民. 以尺量天——中国古代目视尺度天象记录的量化与归算. 济南：山东教育出版社，2008.46
[16] 中国天文学史整理研究小组. 中国天文学史. 北京：科学出版社，1981.96
[17] 王荣彬. 刘焯《皇极历》插值法的构建原理. 自然科学史研究，1994，13（4）：295
[18] 薄树人. 中国科学技术典籍通汇·天文卷. 第1册. 郑州：河南教育出版社，1993.709-715
[19] 曲安京. 一部撰写了40年的著作终于出版了. 中国科技史杂志，2006，27（3）：270
[20] 李俨，钱宝琮. 李俨钱宝琮科学史全集. 第四卷. 沈阳：辽宁教育出版社，1998.142
[21] 曲安京. 中国历法与数学. 北京：科学出版社，2005.178
[22] 张培瑜，陈美东，薄树人，胡铁珠. 中国天文学史大系·中国古代历法. 北京：中国科学技术出版社，2008
[23] 邓可卉. 授时历中的弧矢割圆术再探. 自然科学史研究，2007，26（2）：162

Tentative Discussion on E. Diaz and the Influence of *Tianwenlue* on the Chinese Astronomy

Yao Licheng

(The Inistitute for the History of Natural Science, CAS)

J. Needham (1900-1995) said: "In the history of intercourse between civilisations there seems no parallel to the arrival in China in the 17th century of a group of Europeans so inspired by religious fervour as were the Jesuits, and at the same time so expert in most of sciences which had developed with the Renaissance and the rise of capitalism."[1]. The Jesuits who arrived in China during the Ming and Qing Dynasties opened the gate, which the western and Chinese cultures could mutually exchange, and at the same time their ideas had strongly impacted on the Chinese society, culture and spirit. In the Ming Dynasty, the Confucianism was orthodox, and also took the prime place in the social function. The Jesuits, as Matteo Rcci (1552-1610), realized that they must emphasize common ground between Christianity and Confucianism and pass on the western knowledge of sciences as adjunct work if they desired to stay at China for ever and let the Christianity to root in the hearts of the Chinese[2].

The Jesuits wanted to make a significant contribution to calendar reform and acquire the ruler's confidence by means of introducing the knowledge of occidental astronomy when the Bureau of Astronomy (钦天监) forecast mistakenly the solar eclipse in 1610[3]. The western knowledge of sciences was completely different from the Chinese so that they brought about attention and interest from the Chinese scholars. Faced the new impact, Xu Guangqi (徐光启, 1562-1633) said: "(If the Chinese) want to surpass (the western), (we) must digest (these knowledge); before (we) digest (them), (we) must translate (their books)."[4] When referring to the occidental astronomy, Li Zhizao (李之藻, 1565-1630) said: "The knowledge of astronomy and calendar that they studied were more than that the former Chinese-scholars had done. They did not only show how they calculated, but also they could explain why they did so. They made the astronomical instruments of observing sky, observing sun, and each was exquisite and unsurpassed. Taking their calendar, we should translate it and publish."[5] Many of officials, include Xu Guangqi and Li Zhizao, admired whole-heartedly the occidental astronomy and propagandized it among the scholars. After Xu Guangqi and Li Zhizao acquired the permission from the emperor, they selected and organized people to translate the western books. E. Diaz was one of participants. The astronomical books that were translated had not only an effect on the mathematics and astronomy themselves in China, but also changed the scholars' methods of problem-asking and problem-solving in China.

1. The time of E. Diaz entering into China

Emmanual Diaz (Jr. 1574-1659), Portuguese, his Chinese name is Yangmanuo (阳玛诺), alias Yanxi (演西). He entered into China in the late Ming Dynasty. He was recommended by the Jesuit missionaries in the calendar reform and making fire-weapon, and one of the important Jesuits during the Ming and Qing Dynasties. In 1615 (the 43rd year of the Wanli period), he wrote *Tianwenlue* (*Explicatio Sphaerae coelestis*), which was one of the earliest books the Jesuits introducing the knowledge of occidental astronomy, particularly in this book Diaz first referred to the telescope and Galileo's observations in Chinese.

Though E. Diaz had stayed and done missionary work for more fifty years in China, unfortunately there were a few materials about him. For example, the time of E. Diaz entering into China indeed was not ascertained, and there were a few different dates. Usually it has two dates, 1610 or 1611. So when E. Diaz is mentioned in the Chinese book, the date of his entering into China is always in 1610 or 1611. L. Pfister's (费赖之, 1833-1891) book *Notices biographiques et bibliographiques sur les jésuites de l'ancienne mission de Chine* 1552-1773 (《在华耶稣会士列传及书目》) and J. Dehergne's (荣振华, 1903-1990) book *Répertoire des Jésuites de Chine de 1552 à 1800* (《在华耶稣会士列传及书目补编》) is manual that people study the Jesuits in China. The date of E. Diaz that circulating among the people in China derived from the two books. In fact, both of them didn't tell us the exact time of E. Diaz entering into China in their book, but L. Pfister only mentioned E. Diaz "A. 1610"[6]. "A. 1610" means "arrivé dans la Mission", not "arrive in China". The mistake was owing to the translator. According to L. Pfister and J. Dehergne, and the other information about E. Diaz, we could make sure his date of arrival in China. If E. Diaz had taught the theology for six years in Macao, then he arrivé dans la Mission in 1610, he and G. Ferreira left for Chaozhou together in 1611[7], it is reasonable that he may be enter into China between 1604 and 1605.

According to J. Dehergne, E. Diaz left for Goa in April, 1601 and got to there in October at the same year[8]. E. Diaz had studied in Goa for 3 years, then he, A. Vagnoni (高一志, 1566-1640) and F. da Silva (林斐理, 1578-1614) together left for China by ship in April, 1604. In July, they arrived at Macao in China, and E. Diaz stayed at there since he was ill. In fact, I suppose that E. Diaz arrived in China in July, 1604.

2. E. Diaz and the calendar reform in the late Ming Dynasty

There was an eclipse on December 15, 1610, but Qintianjian (the Imperial Astronomical Bureau) made a mistake in forecasting it, so that their mistake brought out complains and criticisms from many officers and scholars. At that time, Qintianjian had used *Datong Calendar* (《大统历》) to be engaged in astronomical activities. *Datong Calendar* had been applied for about three hundreds years since the Ming Dynasty was established, and it must be fill with inaccuracies. By the Ming Dynasty's law, any people had no permission to study astronomy and take part in astronomical activities in private. If they were found to do it, they would be punished or executed. So *Datong Calendar* fell into disrepair for a few hundred years without amending. People appeal to government for calendar reform. As for this eclipse, the Jesuits, as Diego de Pantoja (庞迪我, 1571-1618) and so on, also forecast it before, and their calculation coincided with the observational result[9]. It was luck for the Jesuits because they and occidental astronomy were regarded seriously with it, and the Jesuits enlarged their influence in China by means of chance of calendar reform. After forecasting incorrectly this eclipse, there were several officials who submitted written statements and asked to translate books of occidental astronomy from 1611 to 1613. As E. Diaz was proficient at astronomy among the Jesuits, Diego de Pantoja invited him together and went to Beijing from Nanjing. The aim of this travel took part in the translating work and made prepare for the calendar reform of the government. Then he wrote *Tianwenlue* during that time, which explained systematically the prime thought and concept of occidental astronomy. In 1615, *Tianwenlue* was firstly published in Beijing.

3. Influence of *Tianwenlue* on the Chinese astronomy

Tianwenlue is written in catechetical form and has 4 chapters, 25 questions and 23 figures. It introduced Aristotelian cosmic theory, namely crystalline spheres, and expounded elementary

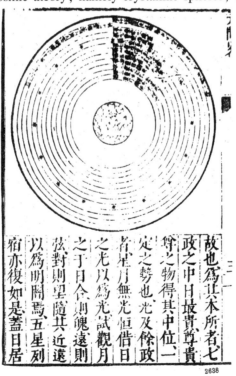

Fig. 1 Aristotelian crystalline spheres
(*Tianxuechuhan*, Vol 5, *Tianwenlue*, p. 3)

knowledge of the western astronomy. The first chapter is *How many layers are there the heaven and what position are the seven planets* (七政); In the chapter, E. Diaz introduced the comic theory, which it has the skies of twelve layers. It had the sky of precession more than the celestial spheres of the eleven skies. The second chapter is *the proper motion of the solar layer and the which degrees are the sun away from the equator*; it explained annual apparent motion of the equator, the Sun and the ecliptic on the zodiac, twenty-four solar terms and so on. It also expounded the irregular motion of the sun because the skies of Qizheng (seven planets) were not concentric with the Earth. The third chapter is *the day or night has changed long or short with the changing of the latitudes*; the fourth chapter is *the sky of moon is the first layer and its proper motion*. In the chapter, E. Diaz firstly introduced eccentric (the truckle system) to show the differences of the eclipse time in Chinese. Particularly, the first reference to the telescope in Chinese is the end of *Tianwenlue* in 1615, and it was the first time that it told the Chinese the discoveries made with Galileo's telescope. *Tianwenlue* expounded the four satellites of the Jupiter, the waxing and waning of the Venus, the rotation of the Sun[10], and so on. *Tianwenlue* was known in China for that. It showed that even if it was in that backward age, disseminate of scientific discovery was very fast. Design of geometrical models of cosmic systems is one of the occidental astronomy traditions, but ancient Chinese astronomy had not this trait, so that when the Jesuits introduced the occidental cosmic models in the end of the Ming Dynasty into China, they promptly arouse the Chinese attention and interesting[11]. Because *Tianwenlue* is one of the earliest books referring to western astronomy by the Jesuits, in fact, it is easy to see that it had made a great impact on the thought and method of the Chinese.

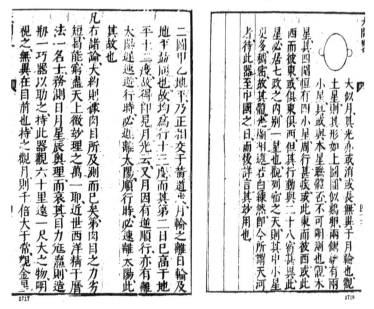

Fig. 2 The last two pages of *Tianwenlue*
(*Tianxuechuhan*, Vol 5, *Tianwenlue*, pp. 46-47)

Although this calendar reform wasn't successful and didn't continue, these books had to have affected on the society in China. *Tianwenlue* was still learned and published until the 19th century, and it was included those prestigious collections in China during the Ming and Qing dynasties, e. g. *Tianxuechuhan* (《天学初函》, *First Collectanea of Heavenly Studies*), *Siku Quanshu* (《四库全书》, *Complete Library of the Four Treasuries*), and so on. Then *Tianwenlue* was introduced into Korea, and we should see that the effect widely spread at that time. The system and theory that it introduced was all new for the Chinese. But we should noted that there was not higher theory and complex calculation of astronomy in it, I think that *Tianwenlue* only played the role of enlightenment of western knowledge in China.

Wang Xichan (王锡阐, 1628-1682) was a famous astronomer and had written many astronomical works, and learned some new astronomical thought and methods from the western astronomy. He published the astronomical book, *Xiaoanxinfa* (《晓庵新法》, *The New Calendar of Xiaoan*) in the Qing Dynasty. In the sixth volume of his book, besides introducing the computing methods of the eclipse, he described the calculation of transit of Venus. The contents were first described in Chinese. How could Wang Xichan think of calculating the transit of Venus? Perhaps, Wang Xichan's method may derive from *Tianwenlue*[12]. In *Tianwenlue*, E. Diaz expounded the transit of the Venus[13]. In the second chapter, E. Diaz accounted for the reason why the Venus and the Mercury do not eclipse the Sun. The Jesuits only referred the phenomena of the inner planet's transit, but they didn't give their computing methods, and it was great that Wang Xichan was able to independently offer in virtue of E. Diaz's describing[14]. *Chongzhen Lishu* (《崇祯历书》) is a large astronomical library, which was compiled in the *Chongzhen* period. Xu Guanqi was the series' general editor and the it was prepare for the calendar reform. The content of *Chongzhen Lishu* was almost introduced from occidental astronomical knowledge, and it contained much new astronomical knowledge, methods and instruments at that times. It mentioned E. Diaz' work and his book *Tianwenlue*[15].

4. Conclusion

J. Needham said: "To seek accomplish their religious mission by bringing to China the best of Renaissance science was highly enlightened proceeding."[16] The Jesuits propagated the western astronomy with the chance of the calendar reform, and translated many books of the western astronomy into Chinese. The occidental astronomical knowledge opened the Chinese views and had great affection on them. In 1615, E. Diaz wrote *Tianwenlue* in Beijing, and it was one of the earliest translated works.

Tianwenlue introduced the basic astronomical knowledge, but the astronomical knowledge, including the figures of cosmos or the Earth, the sizes of planet or cosmos, the reason of the eclipse, etc. were great different from the Chinese traditional astronomy, especially they sounded reasonable, so that it influenced many peoples.

References

[1] Needham J. Science and Civilisation in China. Vol 3. Cambridge: Cambridge University Press, 1959. 437

[2] 顾卫民. 基督教与近代中国社会. 上海：上海人民出版社, 2010. 29

[3] 江晓原, 钮卫星. 天文西学东渐集. 上海：上海书店, 2001. 270

[4] 徐光启. 徐光启集. 北京：中华书局, 1963. 374

[5] 中华书局编辑室. 历代天文律历等汇编. 卷10. 北京：中华书局, 1976. 2538-2539

[6] Pfister L. Notices biographiques etbibliographiques sur les jésuites de l'ancienne mission de Chine 1552-1773. Chang Hai: Imprimerie de la Mission Catholique, 1932. xvi

[7] Pfister L. Notices biographiques et bibliographiques sur les jésuites de l'ancienne mission de Chine 1552-1773. Feng Chengjun tr. Beijing: Zhonghua Book Company, 1995. 110

[8] Dehergne J. Répertoire des Jésuites de Chine de 1552 à 1800. Gengshen tr. Beijing: Zhonghua Book Company, 1995. 185

[9] 徐光启. 增订徐文定公集. 卷1. 上海徐家汇天主教堂图书馆, 1933

[10] 杜升云, 崔振华, 苗永宽等. 中国古代天文学的转轨与近代天文学. 北京：中国科学技术出版社, 2008. 249

[11] 同 [3]. 358

[12] 席泽宗. 试论王锡阐的天文工作. 载：陈美东, 沈荣法主编. 王锡阐研究文集. 石家庄：河北科学技术出版社, 2000. 11-13

[13] 李之藻. 天学初函. 第5卷. 14

[14] 同 [12]

[15] 徐光启编纂. 潘鼐汇. 载：崇祯历书. 上海：上海古籍出版社, 2009. 59
《日躔历指》"推太阳之视差及日地去离远近之算加减之算第八"
"按：天问略等书，皆言地体居天中，止一点。是也。然各重天高下大小不等，各天与地球比例之大小亦不等，惟恒星一重天，比于向下诸天，甚远甚大，以地球较之，极微无数可论，故测候之家，以恒星为求视差之本。"

[16] 同 [1]. 449

An New Explorations of the Origin of Chinese Alchemy*

Han Jishao

(Institute of Religion, Science and Social Studies, Shangdong University)

This paper discusses a new identification of sources and key components of early Chinese alchemy. [①] All the sources recording early Chinese alchemy could be divided into two sorts, the earlier and the later. The earlier show that alchemists of the Western Han Dynasty used cinnabar, mercury and other metals or minerals to make alchemical gold, which was called Huangjin (gold), Huangbai (yellow and white) or Huangye (transforming cinnabar into gold) at that time. Influenced by the idea of Reverted Cinnabar (Huandan, 还丹) which emerged in the middle of the two Han Dynasties, those later sources mistook alchemical gold of the Western Han Dynasty for reverted cinnabar. In other words, there was a theoretical revolution in the early history of Chinese alchemy. So the search for the origin of Chinese alchemy can only be reliably performed with early sources, from which we could find its four key components: the conception of immortality, worship of cinnabar and mercury, worship of gold, rituals. Of which the idea of worship of gold has a close relationship with foreign culture. Chinese alchemy is a wonderful flower of cultural communication.

1. Introduction

During the Warring States period, ideas of immortality prevailed in China. Many people believed there was a special medicine in the place of deities or immortals lived, it could make ordinary people be deathless or immortal. With strong enthusiasm for immortals, activities of searching for this medicine had been practised widely since the late Warring States period. Some Fangshi (方士) gradually came to believe that the medicine could be made from some minerals or metals. This idea is usually regarded as the beginning of Chinese alchemy. When, why and how did it arise? Due to the ambiguity and contradictions between different sources, scholars failed to find the exact time when Chinese alchemy began until now. And little attention was

---

* The present article is a part of my ongoing research project on comparative study between Chinese and foreign alchemy and their communication in history. I am grateful to Professor Sir Geoffrey Lloyd, Professor Christopher Cullen, Dr. Michael Stanley-Baker and the anonymous referee for their comments and revisions.

① In this article, the term "early Chinese alchemy" refers to the first stage of Chinese alchemy. Although different sources gave us different time, I will argue it took place in Western Han Dynasty.

paid to the relationship between Chinese alchemy and foreign culture. Because alchemy appeared in many different civilizations in ancient time, some scholars guessed Chinese alchemy may be migrated from other country in cultural contacts. But these assumptions haven't been proved successfully at all. In fact, what we should pay attention to are not how alchemy migrated from other country to China, but whether Fangshi developed alchemy using of different ingredients including foreign cultures. So identifying the reliability of sources is the urgent need to search for the origin of Chinese alchemy.

The sources about early Chinese alchemy are usually divided into two sorts. One is historical documents, and the other is hagiographies, literary works and the like. The question is that the former show Chinese alchemical practices took place in the reign of the Emperor Wu (Wudi, 武帝, r. 141BC-87BC), whereas the latter tell us an earlier and more detailed history of early Chinese alchemy. In view of the fact that there were so many practices of searching for medicines of immortality before the Western Han Dynasty, some scholars believe it's possible that parts of the latter sources are reliable, and Chinese alchemy should have begun before the Western Han Dynasty. Unfortunately, it is hard to know which parts of those sources are reliable. Subsequently, this uncertainty proves troublesome when searching for the beginning of Chinese alchemy. Thus we should look for a proper way to identify which sources are more reliable. Then we could know the character of the earliest Chinese alchemy, and how it happened.

It is strange that the idea of Reverted Cinnabar emerged much later than early alchemical practices. Usually scholars attribute that to the scarcity of sources. But I notice that those alchemical practices shown by historical documents were quite different from the idea of Reverted Cinnabar, i. e. the idea of Reverted Cinnabar wasn't the earliest theory of Chinese alchemy. Thus, we naturally cannot know the origin of Chinese alchemy. Did the idea of Reverted Cinnabar have any impact on the sources chronologically after it? This question inspires us that it would be helpful to resolve the problem if we prefer the date rather than the property of the sources as a criterion of identifying their authenticity. So taking the period between the end of the Western Han Dynasty and the beginning of the Eastern Han Dynasty as a division, when the first Chinese alchemical texts and the idea of Reverted Cinnabar appeared, all sources will be divided into two sorts, the earlier and the later. We will see that these two different sources depicted different history of early alchemy. The earlier show us a diachronical history, whereas the later show an anachronical history.

Given the discussions above, in this paper I will attempt to look for the original points of Chinese alchemy. I will approach my purpose through four steps: firstly, to discuss alchemical practices of the Western Han Dynasty through the earlier sources; secondly, to introduce the idea of Reverted Cinnabar and its impacts in the history of Chinese alchemy; thirdly, to discuss early alchemy through the later sources; and fourthly, to find core components of early Chinese alchemy, from which we will surprisingly know Chinese alchemy was a fruit of cultural communication between the east and the west.

2. The earlier sources and early Chinese alchemy

There are only a few earlier sources unambiguously referring to alchemy, most of which are historical documents, e. g. *Shiji*（《史记》）by Sima Qian（司马迁）(ca. 145BC-90BC), and *Hanshu*（《汉书》）by Ban Gu（班固）(32-92 AD). These documents show that alchemical practices flourished in the reign of the Emperor Wu. These practices have two branches, one is northern Fangshi, and the other is southern Liu An（刘安）. The accounts about northern Fangshi are as following:

At that time, Li Shaojun（李少君）called on the Emperor with his three arts: making offerings to the Spirit of Furnace, living without eating cereals, and keeping youth forever. And the Emperor honored him... Li Shaojun said to the Emperor, "By making offerings to the Spirit of Furnace one could let the Spirit come. Then cinnabar could be transformed into gold. Make the gold into vessels for eating and drinking, the use of which will prolong one's life. Having attained a long life, he could meet the immortals of Penglai Island（蓬莱岛）in the sea. After that, if he performs the Feng（封）and Shan（禅）ceremonies, he will never die. The Yellow Emperor is just a good example..." So the Emperor began to make offerings to the Spirit of Furnace personally, dispatched Fangshi to the sea to search for the Penglai Island, Anqi Sheng and other immortals, and engaged in practices of transforming cinnabar and other medicines to gold. [1]:1385-1386

He (Luan Da, 栾大) said [to the Emperor], "Your subject has been to the sea, seen Anqi, Xianmen and other immortals. But they didn't believe me because I was an ordinary commoner. And they also thought Prince Kang was not enough noble to be told the formula [of medicine of immortality]. I talked about that to Prince Kang several times, but he didn't make use of me. My master said, 'Gold can be transformed, the breach of the Yellow River's dike can be made up, the medicine of immortality can be found, and immortals can be summoned.' However, all the Fangshi fear to have the same fate with general Wencheng and close their mouths, how do they dare to talk about the formula?" ... At that time, the Emperor was distressed that the breach of the Yellow River couldn't be made up, and gold couldn't be transformed. So he appointed Luan Da as the title of General Wuli. [1]:1389-1391

Emperor Cheng (Chengdi, 成帝) was interested in ghosts and deities very much in his late years... Gu Yong（谷永）persuaded to him, "...Those people—who say that immortals exist in the world, they eat medicine of immortality, fly far away with their light bodies, rise into the highest heaven, look down the hanging garden of Kunlun mountain, float to the Penglai Island, cultivate the five colored cereals which are sown in the morning and harvested in the evening, have lives as long as mountains and rocks, make cinnabar into gold, thaw ice in an instant [by some substance], and practise the arts of visualization and imagination of the Spirit of Five Colors and the Spirit of Five Stores—are all cheats. They use heterodox arts and lies to cheat both common people and the ruler... After the founding of the Han Dynasty, Xin Yuanping（新垣平）, Li Shaoweng（李少翁）, Gongsun Qing（公孙卿）and Luan Da, the last three came from Qi（齐）, were honored by the Emperor for arts of making cinnabar into gold, making offerings to ghosts, and searching for deities and medicine in the sea..." [2]:1260

Among these Fangshi, Li Shaojun's method was to make "gold" from cinnabar and some other minerals. What was this gold? Some ones guess it was gold amalgam, which Li Shaojun used to gild bronze. The procedure is: make mercury from cinnabar, then dissolve gold into mercury, and he would get gold amalgam. Finally daub gold amalgam to bronze vessels for eating and drinking. Unfortunately this assumption is wrong. In China the technique of gilding began from the Warring States period, and became common in the Western Han Dynasty. So it was not so difficult to make gilded bronze for artisans at that time. Contrary to that, no Fangshi succeeded in making "gold", which made Emperor Wu be very sad. So this "gold" should be alchemical gold (fake gold in fact), also called elixir-gold (Danjin, 丹金) in later Chinese alchemy. Other Fangshi had the same aim as Li Shaojun, but we don't know what materials they used.

Liu An, the Prince of Huainan (淮南), also applied himself and his men to make gold. Several books concerned with making gold were attributed to him. However, none of them are fully extant now. His alchemical practices are as following:

[The Prince of Huainan] gathered several thousand of guests and Fangshi, wrote the *Inner Book* (*Neishu*, 《内书》) consisting of twenty-one chapters and the *Outer Book* (*Waishu*, 《外书》) consisting of many chapters. They also wrote the eight chapters book, *Middle Chapter* (*Zhongpian*, 《中篇》), which discussed immortals and the arts of Yellow and White. This book had more than two hundred thousand words. [2]:2145

The Prince of Huainan had *Zhenzhonghongbaoyuan Mishu* (《枕中鸿宝苑秘书》). This book introduced immortals, the arts of making gold, and Zou Yan's (邹衍) methods of prolonging life, but nobody had ever read it. Gengsheng's (更生) father, De (德), got it when he judged the Prince of Huainan in the reign of the Emperor Wu. Gengsheng read it at his childhood, and regarded it precious. Then he presented it to the Emperor, and said gold could be transformed. The Emperor ordered the imperial factory to make gold under the direction of the book. But it cost too much and didn't succeed. So the Emperor sent Gengsheng down to a judge. And the judge charged him with making fake gold, [decreeing that he] should be executed. [2]:1928-1929

Because of the loss of Liu An's alchemical works, we cannot know the details of his practices. What materials did he use? In *Huannanzi* (《淮南子》), a famous philosophy book compiled by Liu An which is still extant, the author said that some metals and minerals could be transformed from mercury over many years. According to that Feng Jiasheng thought mercury was the most important material for Liu An's alchemy. [3] There is another proof in the *Baopuzi Neipian* (《抱朴子内篇》). It quoted an interesting story from *Xinlun* (《新论》) (ca. 16BC-ca. 56AD). It wrote that Chengwei (程伟), a member of the Han Imperial Court, was interested in the arts of Yellow and White (the arts of making artificial gold and silver, 黄白术). He attempted to make artificial gold under the direction of *Zhenzhonghongbao*, but finally failed. When his wife went to see him, he was fanning charcoal to heat the reaction-vessel, in which was just mercury. [4]

As we have seen above, the earlier sources show that all early Chinese alchemical practices were regarded as making [alchemical] gold, and called Huangjin, Huangbai, or Huangye (黄

冶). For fear of confusing them with making artificial gold for the economic purposes which took place in the early Han Dynasty and later alchemy, we should notice their two basic characters. On the one hand, the materials of making alchemical gold were cinnabar, mercury and other metals or minerals, whereas those of making artificial gold for economic purposes were only base metals. Cinnabar and mercury had been used in ancient graves long before the Han Dynasty, and they were often taken for sacred things. Furthermore, alchemical gold was a kind of medicine of immortality, different from artificial gold used for economic purposes or natural gold. So it's hard to draw equivalence between early Chinese alchemy and metallurgy. In fact it had little to do with metallurgy but was concerned with the arts of immortality. To our surprise, why did Chengwei use Liu An's alchemical method to make gold? Of course he wouldn't succeed. Chengwei, however, wasn't the only failure. Liu Xiang (刘向) was actually a more serious victim, and he may be the first man who attempted to make gold for the economic purposes by alchemical arts. On the other hand, the same aim of early alchemical practices was making alchemical gold. It was indeed a kind of elixir, or medicine of immortality, and most of its techniques were probably used in later alchemy, but it was not called Reverted Cinnabar at that time. This is an obvious proof that there was no conception of Reverted Cinnabar in the early Western Han Dynasty.

In conclusion, early alchemy differs greatly from ordinary metallurgy and later alchemy. It's well known that later on, Chinese alchemy has two branches. One is the arts of yellow and white (Huangbaishu), which are mostly concerned with metallurgy; and the other is the arts of golden elixir (Jindanshu). But in early alchemy, the two branches banded together.

3. The Emergence of the idea of Reverted Cinnabar

Before the appearance of the idea of Reverted Cinnabar, the meaning of the Chinese character Dan (丹) was red. Natural cinnabar, which was regarded as a sacred mineral, was also red, so it was usually entitled Dan. Just as what the first Chinese dictionary *Shuowen Jiezi* (《说文解字》) said: "*Dan* is a scarlet stone found in the region of Ba (巴) and Yue (越)." From the Warring States period to the Han Dynasty, there were three main schools of immortal practices: practising breathing, eating medicines, and sexual techniques. For those Fangshi of eating medicines, natural cinnabar was the most important medicine in the beginning. For example, in *Shennong Bencaojing* (《神农本草经》), the first book of Chinese *materia medica* compiled in the early Eastern Han Dynasty, cinnabar stands first on the list of three hundred and sixty-five medicines, and the author wrote: "To eat it persistently can make one communicate with deities and be young forever." From the stories of Daoist hagiography *Liexianzhuan* (《列仙传》) we can see that eating cinnabar was very popular at that time. More interesting, Sima Qian told us a famous family from Sichuan province benefited a lot from the business of natural cinnabar, and the last widow Qing (清) got high respect of Qin Shihuang.[1]:3260

But with the development of alchemy, Fangshi eventually found the secret of Reverted

Cinnabar. There are two pieces of records about this theory in *Shennong Bencaojing*. The first said: "Cinnabar... could turn into mercury." And the second said: "Mercury... can kill (get rid of) poison in gold, silver, copper and tin. When melted, it could turn back into cinnabar." Now we know this is a wrong conclusion. When melted, mercury can be oxidized to red mercury oxide. But it was taken for red cinnabar at that time. From cinnabar to mercury, then back to cinnabar, this was the basic procedure of preparation of Reverted Cinnabar. And this conception became the focus idea of Chinese alchemy from then on. From the late Western Han Dynasty to the early Eastern Han Dynasty was born the earliest Chinese alchemy texts. In *Huangdi Jiuding Shendan Jing* (《黄帝九鼎神丹经》), which is one of a few extant ones now, the author introduces the idea of Reverted Cinnabar in the beginning:

> The Yellow Emperor got the art of Reverted Cinnabar, which was the extreme way to immortality... he compounded and ate it, then rose into heaven and became an immortal. The mysterious Maid told the Yellow Emperor: "All those who want to have longevity but cannot obtain the divine [Reverted] Cinnabar (Shendan, 神丹) and Golden Liquor (Jinye, 金液) merely bring suffering upon themselves. Practising breathing, guiding and pulling (Daoyin, 导引), exhaling old and inhaling new breath, eating medicine of herbs and plants can make one live longer, but cannot free him from death. Eating divine [Reverted] Cinnabar can make one become an immortal and transcend this world. He will be coeternal with the heaven and the earth, and as luminous as the Sun and the Moon. He can see five thousand kilometers away even seated, impel ghosts and immortals, rise into the heaven carrying his whole family, fly with no wings, mount clouds and steer a carriage pulled by dragons, go up and down to the Great Clarity (Taiqing, 太清), and tour the eight poles (Baji, 八极) in a very short time. He will not be restricted by any river, or hurt by any toxicant."[5]

The paragraph above shows that alchemists began to regard Reverted Cinnabar and Golden Liquor as the only ways to immortality, whereas other methods were merely could be used for regimen. It seems that Golden Liquor has a same function as Reverted Cinnabar. However the whole text only introduces how to make Reverted Cinnabar from all kinds of minerals. So Reverted Cinnabar is actually the most important medicine of immortality in this scripture. Apparently, this idea is much different from the alchemical theory of the Western Han Dynasty. Reverted Cinnabar became the real elixir instead of alchemical gold. This change is the proof that there was a theoretical revolution in early history of Chinese alchemy. Cinnabar and mercury can never be transformed into gold, but through many failed practices, alchemists finally discovered the secret of Reverted Cinnabar. Before the revolution, alchemical gold was the unique elixir for the alchemists. After the revolution, however, the alchemists regarded Reverted Cinnabar as the real elixir, which had stronger function than any of the other techniques or medicines, including natural cinnabar.

Here we should point out again, although the term Reverted Cinnabar is not correct—because alchemists had and would never transform base metals or minerals to cinnabar—it was the core idea of Chinese alchemy for more than one thousand years after that. Of course, alchemists gradually knew those elixirs weren't Cinnabar in fait, but they always named them Dan. And the terms Dan and Huandan had more meanings than Reverted Cinnabar in later alchemy. In other words, the word Huandan could be named as Reverted Cinnabar only in early Chinese alchemy.

4. The later sources and early Chinese alchemy

Compared with those earlier sources, the later tell us more details of early Chinese alchemy. Take the Fangshi of the Western Han Dynasty for example. Sima Qian mentioned in *Shiji* that Li Shaojun mastered four arts: making offerings to the Spirit of Furnace, living without eating cereals, keeping youth forever, and making cinnabar to gold. So there is no evidence showing he knew the secret of Reverted Cinnabar. However, in *Shenxianzhuan* (《神仙传》), usually attributed to Ge Hong, the alchemical gold prepared by Li Shaojun was regarded as Shendan (divine [Reverted] Cinnabar) directly:

> Li Shaojun came from Qi. The Emperor Wu recruited Fangshi. Shaojun got the formula of the divine [Reverted] Cinnabar from Mr. Anqi, but was too poor to buy medicines... Then he submitted the formula to the Emperor and said, "Cinnabar can be transformed into gold. Eat the gold then one will become immortal..." Shaojun prepared the divine cinnabar secretly. When got it... He gave the Emperor a part of the formula, and then said that he was ill. [6]:43

The same thing happened to Liu An. He was even said to have had thirty-six scrolls of Reverted Cinnabar scriptures conferred upon him:

> (Liu An) wrote the *Inner Book* consisting of twenty-two chapters. As well as the *Middle Chapter*, of which eight chapters named of *Hongbao Wanbi* (鸿宝万毕) introduced immortals and arts of Yellow and White, and three chapters discussed methods of transformation. The entire book had one hundred thousand words... [Among Eight Masters] one can fry mud into gold, solidify lead into silver, deal with the eight stones by water, distil mercury, mount clouds and steer a carriage pulled by dragons, and fly up the Great Clarity... At last An got thirty-six scrolls [Reverted] Cinnabar scriptures from the Eight Masters. [6]:25-26

According to the *Shen xian zhuan*, many earlier people engaged in alchemy. But to our surprise, it even dated the beginning of Chinese alchemy to the Shang and Zhou Dynasties. Some stories are as following:

> Peng Zu (彭祖)... one of Emperor Zhuanxu's (颛顼) great-great-grandsons, was seven hundred and sixty-seven years old in the late Shang (商) Dynasty... Peng Zu said: "one who want to ascend his body to the heaven and became a deity officer should eat Golden Cinnabar (or Golden Elixir, Jindan, 金丹), by which Jiuzhao Taiyi (九召太一) arise into the heaven."

> Yuzi (玉子)... King You of the Zhou Dynasty called up him but failed... Later he went to the Kongtong Mountain to compound Reverted Cinnabar, and arose into the heaven in day time.

Probably nobody would like to believe such legendary stories. But another famous fiction *Shiyiji* (《拾遗记》) said that Zhao Gao, the chief minister of the Qin Dynasty, mastered the method of preparation of nine vessels reverted cinnabar.[7] Parallel with historical documents, why are these accounts so different? Firstly we will attribute the cause to the nature of these sources. In fact, almost all of confusions lie in the authenticity of these sources. Take Zhao Gao's story for example, is it an entire fiction? According to an account from *Shiji*, Qin Shihuang had ever ordered many Fangshi to prepare the medicine of immortality, "(Qin Shihuang) gathered many literati and Fangshi, and wanted to make great peace and the medicine of immortality"[1]:258. This was the first idea in Chinese history to claim that the medicine of immortality could be made by human being. So Zhao Kuanghua guessed Zhao Gao's story may be authentic in some aspect, and alchemy probably had appeared since the Qin Dynasty.[8] If we ignore the authenticity of the whole story, the idea of Reverted Cinnabar in these later sources should firstly be doubted. Why did later sources regard early alchemy whose purpose was making alchemical gold as behaviors of making Reverted Cinnabar? That resulted from their anachronism. As the theory of alchemical gold was actually the pre-idea of Reverted Cinnabar, it had been taken for behaviors of making Reverted Cinnabar by most sources since its appearance. Moreover, early alchemical gold was made from cinnabar, mercury and other minerals which were also materials of later elixirs, so it was difficult to notice their difference, which may be an unconscious mistake.

5. Origins of Chinese alchemy

Among those diverse discussions about the origin of Chinese alchemy, there is a popular opinion, which connects the origin with Daoism. We know now alchemical gold not Reverted Cinnabar was the first elixir in China, so these opinions are virtually incorrect. Those elements which generated Chinese alchemy should be searched depend only on those earlier sources rather than later Daoist texts. The root ideas of producing Chinese alchemy are undoubtedly many, almost all of which are concerned with ideas of immortality. Dr. Joseph Needham reduced them to three: the pharmaceutical-botanical tradition of the herb or plant of immortality, the metallurgical-chemical tradition of the making of artificial gold, and the medical-mineralogical tradition of the using of inorganic and metallic substances in therapy.[9] However, according to those early sources, we can find four direct components of early Chinese alchemy: the conception of immortality, cinnabar (including mercury), gold and rituals.

The conception of immortality could date to the late Warring States period, before when there were only ideas of longevity, prevention of ageing, ghosts and so on. So if there was no idea of immortality, there would be no Chinese alchemy. Many scholars agree that foreign cultures were an indispensable ingredient in those early ideas of immortality, but we couldn't here regard the first component as foreign culture directly.

Usage of natural cinnabar was a very old tradition in Chinese antiquity. It was reported that cinnabar was found in a Neolithic tomb. During the Shang Dynasty cinnabar was often used as a kind of important mortuary object. The amount for one tomb may surprise many modern people. For example, archaeologists find in Erlitou (二里头), Henan Province, a great deal of cinnabar covers the bottom of some early Shang tombs. The volume of cinnabar in one tomb is

2.4 meters long, 1.5 meters wide, and 1.5-5 centimeters thick. [10] That in another tomb is 1.7 meters long, 0.74 meters wide, and 5-6 centimeters thick. [11] From the Spring and Autumn period onwards, mercury became another important mortuary object. It is well known that there is a great deal of mercury in Qin Shihuang's tomb conformed by archaeological investigation. Therefore we have full proofs to believe that cinnabar and mercury were regarded as sacred objects in antiquity, just like jade. Since the Warring States period cinnabar had become especially important among those minerals and herbs Fangshi used to eat. Moreover, cinnabar has close relationship with gold. Guanzi said, below the place where there was cinnabar there was gold. Maybe this idea inspired those Fangshi who tried their best to make gold from cinnabar in the Western Han Dynasty.

In contrast with cinnabar, gold, however, wasn't considered a sacred material before the Qin Dynasty. Archeological sources have shown that gold appeared in China much later than other ancient civilizations. The Chinese character Jin (金) was found in Shang inscriptions on bones or tortoise shells, but it mostly referred to copper or other metals. Before the Qin Dynasty most gold objects were small adornments, and had nothing to do with immortality. In other words, gold didn't get the same status as cinnabar, jade and the like before the Qin Dynasty. On the contrary, worship of gold was very popular in many other ancient civilizations, e.g. India, Middle and West Asia, Egypt, Greece etc.. Was it possible that they influenced Chinese through Indian or Scythian in the West Region? Joseph Needham attributed Chinese alchemists' worship of gold to Indian culture, he wrote: "Thus it is clear that in ancient Indian liturgical theology of the first half of the -1st millennium the idea of a connection between metallic gold and immortality, even if primarily symbolic, was very explicit… Such then, it would seem, is the background both of the 'plant of immortality' that Chhin Shih Huang Ti was to seek for, and the gold that Li Shao-Chün was to undertake to manufacture. Surely some rumour of persuasion of a most compelling character, reaching China, turned divinity into philosophy, or, to speak more precisely, liturgiology into proto-science."[12] Although we should treat this kind of opinion cautiously, it's probable that the idea of gold in early alchemy was influenced by foreign culture. I have found some other evidences in Daoist Canon now, but let me prepare another article to discuss them. So the second component is a key to know how foreign culture influenced early Chinese alchemy. Of course, there is still much work to be done to know how this phenomenon happened.

Ritual was another essential factor in the naissance of Chinese alchemy, which hasn't had enough attention paid to it before. However, the earlier sources tell us too little except Li Shaojun's making offerings to the Spirit of Furnace. Chinese alchemy has two basic methods, the fire method and the solution method. In fact, the solution method has nothing to do with the origin of Chinese alchemy. For example, the famous Thirty-six Shuifa (三十六水法) were only used to drink for the purpose of being immortal. [13] In Li Shaojun's alchemical procedure, making offerings to the Spirit of Furnace was a prerequisite condition. That undoubtedly refers to the apparatus of making alchemical gold. Yet what I want to emphasize is that worship of the Spirit of Furnace was very common in Chinese antiquity. It evolved from the earlier worship of the Spirit of Fire, who was usually concerned with Chinese deities Yandi （炎帝） and Zhurong （祝融）. So I believe this kind of ritual was a native culture.

When rumors about medicines of immortality appeared in the Warring States period, nobody knew details about it. In the beginning, the essential character of this supernatural medicine was that it wasn't owned by human being, thus has nothing to do with Chinese sacred things (e. g. cinnabar, jade and the like) or religious beliefs (e. g. Yellow Emperor, ghosts, the Spirit of Fire and the like). However, the Western Han Fangshi attempted to make the medicine of immortality by minerals or metals. And their methods were actually a mixture of Chinese traditions and foreign culture. In other words, ancient Chinese compounded those related cultural elements, and created a new method to become an immortal—alchemy. With its development after the Western Han, more cultural ingredients converged into the river of Chinese alchemy, including native ideas of Yinyang （阴阳）, Wuxing （五行, five elements）, Bagua （八卦, eight trigrams）, and minerals from India, Persia and the Arabian Empire.

References

[1] 〔西汉〕司马迁. 史记. 北京：中华书局, 1959
[2] 〔东汉〕班固. 汉书. 北京：中华书局, 1962
[3] 李光璧, 钱君晔. 中国科学技术发明和科学技术人物论集. 北京：生活·读书·新知三联书店, 1955. 129
[4] 王明. 抱朴子内篇校释. 北京：中华书局, 1985. 285
[5] 道藏, 北京：文物出版社, 上海：上海书店, 天津：天津古籍出版社, 1988. 18：795
[6] 〔晋〕葛洪. 神仙传. 北京：中华书局, 1991
[7] 〔晋〕王嘉. 拾遗记. 北京：中华书局, 1981. 105
[8] 赵匡华. 中国古代化学史研究. 北京：北京大学出版社, 1985. 134-135
[9] Needham J. Science and Sivilisation in China. Vol 5. Cambridge：Cambridge University Press, 1976. part Ⅲ, 48
[10] 中国科学院考古研究所二里头工作队. 河南偃师二里头遗址三、八区发掘简报. 考古, 1975, (5)：302-309, 294
[11] 中国科学院考古研究所二里头工作队. 偃师二里头遗址新发现的铜器和玉器. 考古, 1976, (4)：259-263
[12] Needham J. Science and Sivilisation in China. Vol5. Cambridge：Cambridge University Press, 1974. part Ⅱ, 120-121
[13] 韩吉绍. 三十六水法新证. 自然科学史研究, 2007, (4)：507-522

The Making of *Quanti Xinlun*

Chan Man Sing[1] Law Yuen Mei[2]

(1 University of Hong Kong, 2 City University of Hong Kong)

Little is known of the making of *Quanti Xinlun*（全体新论）, a treatise of physiology complied by Benjamin Hobson (1816-1873) and his Chinese assistants, published in 1851, and which went on to become one of the best known translations in the Late Qing era. This paper is intended to fill in the hiatus, drawing on archival materials from the Council for the World Mission and the Wellcome Library, and will be in 2 parts: ① a brief account of its composition, and ② Hobson sources.

1. The composition

Hobson worked with at least three Chinese assistants on the *Treatise*, Chan Apoon, Wong Ping, and Chen Xiutang（陈修堂, possibly Apoon's brother）. As early as in January 1849, Hobson reported to London of his assistant Chan Apoon and their collaboration in "preparing an elementary work on Physiology"①:

> Apoon—This clever interesting young man was formerly instructed by me in Macao and Hong Kong. When I left for England, I left him in charge of the hospital during my absence. He fell into temptation, and disagreeing with Dr. Dike, the Colonial Surgeon who visited the hospital, he left for Canton. He practised surgery a little while, and not finding that profitable he joined himself to a Chinese firm and acted as English interpreter where his profits are considerable. He visits me three evenings of the week and I continue the instruction to him and to mutually assisting each other in preparing an elementary work on physiology bearing on and illustrating natural theology.

The collaboration did not seem to have fructified, however, for two years after, in January, 1851, we find again Hobson reporting on his renewed effort on the book project, this

---

① Hobson's Letter to Dr. Tedman, Jan. 27th, 1849, Council for World Mission Archives, South China, Box 5, 1848-1849, microfiche No. 83. See also Benjamin Hobson's *General Report of the Hospital at Kum-le-fou in Canton from April* 1848 *to November* 1849, pp. 35-37 (MS 5852/42, Morrison-Hobson Archive, Wellcome Library), where the family name of Apoon is given as "Chan"（陈 in Cantonese）.

time with his Chinese teacher, Wong Ping①:

Canton. Jan. 28th. 1851.

Dear Sir,

... I have of course to study Chinese (I wish I could do more) and to read to refresh my memory on the practice of medicine, to enable me to furnish correct information to my pupils. We have met for some months past, three times a week, 2 hours at a time, and have gone through a course of physiology and general anatomy, and we are now occupied with materia medica, and then follow a course of practical medicine and surgery. My teacher, who may also be regarded as a learner in the art, takes down my lecture in running hand, and after writing it out in good Chinese, brings it to me for correction. In this way, a book is ready for the press on physiology, for which a number of drawings are prepared by the lithographic press. They are copied by a friend of mine from drawings in my possession, and when lithographed, are printed off by a Chinese whom I have taught to use the press.②

The book in question came out as *Quanti Xinlun* later, in October of the same year.③ So it began life, as was explained by Hobson himself, in a pedagogical setting, with Hobson translating orally from some western medical work and lecturing on its content, presumably in the local dialect of Cantonese④, and his Chinese teacher wrote it down, and later fleshed it out in the Wenyan (文言) style. This must have gone on very well, so that by the time of his reporting to LMS (London Missionary Society) in January, 1851, "a book is ready for the press". But the two fell out for some unknown reasons in the middle of the year, and by August, we know from Hobson's report that Wong had left and Apoon's brother was helping him with preparing the book for the press⑤:

> Some of whom I thought well have turned away from the truth, especially the Teacher of whom I spoke in a former letter... Circumstances have transpired that lead me believe that he was a hypocrite. He is now at Hong Kong, assisting in Dr. Smith's school and has applied twice to Dr. Legge for baptism... His [Apoon's] brother is assisting me in preparing my work on Physiology which is now nearly ready for the press, and will be illustrated with numerous lithographic drawings.

This brother of Apoon's was possibly Chen Xiutang, whose name appeared in the Preface as the "Co-author". He too seemed to be actively involved in its composition, even at this

---

① Hobson's letter to Dr Tedman, Jan. 28th, 1851, in Council for World Mission Archives, South China Box 5, 1850-1851, microfiche No. 88. See also Hobson's letter to LMS, July 18, 1850, from the same archive, Incoming Letters, South China, Box 5, 1850, microfiche No. 85. where we are given his family name as "Wong" (王 or 黄 in Cantonese).

② The "friend" is Dr W. G. Dickson. See *Brief Report of the Hospital at Kum-le-fou in Canton, during the Year 1852*, by Benjamin Hobson, in Morrison-Hobson Archive, MS 5852, Wellcome Library. Hobson had another illustrator, a young Chinese Zhou Xue (周学), who, according to Leung Ah-fat (梁亚发), was specially asked by Hobson himself to execute the drawing. See "Zhonghua Zuizao budaozhe Liang Fa" (中华最早布道者梁发) in *Jindaishi Ziliao* (《近代史资料》), 1979, 2: 211.

③ "A Note to the Foreign Reader" (in English) was appended to the 1851 edition and dated "October, 1851". This edition can be accessed from the Australian National Library website at http: //nla. gov. au/ nla. gen-vn1869894.

④ The published version still contains distinct marks of its dialectal origin, which will baffle many non-Cantonese speakers. See Zhang Shanlei (张山雷), *Quanti Xinlun Shuzheng* (《全体新论疏证》) (Lanxi: Zhejiang Lanxi Zhongyi Xuexiao (浙江兰溪中医学校), 1927) in 2 *juan*op. cited, pp. 6b, 15b, 16a (juan 1) and p. 4b (juan 2).

⑤ Hobson's letter to Dr. Tedman, dated August 20th, 1851, Council for World Mission Archives, South China, Box 5, 1851, microfiche No. 89.

stage "writing and translating, extracting the essentials and trimming away the over-elaborations" (删繁撮要，译述成书), as the Preface informs us. As for the "Teacher" Wong himself, James Legge wrote of him briefly in a letter to the LMS dated 22nd July, 1852①:

> For some years he was employed as a teacher by Dr. Hobson and he applied to the Church in Canton for baptism nearly three years ago. He first came to Hong Kong last year, as a teacher in Bishop Smith's School, and shortly after his arrival, wrote to me requesting that he might be baptized. I had then several interviews with him. Being a scholar and having read much of the *Bible*, and enjoyed Dr. Hobson's instruction, his knowledge was of course very considerable... Now and then in the course of our conversations, sparks were thrown out from the proud, unsubdued heart of a Chinese professor of literature.

Given the rich, highly literary language in *Quanti Xinlun*, and also the fact that it was "ready for the press" even as early as in January, 1851, when Wong was still in Hobson's employ, it is reasonable to assume that the book came largely from the pen, or the ink and brush for that matter, of this "Chinese professor of literature". Hobson's failure to acknowledge his contribution is understandable, given the acrimonious and of their relationship.

2. Hobson's sources

Hobson did not mention his textual sources in *Quanti Xinlun*, but in a long essay on medicine in China (1861) where he wrote at length of his own Chinese works, five full treatises in all, on subjects ranging from Anatomy and Physiology, Surgery, Midwifery, to General Medicine and Science, he briefly touched on his sources and method②:

> These five works were not translations of any oneEnglish work in particular, but were rather selections from many works on the same subjects, adapted to use. Those, however, more specially used were Dr. Carpenter's Physiology, the Surgeon's *Vade Mecum*, Dr Churchill's Midwifery, Dr. Watson's Practice of Physic, Dr. Arnott's Works, and Dick's Solar System.

For *Quanti Xinlun*, we have further details from an anonymous review of Hobson's works, published in *Medical Times and Gazette* of May 28, 1859 (pp. 555-556):

> This work (i. e. *Quanti Xinlun*, my note) is mainly a translation of Dr. Carpenter's admirable treatise on Animal Physiology, with extracts from Quain's, Wilson's and Paley's works.

The rather precise nature of the intelligence suggests it might most likely also come from Hobson himself. In any case it corroborates with what he wrote earlier of his pictorial sources in the *Canton Hospital Report* of 1852, when he was busily preparing a new edition of *Quanti Xinlun*③:

> In order to furnish a larger supply of suitable drawings to illustrate the work, the author is now

---

① Legge's letter to Dr. Tedman, dated 22nd July, 1852, Council for World Mission Archives, South China, Box 5, 1852, microfiche No. 95.
② B. Hobson. The History and Present State of Medicine in China. Medical Times and Gazette, Jan. 12, 1861. 35
③ *Brief Report of the Hospital at Kum-le-fou in Canton, During the Year 1852*, by Benjamin Hobson, in MS 5852, Morrison-Hobson Archive, Wellcome Library.

having them cut on wood by a native workman under his direction. The number will not be far short of one hundred, and will be selected from Quain, Wilson, Cruveilhier, Bell, Paxton, Carpenter and Mouat.

This new edition came out in 1853, completely re-illustrated, but textually almost identical to the first.① The pictorial sources, we may assume, should naturally serve to be the textual sources as well, since these books—Quain, Wilson, and Carpenter etc. —were in fact the most wisely used textbooks in English on anatomy and physiology in the first half of the nineteenth century. And indeed the two lists of sources, one textual and the other pictorial, basically overlap. (William) Carpenter, (Jones) Quain, and (Erasmus) Wilson appeared on both, and William Paley's *Natural Theology* (1802) was, as was well known, illustrated by Paxton, himself an anatomist. On Hobson's textual sources, therefore, it will be useful to consider also the pictorial list of 1852, if we hope to be exhaustive.

To prove our point, and to give some idea as to how the translation, or "adaption" as Hobson would sometimes have it, was actually done, let's examine an example from the first section of *Quanti Xinlun*, "A Brief Discourse on the Human Body (身体略论)", which, as we shall soon demonstrate, is based on Cruveilhier's:

> *Quanti Xinlun*: *Our body is enveloped with a covering, like a garment, both on the outside and the inside. It covers all parts of the body on the outside, then from the apertures such as the mouth*

---

① The Harvard Yenching Institute keeps a copy of this second edition (Shelfmark: TA7850 39). A microfiche copy is also available from *Protestant Missionary Works in Chinese. J, Physiology and Medicine*, CH-1234; J2. [Microfiche, Zug, Switzerland: IDC 1982]).

A short note is here called for on the early editions of *Quanti Xinlun* when Hobson was reporting back to LMS in August 1852, he was already undertaking the preparation for a new edition of book. Thus he wrote:

I am preparing a second edition of plates for my work on physiology. I am encouraged in this by the demand for the work, and the good opinion everywhere expressed in its form both here [ie. Canton; my note] and at Shanghai and Ningpo... Very lately a rich and influential Chinese has recut and re-published the book at his own expense, and added it to a collection of native authors which sells for about \$30. He did not however ask my permission, and has thought proper to expunge all that referred to the Jesus and God. He has also condensed several of the plates to such the size of the book and consequently has made frightful things of them.

(Letter to Dr Tedman, Canton, August 21st, 1852. Incoming Letters (South China), Box No. 5 (1851-1852) Microfiche No. 92, Council for World Mission Archives.)

The "rich and influential Chinese" was Pan Shicheng (潘仕成) (1832 provincial degree), from a Cohong merchant family, and a bibliophile. This new edition came out in the year fouowing. In the hospital report for 1853, Hobson wrote:

The Tract Committee has also printed 200 copies of the work on physiology, divided into five parts... A work on surgery, to be illustrated with woodcuts, has been begun, but want of time and other circumstances will prevent its being finished this year.

(*Report of the Missionary Hospital in the Western Suburb of Canton*, in MS 5852 Morrison-Hobson Archive, Wellcome Library. pp. 8-9)

The physiological work in question was the newly revised *Quanti Xinlun*, and in the same year Hobson moved on to work on his next medical treatise, *Xiyi Luelun* (《西医略论》) (*First Lines of the Practice of Surgery in the West*).

This new edition was completely reillustrated. Textually, however, apart from a rewritten introduction "身体略论", and the addition of a new section at the end "On the wonderful usefulness of the Soul" (灵魂妙用论) the two editions resemble each other closely.

Between 1855 and 1858, a third edition came out, which differs only slightly, both textually and pictorially from the second. See 陈万成:"《全体新论》的撰译与早期版本" and "《全体新论》插图的来源—兼评松本秀士、坂井建雄文"「全体新论」に揭載される解剖図の出典について (soon to be published).

and nose, turns inward to become the inner covering, enveloping the viscera therein... Under the outer skin there are adipose tissues, which are in the shape of many small hollows linked to one another, like the eyes of a fishing net. Its use is to elevate the skin and to give the body the roundness of form... Beneath the adipose tissues, there are muscles in several layers, bright red in colour, in a flat round shape, and joined together to hold firm onto the hard bones, which lend them support, and on which they are attached. Hidden between the muscles and the bones are the main blood vessels and nerves, thus removed from possible injuries. There are also tendons and strong fibrous membranes, some for binding layers of muscles together, others for holding the muscles and bones in place, and to prevent the muscles from slipping out of place. ①

(人身内外，咸有皮肤，如衣服一般，外则遍布全体，从口鼻入腹，成内皮，以裹脏腑……外皮里俱有肥网膜密遍，此膜状如小孔联结，若网罟之眼，其用所以使浑身连贯圆满……人身肥网膜下，有动肉或曰肌，即字典所谓附骨之肉，数层，其形圆扁，其色鲜红周围裹合，坚骨在其中，以辅佐之，使之有所附丽，在动肉与骨之间，有最要血管及脑气筋藏聚之处，所以免其易于被害也，此外另有筋带筋包不少，或以束数层之肌，或以裹肌及各骨节，皆所以保护各肌，不致乖离。)

Cruveilhier: There is one general covering, which, like a garment, envelops the whole body, and is moulded, as it were, round all its parts. This covering is the skin, of which the nails and hair are dependences. The skin presents a certain number of apertures, by means of which a communication is established between the exterior and the interior of the body... Under the skin there is a layer of adipose cellular tissue, which gently elevates it, fills up the depressions, and contributes to impart that roundness of form which characterizes all animals, and particularly the human species... Below the cellular tissue are the *muscles*, red, fleshy bundles, arranged in many layers. In the centre of all these structures are placed the bones, inflexible columns, which serve for a support to all that surrounds them. The vessels and the nerves are in the immediate neighbourhood of the bones, and, consequently, removed as much as possible from external injury. Lastly, around the muscles and under the subcutaneous adipose tissue are certain strong membranes, which bind the parts together, and which, by prolongations detached from their internal surface, separate and retain in their situation the different muscular layers. ②

Here, the translator apparently follows Cruveilhier's "Introduction", retaining even the figure of "garment" (如衣服一般). He also exhibits a general tendency to simplify and to keep the message easily digestible, though, we may add, he very occasionally amplifies to explain an unfamiliar concept or a point of considerable difficulty (e. g. "此膜状如小孔联结，若网罟之眼"). Overall, it conforms to what one will expect from the Late Qing practice of "dual translation" (合译). The foreign translator tended to abridge, possibly because of his limited facility in the local language on one hand, and on the other, that he found it simply more

---

① *Quanti Xinlun*. first edition. 1851. 1a-b. Between 1855 and 1858, a third edition came out, which differs only slightly, both textually and pictorially, from the second. See 陳萬成：" 《全體新論》的撰译与早期版本" and "《全體新論》插圖的來源—兼評松本秀士、坂井建雄文「全体新論」に揭載される解剖図の出典について" (soon to be published).

② J. Cruveilhier. An Anatomy of the Human Body. trans by Dr Madden, edited by G. S. Patterson. New York: Harper&Brothers, 1844. 2.

profitable to the audience to keep it elementary, paring off the inessentials, Occasionally, however, he also amplified, specially when the original appeared terse or too difficult for the Chinese assistant, himself with little knowledge of the subject in question. The Chinese assistant, translating from the foreigner's dialectal delivery into Wenyan, would also tend to simplify. Limited by his understanding, he would most likely further reduce the message rather than increase it. The resulting translation is, as we have seen here, a much pared-off version of the original, but still recognizable as derived from a source text.

This first section was drastically re-written in the 1853 edition, Consequently much further skewed from the original away from the original: Here is its opening portion, which should suffice for illustration:

> Of all things on earth the human body is the most wonderful. There is a general covering both inside and out, forming a cutaneous layer on the outside, and on the inside, a canopy enfolding the viscera and the various organs... Beneath the outer skin are the adipose tissues, in the shape of small hollows linked together, like the eyes of a fishing net. They serve to elevate the skin and to give the body the roundness of form.

（世上万类，以人身为最奇，不论内外，皆有皮肤，遍布外体为外皮，分布脏腑为内皮。……外皮之里，俱有肥网脂膜，狀如小孔联结，仿类网眼，其用所以使浑身连贯圆满。）

The rewriting apparently aims for stylistic felicity, with better poise and balance in the cadences (e. g. "俱有肥网脂膜" in place of "俱有肥网膜"; and "仿类网眼" for "若网罟之眼") and parallelisms (e. g. "遍布外体为外皮，分布脏腑为内皮" in place of "外则遍布全体，从口鼻入腹，成内皮，以裹脏腑"). Consequently the garment figure is suppressed, the inner-outer（内外）counterpoint mow much more salient than before, and Cruveilhier's original utterly transformed beyond recognition.

Such is the nature of translation in *Quanti Xinlun*, and perhaps also in most other works of *xixue*（西学）in Late Qing, produced through collaborative efforts by western missionaries and their local assistants. Communicative efficacy almost always overrode all other concerns, such as fidelity, a fact certainly not to be taken lightly in studies of China-West translations, scientific or otherwise in the nineteenth century.

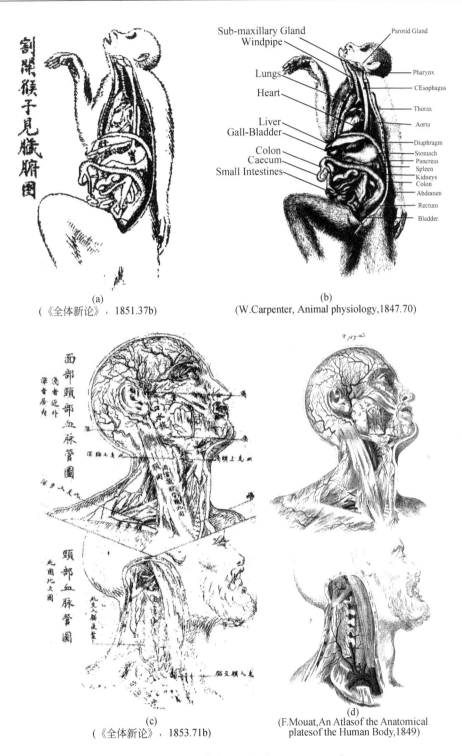

Fig. 1 Sources of *Quanti Xinlun*: two examples

The Jesuit João de Loureiro (1717-1791) and the Medicinal Plants of China

Manuel S. Pinto¹   Wang Bing²   Noël Golvers³
Rui Manuel Loureiro⁴   Rosa Pinho⁵

(1 University of Aveiro, Portugal; 2 Institute for the History of Natural Science, C. A. S., China; 3 University of Leuven, Belgium; 4 New University of Lisbon, Portugal; 5 University of Aveiro, Portugal)

1. Introduction

In 17th and 18th centuries the Jesuits contributed to the scientific knowledge of the plants of China, namely the medicinal ones, and made Chinese botany better known to the West①. Among the fathers of the Society of Jesus who performed such studies in the Far-East, João de Loureiro (1717-1791) stands out because he is the author *Flora Cochichinensis*[1,2], in which mention is made to many plants from China and their uses.

The present paper is about some Chinese medicinal plants referred to in that book. An excellent paper on such plants as used in Portuguese medicine in 18th century and on the important role of the Jesuits in introducing Chinese drugs in western medicine has been recently published by Prof. Ana Maria Amaro[3]. Previously the same author had written about the introduction of western medicine in Macao[4].

As Loureiro, when he returned to Portugal, brought with him a printed copy of Li Shizhen's (1518-1593) *Bencao Gangmu*, dated 1655[5], a tentative comparison is made between the two books, using for that purpose just some of the plants described in them.

2. Short bio-blibliographic note on Loureiro

Tab. 1 is a biographic sketch of Loureiro compiled from Bernardino A. Gomes, his main biographer, who published Loureiro's biography in 1865[6].

---

① See, for instance, several papers about "European Travellers and the Asian Natural World" in *Revista de Cultura—Review of Culture*, International Edition, 20, 21. Macau: Instituto Cultural, Governo da R. A. E. de Macau, 2006, 2007. Not only travellers but also the uses of plants, like the Socotra aloe (by Z. Biedermann[16]) betel palm (by R. M. Loureiro[17]) and the Brazilian tobacco (by A. M. Caldeira[18]) are subject to research studies in those editions of the *Review of Culture*.

The Jesuit João de Loureiro (1717-1791) and the Medicinal Plants of China

Tab. 1　João de Loureiro's biographic sketch

1717	Born in Lisbon; he studied there in the religious Jesuit college of S. Antão.
1732	Entered the Society of Jesus.
1735	Went to Goa as a missionary.
1738	Went to Macao as a missionary.
1742	Went to Cochinchina as a Jesuit astronomer-missionary.
	He became interested in medicinal plants used by local people. Botanical studies based successively on Dioscorides' and Linneus' works; became official naturalist and astronomer of the Court; mounted an astronomical observatory in the royal palace. He got the botanical books from a friend (T. Rieddel) who was the captain of a British ship and brought them from Europe.
	The time was of disputes between the Portuguese and the French missionaries in Southeast Asia.
1750-1752	He was out of Cochinchina due to religious persecutions. Botanical studies in the Philippines and Sumatra.
~1753	Returned to Cochinchina where he conducted more botanical studies.
1759	Expulsion from Portugal of the Jesuits; they were expelled from Macao in 1764.
1767	The Society of Jesus is suppressed in the Catholic countries (Pope Clement XIV brief of 1773).
1777	Went to Canton where he made botanical studies in the area.
	When in Asia he was invited in 1780 to go to London by Joseph Banks (but declined the invitation) and was made a member of the Royal Society of London
1781	Returned to Portugal. Spent 3 months based in the Mozambique Island; carried out botanical studies in East Africa.
	In Lisbon was made a member of the Lisbon Academy of Sciences.
1782	Arrived at Lisbon.
1790	*Flora Cochinchinensis* (finished in 1788) published by the Lisbon Academy of Sciences.
1791	Died in Lisbon. He left a herbarium, drawings, letters, etc. to the Academy.
1793	German edition of *Flora Cochinchinensis* published in Berlin.

Besides *Flora Cochinchinensis*, his *opus magnum*, Loureiro also the author of nine papers (six on botany, one on palaeontology, one on anthropogenesis and one on astronomy) published in several memoirs of the Lisbon Academy of Sciences, either before or after his death. He is also the author of the manuscript *Nova Genera Plantarum in Cochinchinâ...*[7] According to Bernardino Gomes, Loureiro sent the manuscript to London asking for it to be published there, but he was dissuaded from that by friends, particularly Joseph Banks, president of the Royal Society.

In the preface of *Flora Cochinchinis* Loureiro let the reader know about the difficulties that he had to face when working in the Canton area because foreigners were not allowed to trespass the limits of the settlements and he had to rely on information that he got from a peasant who worked for him collecting specimens and telling him the local names of plants. China at the time was under emperor Qianglong's (1711-1799) rule that lasted from 1735 to 1796. On the other hand he would not be able even to visit Macao given the expulsion of the

Jesuits from there in 1764.

3. Flora Cochinchinensis

Flora Cochinchinensis is a 744 page book, plus errata, in Latin, which begins with an address made by the author to the members of the Royal Academy of Sciences of Lisbon where he thanks the Queen of Portugal who had founded the Academy in 1779. The address is followed by the presentation of the book in which Loureiro describes the way it had been prepared: the general areas where he had collected the plants, the two main reference works that he had used for classifying the plants—*De Materia Medica*, by Pedanius Dioscorides (ca. 40AD-ca. 90AD)① and *Genera Plantarum*, by Carl Linnaeus (1707-1778)② —the difficulties that he had to face, mainly in Cochinchina and in China, etc.

According to E. D. Merrill (1876-1956), a well known American botanist who published several studies about *Flora Cochinchinensis*, the book describes a total of 1292 plants, 697 bring from Indochina, 254 from China, 292 both from China and Indochina, 5 from India, 4 from Malacca, Sumatra and Philippines, 29 from tropical East Africa and 18 from Mozambique (9), Zanzibar (8) and Madagascar (1)③. [8]

The description of the plants follows Linnaeus' system: the species belong to 672 genera in turn included in 24 classes each genera being subdivided in orders. The Latin names of the species are followed by a Greek letter indicating general geographic provenance (α for plants from Cochinchina, β for plants from China, γ for plants from Africa and δ for plants from Cambodja, Champa, Bengal and Sumatra—the "Indias" as he calls these areas and by the local name of the plant in either in Chinese or Annamite, if not in both languages. The areas and locals in China where they occur and their medicinal and other uses are described.

The bibliographic list is a long one, comprising more than 90 references[9].

Loureiro prepared all along his life in the Far East several herbaria that he kept conveying to

---

① In referring to the Dioscorides' book, Gomes (Loureiro's main biographer) wrote that the Jesuit had used an edition commented by "Laguna, Rajo, e Tournefort"[6]:18 to carry out his botanical work in Asia. Gomes was referring to the 1555 translation into Spanish of *Materia Medica*, made and commented by Andrés de Laguna (1499-1559); to *Institutiones rei herbariæ* by Joseph P. de Tournefort (1656-1708), published in 1694, and to *Historia Plantarum*, by Joannes Rajo (1627-1705), published in 1686. The works by Tournefort and by Rajo (Ray) are in the list of references of *Flora Cochinchinensis*, but not the work by Andrés de Laguna.

② In the list of references Loureiro mentions the 1764 edition of Linneus' Genera Plantarum, published in Stockholm, plus nine works authored by the Swedish botanist, among them *Systema Natura* (1767), *Species Plantarum* (1763), *Philosophia Botanica* (1770) and *Systema Plantarum* (1779). If Loureiro might have consulted the first four books when in Cochinchina, the last being dated after 1777, when he went to China, could only have been consulted in Canton.

③ Different authors give different numbers for the genera and species found in *Flora Cochinchinensis*. For instance, and apart from Merrill[8], Gomes[6] mentions 1400 species for Cochinchina out from a total of 2000, and a total of 672 genera (489 Linnean and 183 new ones); Fernandes[15] mentions a total of nearly 1300 species, 564 out of which found in China, 697 found in Vietnam and in SE Asia and the remaining ones found in Zanzibar, Philippines, Sumatra, Madagascar and tropical Africa.

several botanists in London, Sweden and Paris with whom he corresponded. To the Lisbon Academy of Sciences he left also a herbarium. So several medicinal plants used in China reached Europe. These are the cases of *Curcuma longa* L. , a plant with several medicinal applications, that was in the herbarium of the Academy of Sciences of Lisbon and also of *Liriope spicata* Lour. sent to Paris to the Muséum National of Histoire Naturelle.

Braga[9] summarizes as follows the fate of the Loureiro's herbarium: ① it consisted possibly of more than 400 specimens; ② about 60 were conveyed to England and Sweden in 1774 (from Cochinchina); 270 specimens were sent to England in 1779 (from Canton), being kept at the British Museum; ③ 87 reached Paris in 1807 or 1808, taken from Lisbon; ④ 37 were kept in Lisbon. Burns-Balogh[10] recently listed 92 specimens belonging to the "Herbarium Loureiro" kept in Paris in the Laboratoire de Phanérogamie du Muséum National d'Histoire Naturelle.

Controversy has occurred about the correct classification and nomenclature of several genera and species described by Loureiro in *Flora Cochinchinensis*, as well as about the chronological priority of some of his descriptions. This is one reason why so many genera and species names in that book cannot be found in the *Dictionnaire Ricci des Plantes de chine*. A small number of criticisms mostly related to genera and species not accurately determined were made in 1793, in the Germany edition of *Flora Cochinchinensis*, annoted by Carl L. Willdenow (1765-1812)[2]. Wildenow refers to a large number of findings of plants made after the year *Flora* had been published and that in places the book was not consistent in terms of nomenclature, facts that were not mentioned by the German editor. He (Wildenow) hoped that the editor would publish illustrations of the plants in a future edition. Bernardino Gomes in Loureiro's biography[6], studied the genera and species names of 37 specimens of the herbarium that the Jesuit had left to the Lisbon Academy of Sciences by comparing the names written by Loureiro on the pieces of "Chinese" paper wrapping the specimens with names in *Flora Cochinchinenesis* and gave them the correct ones used at the time. The well known French botanist Antoine Laurent de Jussieu (1748-1836) did the same thing with the 87 specimens that had been taken to Paris. Merrill[11] raised several questions about how several plants had been described, in terms of taxonomy and nomenclature, about geographic provenance of some of them, about the chronological priority of Loureiro in classifying some species, about lack of information relative to several common plants found in Cochinchina, etc. From all this, changes in the systematics of many medicinal plants classified by Loureiro are abundant. Just to cite two examples: *Areca hortensis* Lour. is now *Areca catechu* L. and *Amomum zingiber* Lour. is now *Zingiber officinalis* Roscoe.

In the preface of *Flora Conchinchinensis* Loureiro discusses also problems related to several classes, namely *Gynandria*, *Monoecia*, *Dioecia* and *Polygamia*.

4. A tentative comparison with Li Shizhen's *Bencao Gangmu*

Some 180 species belonging to some 140 genera of medicinal plants from China are listed in

Flora Cochinchinesis (*FC*). However, for several reasons, only 68 species out of the 180 could be found having a complete and accurate description in the reference work *Dictionnaire Ricci des Plantes de Chine*[12]. Based on eight selected medicinal plants that are found both in *FC* and in Li Shizhen's *Bencao Gangmu* (*BCGM*), a short and tentative comparison between the two books is made in Table II aiming at finding what differences and similarities may exist in terms of their medicinal properties. It is hoped that in the near future such type of comparison may be made for all the 68 plants.

Written by Li Shizhen during 1552-1579 the first edition of *BCGM* was published in Nanjing in 1596. Afterwards it was reprinted many times in China. The Jesuit Michel Boym (1612-1659) dealt with some contents in his *Flora Sinenesis* (1656); then several decades later J. B. du Halde (1674-1743) in his famous book *Descriptions... de la chine et de Ca Tartarie chinoise* (1735) dealt with the book, too. Parts of *BCGM* have been published in various languages (Russian, German, French and English) in Europe since 1656. The herbs' entries were roughly translated into English by the end of the 19th century and an abridged English version was also published in 1932, as referred to in www.medarus.or/Medecins/Medecins/Textes/lishizhenhtml.[13] According to www.china.org.cn/english/2006/sep/180652.htm [14] as recently as 2003 the first complete English translation of the Chinese book has been published.

The copy of *BCGM* brought to Portugal by Loureiro was a revised edition by Wu Yuchang and published by Taihetang in 1655, almost 60 years later than its first edition. However, it needs to be fully studied in order to determine when and where the copy was printed, and how many volumes of the copy were brought back by Loureiro to his motherland.

Not referring specifically to the book by Li Shizhen, in the preface of *FC* Loureiro wrote that he had used Chinese books to confirm the medicinal properties of many plants that grew in China and in Cochinchina; and in fact in the text he made mention to some parts of *BCGM*.

Tab. 2 Medicinal properties of some plants described in *Flora Cochinchinensis* and in *Bencao Gangmu*

Scientific name in *Flora Cochichinensis*	Chinese name	Pinyin	Medicinal properties in *Flora Cochichinensis*	Medicinal properties in *Bencao Gangmu*	Common names in English
Areca catechu L.	槟榔	Binglang	Nuts: expectorant, laxative, beneficial in tenacious rhinitis. taenifug Fibrous bark: diuretic	子苦辛温涩无毒。消谷，逐水，除痰，辟杀三虫。治泻痢后重，心腹诸痛，大小便气秘，痰气喘急，疗诸疟，御瘴疠。	betel Palm and others

Scientific name in *Flora Cochichinensis*	Chinese name	Pinyin	Medicinal properties in *Flora Cochichinensis*	Medicinal properties in *Bencao Gangmu*	Common names in English
Carthamus tinctorius L.	红蓝花、红花	Hong Lan Hua、Honghua	Seeds: purgative. Flowers: laxative, resolutive, emmenagogue, treatment of abdominal pain and births' inflammation	花温辛无毒。主治产后诸疾，活血润燥，止痛散肿，通经。子功效与花同。	safflower and others
Coriandrum sativum L.	芫荽、胡荽、香荽	Yansui、Husui、Xiangsui	Seeds: soporific; carminative	根叶辛温微毒。主治消谷，治五脏，通小腹气，止头痛，辟蛊毒恶气。子辛酸平无毒。主治消谷能食，蛊毒五痔。	coriander and others
Croton tiglium L.	巴豆	Badou	Seeds: purgative; emetic; emmenagogue; treatment of womb obstruction, of dropsy and of the alteration of the body humours	辛温有毒。主治伤寒温疟寒热，积滞泻痢，破血排脓消肿。	croton and others
Datura metel L.	［白］曼陀罗（花）	Bai Man tuoluo	Seeds: soporific; inebriant, antispasmodic; treatment of asthma crises. Leaves: treatment of haemorrhoidal pain, burns and corrosive ulcers	花、子，辛温有毒。主治诸风及寒湿脚气。又主惊痫及脱肛，并入麻药。	angel's trumpet and others
Dichroa febrifuga Lour.	［黄］常山	Huang Changshan	Roots and leaves: febrifuge, treatment of tertian and quartan fevers; emetic	苦寒有毒。主治伤寒寒热，温疟鬼毒，胸中痰结。治诸疟，吐痰涎。	Chinese quinine and other
Kaempferia galanga L.	山奈	Shannai	Stomachic; good for headaches; diaphoretic; effective against contagious diseases	根辛温无毒。主治暖中，辟瘴疠恶气。治心腹冷痛，寒湿霍乱，牙痛。	galanga and others
Stemona tuberosa Lour.	［对叶］百部	Duiye Baibu	Roots: expectorant; refreshing; demulcent; treatment of lungs' illness, of phthisis, of chronic cough	根甘微温无毒。主治咳嗽上气。治肺热润肺。	wild asparagus and other

Tab. 2 allows for some similarities and differences to be found between the two books: ① for classifying plants, different systems were used in *BCGM* and *FC*. In *BCGM*, Li Shizhen used the Chinese traditional method to classify plants as five categories, i. e., grass, grain, vegetable, fruit and wood. This classification originated from different sorts of ordinary food, and was used from ancient times. Indeed in China there is a common saying "medicine comes from some food". In *FC*, Loureiro used Linnaeus' system founded in 1750s, which greatly effected the development of plants classifying in the history of modern biology. So in fact the two authors followed entirely different system of knowledge. ② As for the 8 kinds of plants in Tab. 2, the parts of plants for medical uses are more or less alike: *Coriandrum sativum* (seeds), *Croton tiglium* (seeds), *Kaempferia galanga* (roots and stems) and *Stemona tuberose* (roots), are used with the same purposes in both books. However, there are some differences: flowers of *Datura metel* and roots of *Dichroa febrifuga* are used in medicine in *BCGM*, while seeds and leaves of the former and roots and leaves of the latter are used in *FC*. ③ The medical properties are roughly the same, for examples, *Croton tiglium* (as purgative), *Dichroa febrifuga* (as febrifuge, treatment of tertian and quartan fevers), *Kaempferia galangal* (stomachic, headaches; effective against contagious diseases) and *Stemona tuberose* (treatment of lungs' illness, of chronic cough). Meanwhile, there are some differences: for example *Coriandrum sativum* for expulsion of stagnant air, in *BCGM*. However, it is also used as soporific, and *Datura metel* for the treatment of burns and corrosive ulcers, in *FC*. One wonders whether the differences in ② and ③ are related to the history of medical uses. ④ Since ancient times many Chinese books titled "Bencao" (dealing mainly with Chinese herbs, and animal and mineral medications as well) have given detailed descriptions of every medical plant. Li Shizhen's *BCGM* is a comprehensive expression of them up to his times. The Chinese traditional knowledge of medicine plants is very specific and professional. The knowledge system on *materia medica* in ancient China is much different from that in other peoples and countries. Moreover, greatly difference exists in the past and nowadays in the concepts and understanding of disease, treatment, medication and curative effect.

5. Concluding remarks

The importance of Loureiro's botanical activities may be summarized as follows: ①they were among the pioneering ethno-botanical studies in China made by Westerns; ②they included the preparation of herbaria, drawings and written descriptions that made known to Europe for the first time many Chinese plants; ③the medicinal uses of many plants were also made known to Europe; ④they promoted the transfer of some plants from China; ⑤they contributed to the progress of botany in general and of Chinese botany in particular; ⑥apart from the uses in medicine, they made known to Europe many other kinds of uses; ⑦having corresponding with western botanists, Loureiro played an important role in the West-East scientific exchange in the 18th century in the field of botany; ⑧his work should be seen in the framework of the physiocratic ideas of European Enlightenment; ⑨as scholars, whether in China or in Europe, have hardly investigated the relationship between *Bencao Gangmu* and *Flora Cochinchinensis* and as one of the important figures and events of the cultural exchange between China and Portugal, Loureiro and his book are worth much more studies.

A final remark: the role in China of the Jesuits as naturalists (not as astronomers or mathematicians or just geographers) needs much more attention that it has deserved so far.

References

[1] Loureiro J de. Flora Cochinchinensis: Sistens Plantas in Regno Cochinchina Nascentes. Quibus Acceduntaliæ Observatæ in Sinensis Imperio, Africa Orientali, Indiæque Locis Variis... Tomus I, Tomus II. Ulyssipone Typis, et Expensis Academicis. Lisboa: Real Academia das Sciencias de Lisboa, 1791

[2] Loureiro J de. Flora Cochinchinensis: Sistens Plantas in Regno Cochinchina Nascentes. Quibus Acceduntaliæ Observatæ in Sinensis Imperio, Africa Orientali, Indiæque Locis Variis... Denuo in Germania edita cum notis Caroli Ludovici Willdenow... Berlin: J J Fuchs, Bibliopolam in Ripa Augustinorum, 1793

[3] Amaro A M. Fármacos Chineses Usados na Medicina Portuguesa no Século XVIII. Revista de Cultura—Review of Culture, International Edition, 2007, 21: 121-136

[4] Amaro A M. Introdução da Medicina Ocidental em Macau e as Receitas de Segredo da Botica do Colégio de São Paulo. Macau: Instituto Cultural de Macau, Colecção Macaense, 1992, 2

[5] Li S Z. Bencao Gangmu (Kept in the Rare books section in the Lisbon Academy of Sciences). Revised Edition, 1655

[6] Gomes B A. Elogio Historico do P^e João de Loureiro. Lisboa: Academia Real das Sciencias de Lisboa, 1865

[7] Loureiro J de. Nova Genera Plantarum in Cochinchina nuper inventa et Linnæanâ Methodo descripta, Cantão 20-2-1779. Lisbon: Archives of the Lisbon Academy of Sciences, Mss Azul 386, 1779

[8] Merrill E D. Loureiro and his Botanical Work. Proceedings, American Philosophical Society, 1933, 72: 229-239

[9] Braga J M. Um Missionário Português Botânico Padre João de Loureiro. Boletim Eclesiástico da Diocese de Macau, 1938, 408: 618-634

[10] Burns-Balogh P. Guide Loureiro Herbarium. Leiden: Brill - IDC Publishers, 1987

[11] Merrill E D. A Commentary on Loureiro's Flora Cochinchinensis. Transactions, American Philosophical Society, II, 1935, 24 (2): 1-445

[12] Fèvre F, Métailié G. Dictionnaire Ricci des Plantes de Chine. Paris: Association Ricci—Les Éditions du Cerf, 2005

[13] www. medarus. org/Medecins/Medecins/Textes/lishizhenhtml. [2010-09-30]

[14] www. china. org. cn/english/2006/sep/180652. htm. [2010-09-30]

[15] Fernandes J. João de Loureiro, um Português Esquecido. http://dererummundi. blogspot. com/2008/12Joo-de-loureiro-um-portugus-esquecido. [2010-09-30]

[16] Biedermann Z. Uma Erva de Muitas Virtudes. O Aloé Socotorino na Mira de Botanistas e Viajantes desde a Antiguidade até à Idade Moderna. Revista de Cultura—Review of Culture, International Edition, 21: 30-48. Macau: Instituto Cultural, Governo da R A E de Macau, 2007

[17] Loureiro R M. A verde folha da erva ardente. Betel Chewing in 16th Century European Sources. Revista de Cultura—Review of Culture, International Edition, 21: 49-63. Macau: Instituto Cultural, Governo da R A E de Macau, 2007

[18] Caldeira A M. A Divulgação do Tabaco Brasileiro na China. A Miragem de um Mercado. Revista de Cultura—Review of Culture, International Edition, 21: 64-81. Macau: Instituto Cultural, Governo da R A E de Macau, 2007

The Feuds of the Medical Sects in Republic of China and Colonial Modernity

Xia Yuanyuan

(School of Health Policy & Management, Nanjing Medical Unversity)

The article examined the formation and influence of German-Japanese medicine and Anglo-American medicine, described the German-Japanese dispute with the Anglo-American sect of the process, compared the similarities and differences between the two medical schools, analyzea the trait of the medical sects, and point out the root cause of the feuds lying in the colonial modernity in the political, economic, and cultural, as well as the complexity of the construction of medical education system.

民国时期的医派纷争与后殖民现代性

夏媛媛

(南京医科大学医政学院)

虽然自民国建立始,中国已成为独立的国家,教育主权在20年代的教育独立运动中逐渐收回,但殖民的影响却渗透在各个方面,远未完成解殖的过程。在医学上殖民利益争夺的表现尤为明显。德日医学派与英美医学派的斗争从根本上就是这种殖民利益斗争在医学上的反映,深具权力垄断的性格,同时医派纷争也表现为对医学标准化权力的争夺。而正是在这样的后殖民的权力斗争中,中国的西医教育得以曲折而艰难地建立。

1. 英美派与德日派的形成与发展

1.1 英美派的形成与发展

1866年,嘉约翰设立附属于博济医院的"博济医学校",这是中国近代第一个教会医学校。从1875年起,由于西方国家在华开办的企事业机构日益增多,在这种大环境刺激下,教会学校急剧发展。到1896年,医学传教士已在39个地方进行医学教育,共培养

268 名男、女西医医生[1]。辛亥革命以后，除原有的一些教会医学校得到发展外，还建立了一些新的教会医学院。据统计，1915 年美英教会医学校有 23 处，护士学校 36 处[2]。十数年间教会医学院校迅速发展，对中国医学教育产生了一定影响。

除此以外，留美归国的医学生也对医学教育的进一步发展起到了促进作用。另外，1918 年美国石油大王洛克菲勒出巨资改组北京协和医院为北京协和医学院，每年招收中国各地的医师入该院各科进修，并选送一批中国医师赴美深造。协和医学院培养出来的医学生接受的完全是美国式的医学教育，毕业后或从事医疗，或成为卫生官员，或进入各医学院校成为教员，进一步培养了亲英美的医学生。据 1935 年的统计表明，全国医学院校的教员 964 人中曾留学美国的有 142 人，留学英国的有 87 人，协和医学院毕业的又有 70 人，仅这三项，在医学校中亲英美的人数就多达 299 人，占全部教员数的 31%[3]。

英美教会医学院校加上留学英美学医归国的毕业生再加上协和医学院毕业的学生，他们共同形成了英美派西医，这些医学校则被称为英美派医学校。而随着英美派医学团体——中华医学会的建立，英美派西医的影响则日益广泛。

1.2 德日派的形成与壮大

1896 年唐宝锷、胡宗瀛、朱忠光等 13 名学生赴日留学，标志了中国近代留日运动的开始。而 1907 年清政府规定，凡官费留学生回国后，皆需充当专门教员五年，以尽义务；在义务期未满之前，不得调用派充其他差使。这使得全国各地各级各类高校，出现了一个数量比较可观、分布十分广泛的留日学生群体。在他们中间，留学日本学医的毕业生，归国后有的直接从事医学教育，有的成为卫生官员，有的开业行医。他们在开办医学校或任用卫生人员时均偏好用一些同样有日本留学背景的人，于是便有了日派西医的称谓；再加上日本的医学又学自德国，于是他们和德国留学归国的学医者合称为德日派西医。这些留学日本、德国归国的毕业生回国后办的一些医学校就被称为德日派医学校。

德日派西医利用北京政府奉行仿效日本明治维新的方针，在北京、浙江等地兴办七所医学校，皆聘用日本人或日本留学生充当教员，所用教本多译自日本。一些德日派系的学校如德国人开办的同德医学院、上海德文医学校、日本人开办的满洲医科大学等都享有时誉。日本人在北京、济南、汉口等地所设的日华同仁会医院及其所设的医学校也训练出不少中国西医[4]。据 1935 年的统计，当时全国 27 个医学院校的教员共 964 人，其中曾留学日本者共 106 人，占全数的 11%[5]。留日的医学生创立的学校有些绵延至今，成为我国著名的高等医学学府。德日派创办的中国药学会与中华民国医药学会也成为他们发挥自己影响力的阵地。

2. 德日派与英美派的纷争

2.1 争夺话语权的斗争

英美派与德日派的对立从民国开始一直就存在，但表现并不明显。中华民国医药会与中华医学会在医学教育、公共卫生、国家医学体制以及对待传统中医等问题上存在着分歧，但两会之间的关系总体上还是较为融洽的[6]。后因两派对于医学标准化及其话语权的争夺，医派斗争则越来越趋于白热化，逐渐脱离了对学术观点的争论，并在 30 年代中期表现得最为明显。

导致这一争斗白热化的导火索就起因于《中华医学杂志》于 1933 年第 19 卷第 2 期的一篇文章"我国的医学教育",其中一段写到了日本与中国医学教育:

> 民国初年所办之医学校:如陆军军医学校及北京,江苏,浙江,直隶等医专学校,皆以毕业日本之人充当校长及教员。此毕业生留学时,日本医学尚未发达;而日本学校当局对于中国留学生,又向取放任主义;是以多数皆学无专长。回国后,仅一普通医学士,并无所谓专门。当时因人才缺乏,故荣任教授,主讲大学。当讲书时,仅以自己之讲义,向学生背诵。"讲"之一字,已谈不到;其不称职,可想而知。……此辈主办之医学校,教员及设备,皆极不良,应加改革。但若辈势力,根深蒂固,改革实不易言[7]。

这一文章发表后立刻招来了猛烈的炮火,《医事公论》、《医事汇刊》上接连刊发了多篇反击和批评的文章,纷纷谈及医界团结问题,一直到 1936 年后方渐渐平息。特别是《医事公论》杂志,由中国医事改进社创办,在其 1933 年 10 月 10 日的创刊号中刊登了医事改进社的章程,第一章总则的第三条本社职务的第一款内容就是:"彻底谋中国医事之整理与改进,消除医学派别上之偏见以使医药界同志互助团结,共趋于为国家为致用为学术努力之途径。"[8]足见当时医派矛盾之激烈已引起了医界人士的高度重视。甚至可以认为,医事改进社的创办本身也和医派的斗争有很大关系。因为在《医事公论》的创刊号中第一篇文章便是《中国医事改进社创立旨趣书》,鲜明表示了改进社创办的主要目的之一就是谋"全国医界的精诚团结"[9]。从 1933 年到 1936 年《医事公论》中就有大约十几篇文章谈及医派纷争。另外,《医育》杂志创办于 1935 年,创办人汪元臣毕业于德国柏林大学,按当时的标准他也属于德日派,而在《医育》的第一卷的一、二期详细附载了世界各国医学校的课程概要,第三期则重点介绍了日本近代医学教育的沿革以及日本九州帝国大学医学部的规程[10]。其立场可以说是不言自明的。

从当时医界人士发表的相关文章来看,德日派与英美派的争斗主要表现在几个方面:一是学会的发展,二是权力机关的任职,三是医学校课程的制定,四是医学校的语言。

从学会的发展来看,中华民国医药学会与中华医学会同在 1915 年先后成立,德日派的中华民国医药学会在北洋政府时期影响力颇大,到 1930 年会员已达 800 多人;英美派的中华医学会在南京国民政府时期的影响力逐渐加强,到 1930 年会员也达 775 人[11]。1932 年中华医学会与博医会合并后,财力规模都空前壮大,会员达 1500 多人[12]。面对英美派的中华医学会如此迅速的发展,再加上言辞上对德日派又有蔑视之意,德日派的中华民国医药学会是颇有意见的,在言辞上也针锋相对,认为"某某医学会的发展,何尝是自己的能力,还不是依赖外国人的势力吗。……中华民国医药学会,是中华民国医药同志所组成,并非外国人在中国所开办的一个会,不愿卖身,也就是不愿卖国。……留日的学生,有一个特别的性质,是能够认清国家……吾想一个人,连自己的国家,遂不能认清,倒是旁观的人认得清楚,认贼作父,世果有其人耶……"[13]其措辞相当尖刻。由于言语上的交锋甚为激烈,导致原本两会合并的讨论也暂告终止。

从权力机关的任职来看,德日派与英美派的矛盾也很突出。北京国民政府时期,中华民国医药学会成为政府在医政方面的主要咨询机构,该会的创始人汤尔和同时也成为炙手可热的政要。南京国民政府成立后,中华医学会对政府的影响力逐渐加强。长期担任的"卫生部(署)"长的刘瑞恒及继任者颜福庆均曾任中华医学会的会长。而德日系的侯希民、陈方之、方石珊等人仍在卫生行政部门担任重要职务,但多是技术性强的职务,对卫生决策的影响力已经远不如英美系了[14]。1930年"卫生部"取消,改设"卫生署",编制缩小,日、德、法各派西医的高级职员都自动去职。此后,"中央"卫生行政权一直都掌握在亲美派西医的手里[15]。

对于医学校的课程制定与教学语言的确定,也一直是德日派与英美派角逐的重点。1935年6月"教育部"公布了大学医学院及医科暂行课目表[16]。而这一课目表的制定也引起了德日派的质疑。其一是质疑委员会的组成,认为"这次政府所组织的委员会的委员据我们所闻不到十人,而一大部是"卫生署"和京沪一带几位医界的人物。这种少数力量不平均的委员集合成的委员会试问能代表中国全国吗?而由这少数的委员所造成的草案试问能适合于全国的各学校吗?"其二是质疑课目表制定所参照的对象,"这次由该会拟就的医学院课程大纲是参考北平私立协和医学院的课程大纲作的。关于抄写协和的课程的确有价值我并不反对,不过仔细考虑一下,他并不是十分完善的课程"。其三是质疑其课程设置第一、第二外国语与中国国情不符,认为"该委员会是想用着这种巧妙的手段,拥护中国纯用英文学医的学校,而压迫纯用其他外国文学医的学校"。他们还认为美国的医学教育在当时还比较幼稚,从课程上看更注重实用而不重理论,应该参考德、奥、瑞、荷、丹麦及日本等国的课程,并另拟了一份作为研究的参考。

很明显,到20世纪30年代中期,英美派在以上四个方面均取得了优势地位,话语权基本掌握于英美派手中。但这是否意味着德日派就一无是处,真的如英美派所认为的"教员及设备,皆极不良"?到底这种说法有多少真实性,德日派医校与英美派医校究竟有无本质分歧与差异,下面可以做一个比较来说明。

2.2 德日派医校与英美派医校的异同

在德日派医学校与英美派医学校中,"国立北京医学专门学校"和私立北平协和医学院是最具有代表性的,并且"国立北京医学专门学校"于1928年改为"国立北平大学医学院"[17],从医学专门学校升格为独立医学院,与协和医学院也有一定的可比性。下面不妨用这两个学校在30年代初的情况来做个比较。

2.2.1 基本概况(表1)

表1 北平大学医学院与协和医学院基本情况比较

学校	办学宗旨	入学程度	毕业年限	教师人数[18]	学生人数[19]	经费数[20]	年学费/元[19]
北平大学医学院	教授高深医学,养成硕学宏才,应党国需要[19]	中学六年毕业	六年	34人	185人	28.36万元/年	20
协和医学院	提倡医学公共卫生,造就医士,培植医学教员[20]	大学三年	五年	71人	103人	350万元/年	100

从基本概况来看,这两校的确有很大的不同,首先是办学宗旨,协和医学院明确提出培植医学教员,这与培养医师的实用目的有较大差异;其次是学制,协和医学院实际的学制较长,从中学毕业后需要8年的时间;第三是师生比,一个是1:5.4,一个是1:1.4;第四是经费,协和医学院更具优势;最后从学费上来看,相对而言协和医学院也是较贵

的。两者比较，协和医学院精英化的趋势更加明显，北平大学医学院则更趋于大众化，但这种办学宗旨的区别并不能说明教学质量的差异。

2.2.2 师资情况（表2）

表2　北平大学医学院与协和医学院师资情况比较

学校	教员总数	教授	副教授和助教授	讲师	助教
北平大学医学院[21]	75	20（26.7%）	0	20（26.7%）	35（46.7%）
协和医学院[20]	109	16（14.7%）	50（45.9%）	2（1.8%）	41（37.6%）

师资情况表中可以看出两校人师资构成的最高级别和最低级别相差不大，而副教授和讲师这两个中间级则相差较多，协和医学院的讲师只占1.8%，而北平大学医学院的副教授数为0。这可能与协和较为重视师资的培养、晋升有关。

2.2.3 课程情况（表3）[22]

表3　北平大学医学院与协和医学院课程情况比较

学校	医科课程门类	总课时数	基础课程时数	临床课程时数	讲授语言
北平大学医学院	13	3453	1617	1836	中文
协和医学院	13	4000	1642	2358	英文

对于德日派与英美派的课程，以往最多的评论是英美派的课程注重实用而不重理论，从实际情况来看，两者总课时相差547学时，而基础课程课时只有25学时之差，主要是在临床课程上拉开了差距，相较而言英美派的确更加注重实用，但并不能得出其不重理论的结论。只是有一些课程英美派医校没有开出，如"处方学"、"局部解剖学"；还有一些课程名称不同，如德日派称"卫生学"而英美派则称"公共卫生学"，德日派称"临床讲义"而英美派称"讲义与临床"，引发了德日派认为英美派不注重理论，"讲义与临床"实则只有临床没有讲义的指责[23]。对此，英美派则认为"以时数之多寡，而定课程之优劣，加增时数，即所以改善课程，此种趋势，殊为可悲"[24]。况且，偏重理论与偏重实用是否就能成为决胜优劣的关键也值得商榷。

从以上的比较可以看出，所谓德日派与英美派的不同并没有什么学理上的真正分歧，学校实力的强弱与经费的投入有很大关系，与所属的派别也没有根本关联，学校宗旨也本来就不尽相同，以适应社会不同的需要，唯一可称为差别的只是教学语言的不一致，所以形成医派并导致医派对立的背后深层原因则更值得深思。

3. 医派纷争与后殖民现代性

中国虽然已摆脱了"殖民地国家"的称号，在政治上取得了独立，但后殖民的影响却渗透到了各个方面，在医学科学领域更是如此。医学精英们对现代化的强烈向往，使其成员常在国家建设的过程中扮演重要的角色。各个帝国主义国家对于这一类民族资产阶级专业精英的争夺是最为激烈的，因为这是他们掌握实质利益与文化霸权的有效途径，因此不同国家医学后殖民的结果则是导致不同西医派别在西医内部分化。医派斗争即可以看做是这一后殖民现代性的表现。客观上来看，纷争本身也促进了中国西医高等教育的发展，并促使医学教育者们更好地对医学教育规律进行了探讨和反思。

3.1 医派纷争的后殖民根源

3.1.1 政治殖民的医学表现

派系斗争在蒋介石政权内始终存在，使民国政局出现变幻奇巧的局面，官随"派"换。从1927年到1937年，内政部长换了12人，平均任期不足8个月[25]。这种局面自然也延伸到卫生与教育部门，因此可以认为医派的斗争充其量只是国民党派系斗争的表现之一。对于这种情况，当时的医界已有所认识，正如当时的"南京卫生局"局长及"省立江苏医政学院"院长胡定安所言："今日中国医界派别之开端也，吾言不曰在社会服务之医师，而曰在政之医事当局，……当以科学大同国家应重消除派别捐弃成见为主要条件，使全国医界只知以促进中国医药学术为唯一职责，庶几医事前途得留一线之曙光，不致如政治潮流彼此倾轧无已时也……"[26]

3.1.2 经济殖民的利益争夺

由于当时中国医学的不发达，从药品到器械都是国外的好，于是中国便成了国外药品与器械最好的推销地，各国为了本国的药品和器械能够推销到中国，最好的办法便是在中国设医院、办学校、选送留学生，通过这种种办法来培植倾向自己的中国医生，最终达到其经济上的殖民目的。正是这些国家在幕后的扶植，助长了当时各派西医之间的斗争，其目的明显在于排挤他国在华的势力，以便达到经济上的独家获利。于是当时医界的有识之士发出了这样的警示："我国医学界中之所以分出派别来，其最后的原因，还是由于几个资本主义的国家在背后操纵的缘故，他们为要推销他们的物品，自会放下这许多线索来，使我国的医学界中人无意间受到他们的牵引。所以我们要设法能够尽量的自制新药和各种医学的器械，也是目前最要紧的事！"[27]

3.1.3 文化殖民与身份认同

对医派斗争影响最大的是文化的殖民性，因为现代生物医学本身就具有"帝国主义"或"殖民主义"的特性。殖民医学史家大卫·阿诺认为"在某种意义上，所有的现代医学都在进行一种殖民的过程"，现代医学在欧美透过与国家的"共生"关系排除民俗医学而取得垄断地位的历史，可以说是一种殖民的过程[28]。在中国，这种殖民过程引发了对中西医之间的斗争。而由于引入的医学渠道不同，它们又各自扩张，相互排斥，终于引发了西医内部的医派斗争。

另外，近年来科技史、医学史以及STS的研究指出，现代医学与科学这种"帝国扩张"过程主要推动力量之一，来自于"标准化"，包括器械与仪器的标准化，也包括人员训练过程与专业资格检定的标准化[29]。在当时的中国，这种源于对来自不同国家医学标准化地位的争夺自然也就导致了医派之间的不和。

再有，不同医派的斗争还源于身份认同的需要。威廉·布鲁姆（Williams Bloom）曾指出："身份确认对任何个人来说，都是一个内在的、无意识的行为要求。个人努力设法确认身份以获得心理安全感，也努力设法维持、保护和巩固身份以维护和加强这种心理安全感，后者对于个性稳定与心灵健康来说，有着至关重要的作用。"[30] 当时的英美派与德日派人士，留学海外后经历了不同的文化碰撞，更由于当时的中国正是全面落后之时，他们回到祖国后便产生强烈的传播海外文化的使命感。他们虽是中国人，但在身份、心理上都倾向于留学国。留学同一国家的人士自然抱成一团，以巩固这种身份的确认；留学不同国家的自然就相互排斥，避免被对方同化。

3.2 医学殖民的特点与作用

一方面，殖民国利用医学达到了巩固殖民统治，破除文化对抗的目的；另一方面，先进医学技术也随之在中国得到了发展。正如医派的纷争虽深深打上了殖民权力争夺的烙印，但客观上却促进了医学的进步及对医学教育的思考。

无论英美派还是德日派，他们留学国外，学习他国先进的医学技术，并将这些带回国内，从而在一定程度上改变了之前中国医学落后的状态。首先是西医教育体系的引入给中国传统中医带来了强有力的冲击，也迫使中医思考自身的生存与发展问题，为中医教育提供了一个学习的样板。其次在医疗卫生方面，各派人士最初均抱着学成归国报效祖国的想法，为祖国的医疗卫生事业做出了不少贡献。开业医师人数的增加，为百姓的疾病诊治提供了有效的途径；公共卫生人才的养成，为城市的卫生防疫提供了有力的保障。最后从医学教育方面来看，各种模式的医学院校为当时的中国培养了不同层次的医学人才，既有独立开业医师，也有医学科研精英，还有卫生管理专才。

最重要的是，在医学教育领域，这种派系纷争引发了人们对医学教育规律的思考。英美派与德日派的教育宗旨各不相同，但这种不同目标培养出的医学生恰恰适应了不同层次的需要。可见，当时中国的国情决定了医学培养宗旨的多样化，并无所谓哪一派的是标准或优于对方。英美派与德日派尤其在课程设置方面有所争议，德日派重视理论基础，英美派重视实践操作，今天看来这些争议对医学教育的发展均是有益的。因为两者对于医学而言同样重要，况且重视理论并不一定意味着忽视实践，反之亦然。由此看来，英美派与德日派在课程设置上是各有侧重，形成了互补而非对立的关系。

归结起来，医派的纷争还是缘于当时中国医学的不发达，各方面还不能独立自强，对外来的各类事物缺乏正确的评价，更由于未能形成适应于自己国情的医学教育体系。所以建立起中国自己的医学教育体系便成为当时医界有识之士的共识，最终西医教育体系的建构便是在与帝国主义及西方现代医学纠缠拮抗的复杂关系及其所涉及的利益与权力中完成的。

参 考 文 献

[1] The China Medical Missionary Journal, Vol XI, No 2, 1897.91
[2] 邓铁涛，程之范. 中国医学通史（近代卷）. 北京：人民卫生出版社，2000.490-498
[3] 武文忠. 我国医药学院校之初步统计. 医育周年纪念刊，1936，72
[4] 金宝善. 金宝善文集（样本）. 北京医科大学公共卫生学院编.1991，106
[5] 武文忠. 我国医药学院校之初步统计. 医育周年纪念刊，1936，72
[6] 刘远明. 中华医学会与民国时期的医疗卫生体制化. 贵州社会科学，2007，6：165
[7] 中国的医学教育. 中华医学杂志，1933，(19) 2：205
[8] 中国医事改进社章程. 医事公论（创刊号），1933：36
[9] 中国医事改进社创立旨趣书. 医事公论（创刊号），1933：2

[10] 九州帝国大学医学部规程．医育．第一卷第三期：30-32
[11] 陈清森．中华医学会80年发展历程．中华医史杂志，1995，(25) 1：1
[12] 陈清森．中华医学会80年发展历程．中华医史杂志，1995，(25) 1：2
[13] 显祖．中华民国医药学会何以不像人家的发展．医事公论（创刊号），1933：27-28
[14] 尹倩．分化与融合——论民国医师团体的发展特点．甘肃社会科学，2008，2：22-26
[15] 金宝善．金宝善文集（样本）．北京医科大学公共卫生学院编．1991，112
[16] 陈邦贤．近十年来医学教育大事记．医育，1939，3 (2-3)：30-44
[17] 李涛．民国二十一年度医学教育．中华医学杂志，1933，(19)：681
[18] 李涛．民国二十一年度医学教育．中华医学杂志，1933，(19)：689
[19] 全国中西医药学校调查报告．中西医药，1936，(2) 1：60-61
[20] 私立北平协和医学院简章．北平：私立北平协和医学院编，1930
[21] 二十三年度教职员调查表．北京大学医学部档案馆，案卷号 11-1934-1
[22] 李涛．民国二十一年度医学教育．中华医学杂志，1933，(19)：685
[23] 西京医师公会对于南京医学教育委员会所拟医学院课程大纲评议点大要．医事汇刊，1935，23：164-167
[24] 狄瑞德．医学教育与北平协和医学院之课程．中华医学杂志，1934，(20)：1530
[25] 王国君．论国民党内的派系斗争．松辽学刊（社会科学版），1994，2：42
[26] 胡定安．中国医事前途急待解决之几个根本问题．医事公论，1934，6：17
[27] 柯士铭．向第三届全国医师联合会代表诸君进一言．医事公论，1934，6：6-7
[28] 李尚仁．医学、帝国主义与现代性：专题导言．台湾社会研究季刊，2004，54：11
[29] 同［28］
[30] 乐黛云．文化传递与文化形象．北京：北京大学出版社，1999：332

Investigation on Traditional Spinning Wheels and Looms in Ze Zhou Region

Lu Wei Yang Xiaoming

(College of Textiles, Donghua University)

Lu silk was a kind of traditional one, which prevailed during the Ming and Qing Dynasty in the southeast of Shanxi province. It embodied the combination of the local traditional culture with its unique skill. Field archaeology and oral history both were used to investigate the existing spinning wheels and looms in Ze Zhou region. This paper discuses the traditional silk-weaving skill, its profound historical and cultural background and realistic significance in the southeast of Shanxi province.

Lu silk, is the most representative and impact textile of Shanxi province as well as of northern China in the Ming Dynasty, enjoys "Songjiang district in south, Lu'an region in north, clothing the world", and "Lu silk over the world" as a reputation, but since the late Ming and early Qing craftsmen of Lu Zhou region uprising destroyed almost all the looms, the reputation of "Lu silk" disappeared, not only a small amount of spinning wheel and loom, but also its technical process in some extent were retained and transmitted in Ze Zhou region. By field investigation and document, this thesis for the first time makes a preliminary study on the spinning wheel and loom in Ze Zhou region, its historical evolution and transmission way, with a view to make a necessary foundation for exploring the weaving technology of Lu silk.

1. Historical Origins of Lu silk

Shanxi province has always been the place for mulberry planting and sericulture. Southeast areas of Shanxi province, due to its unique geographical environment, became the center of silk industry of Shanxi province even the Yellow River basin, and Lu silk was one of the most representatives of textiles.

Southeast areas of Shanxi province mainly include Ze and Lu regions. Lu, that is, Lu Zhou state, includes present Changzhi, Huguan, Zhangzi, Tunliu, Lucheng, Xiangyuan, Licheng etc., anciently called Shangdang County, which was changed into Lu Zhou and promoted to Lu'an House respectively in 1408 and 1529; Ze, that is, Ze Zhou state, including Jincheng, Gaoping, Yangcheng, Qinshui, and Lingchuan six cities and counties, along with Lu Zhou state became a major center of silk industry in the Ming and Qing Dynasties in China.

In the Ming Dynasty, there were four major silk producing industry areas: Zhejiang province, Jiangsu province, Sichuan province, Shanxi province, Fujian and Guangzhou. The main textile product of Shanxi province just was Lu silk. [1] Lu Zhou should be the central region in the north, which was well known in the world for its product-Lu silk. [2] Lu silk was named after Lu Zhou state, the main producing areas are the present Changzhi county and Gaoping city. From the view of geographical location, Gaoping city and Lu'an House were linked, but it belonged to Ze Zhou state from the aspect of administrative division. Thus, both Lu Zhou, and the neighboring Ze Zhou state were the Lu silk origin. The late Ming and early Qing Dynasties craftsmen uprising destroyed almost all of the looms of Lu Zhou state and Lu silk disappeared in Lu Zhou state. Although the name of Lu silk disappeared, the skills was retained in Ze Zhou state and be passed on from generation to generation. Subsequent Lu satin, Ze silk, and Ze Zhou handkerchief were all the continuation and development of Lu silk skills and culture, which has written a splendid touch in Chinese history. The following document literature provides reliable evidences:

> The fourth day of Lunar New Year in April in 1765 (The 30th year of Qianlong in the Qing Dynasty), silk fabrics for trade including Lu satin and Ze silk in Kashgar were from Shanxi province... fifteen bolts blue silk, fifteen bolts gray silk and ten bolts bronze silk were from Gaoping county... [3]

Other historical records documented, from the Emperor Qianlong to Jiaqing in the Qing Dynasty, the amount of Ze silk sold to Xinjiang each year is between 100-300 bolts, which all proved the development of local silk industry.

Ze Zhou region, one of the cradles of Chinese agricultural civilization, was also the critical areas for sericulture and silk culture. The earliest record of Ze Zhou textile tools and textile scene image is the fresco "View of Weaving" in Kaihua Temple of Gaoping city, which describes the scene of the charity Prince's view on women's weaving painting. From another perspective, as the carrier of historical events account, fresco art reflects the traditional textile machinery of Ze Zhou region in the Northern Song Dynasty, and provides evidence for comprehensively study textile industry during the Northern Song period. "View of Weaving", depicts women spinning and weaving. The loom in the painting is a vertical loom with single heald and double treadle, which is popular in the *Zirenyizhi* written by Xue Jingshi, a Shanxi nationality craftsman in the Yuan Dynasty, also shows that the development course of textile machinery in Ze Zhou region in the Song Dynasty.

Zirenyizhi also recorded the materials and function of a small cloth laying machine and jacquard machine, just as Zheng Juxin notes:

> The Song and Yuan Dynasties, Lu'an region in Shanxi province, jacquard machine is a universal machine model... the prosperity of the textile industry of Lu'an region won a reputation as "Songjiang district in south, Lu'an region in north, clothing the world", "Lu silk over the world". [4]

These precious image materials support the studying on the development of silk weaving technology in Ze Zhou region.

So far, there still exists an integral relationship between the local textile industry development and the glorious tradition of Lu silk, based on which, it continuously gets more development and innovation.

It is the reason for selecting Ze Zhou region as our research origin, while Lu silk originated from Lu Zhou region.

2. Main types of traditional spinning wheel in Ze Zhou region

2.1 Investigation sites

The selection of investigation sites, which include the traditional places of mulberry planting and sericulture cultivation, silk and weaving in Ze Zhou region, namely, Feng village, Wang village, Runcheng village, Nanyang village and Jincheng city and so all. There are two types of spinning wheel and three kinds of weaving machine. Three types of weaving machine are respectively called small machine, large machine and spinning rope machine in local areas, distribution as the following Tab. 1:

Tab. 1 Distribution of spinning wheel, spindle-whorl and spinning rope machine

type / village	spinning wheel	Spindle-whorl	small loom	large loom	spinning rope machine
Wang village	√	√			
Feng village			√	√	
Runcheng village	√		√	√	
Baofu village	√				√
Nanyang village	√	√	√	√	√

2.2 Main types

Spinning wheel was produced in ancient China; its original shape might occur in the Shang Dynasty, the forming spinning wheel may appear in the Warring States period, and generally used in the Han Dynasty. Portrait of a large number of Han stone spinning wheels provides the evidence for this. Spinning wheel is a tool specially used to twist and stretch cotton, linen and wool fiber, mainly used for short-fiber drawing. Though somewhat different in shape of the spinning wheel, the basic principle and purpose are the same.

Through field visits, hand operated single-spindle spinning wheel was generally used in Ze Zhou rural areas, and according to local elderly descriptions, they only used this one, which also showed large-scale cotton textile industry did not appear in the local areas, cotton industry was only in self-sufficient state. There were two main types, the most common shown in Fig. 1:

Multi-cultural Perspectives of the History of Science and Technology in China

Fig. 1　Hand operated single-spindle spinning wheel

From Fig. 1, the spinning wheel consists of four main parts: frame, spindle, pulley and handle. It is a transmission unit of rope and wheel, which consists of two groups of "米" character shaped spokes fixed on the axle, generally 20-25cm apart, handle mounted on one end of the axle, wheel and spindle connected by a rope. To turn the handle clockwise, the spindle rotate and twist yarn. On the contrary, turn the handle counterclockwise, the spindle rotate in the opposite direction. Trough this way repeatedly, spinning yarns are fine and evenness.

The main difference of the existing two types of spinning wheel in Ze Zhou region is the width of the wood chip of the rope-wheel unit, and the more common is the narrow wood chip. The under frame length 78cm, maximum width 48cm, distance between the two groups spoke 20cm, spinning wheel height 45cm, diameter of rope-wheel about 67cm, the length of wood chip on spokes 30cm, spindle for the iron.

Cotton, linen and wool fibers must be spun by the spinning wheel and be used to weave the fabric, the spinning process is called twisting. For silk fiber, the silk thread can be woven after reeling; spinning wheel strands the single fiber, known as "doubling". Ze Zhou region is one of the agriculture origin in China, has rich natural resources, hemp and silk fiber are the first textile fibers. During the Song Dynasty, with the expansion of cotton cultivation, people in Ze Zhou region also started to cultivate cotton, but hemp and silk textiles were still retained as traditional products. Local traditional spinning wheel, as a tool, is used in different types of textile fibers, mainly cotton, linen and silk. Cotton spinning, then woven into the traditional old coarse cloth; silk fiber after spinning is used for hand embroidery, made into insole, stomachers and hats.

Traditional spinning tools, spindle-whorl, still exist. Spindle-whorl is the composition of whorl and spindle which is made of wood or iron, the same role as the spinning wheel. As one of the earliest human spinning tools, spindle-whorl emerging can be traced back at least the Neolithic. In more than 30 provinces, municipalities and autonomous regions, the whorl, major textile spindle-whorl parts, have been excavated in the site of large-scale early inhabitants. [5]

The differences of spindle-whorl are the size and largely depend on the materials of the whorl. Folk Ze Zhou region, there is a lot of spindle-whorl, which is called Nianxiantuo. Whorl with hexagonal iron whorl, there are drum-shaped wooden whorl, shown in Fig. 2:

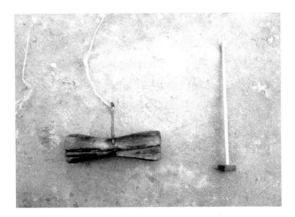

Fig. 2 Spindle-whorl

According to research, spindle-whorl is primarily used for twisting hemp fiber and silk fibers, after which can be used to make shoe soles, the main feature is comfortable, sturdy, especially for summer wear. It is more common in rural areas, today still exist.

Nianxiantuo shown in Fig. 3:

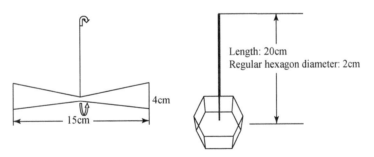

Fig. 3 Nianxiantuo

3. Traditional looms and development in Ze Zhou region

3.1 Laying loom

In Ze Zhou region, there are three types loom: large loom, small loom and spinning rope machine. The most commonly used are small loom and large loom. Both small loom and large loom are used to weave silk and cotton at local family, but small one is most common. As can be seen from the picture, it is similar to the laying loom described in the *Zirenyizhi* written by Xue Jingshi.

According to Zhao Feng's research on ancient laying loom:

From the description on laying loom in Japan and interpretation of the two works of the Yuan

Dynasty, laying loom is a pedal-operated side loom with single-track and single-treadle. There are many different titles for this laying loom later, such as side loom (*Tiangongkaiwu*), "Dahua loom" (Hunan Province), grass cloth loom (Jiangxi Province), Luoji machine (Jiangsu Province), loom machine (Shanxi Province), etc., which are widespread in china with a variety of types. [6]

Obviously, there is s a wide range of laying loom, very popular in villages, but the basic operating principles and components are similar.

In Ze Zhou region, existing small regional loom is 1.6 m in length, 1.7 m in height and 0.75 m in width, which is a single-track and treadle loom, and also known as side loom. From the shape, it belongs to pedal-operated pressure laying loom. Standing frame and laying frame composite the major frame of loom, this is supported by the four foot column. There is a group of Ya, ermu on the standing frame, which connected track with the front-end and treadle with another end. In the middle of treadle and Ya, ermu is pressing warp bar. Matau is fixed in the rear of loom; reed hanging from the two bamboo poles with a rope up, two bamboo poles is fixed next to the upper standing frame. The waist belt and cloth beam connected to the waist of weaver to adjust the warp tension. Reed is also a small tool with different sizes, which has two kinds of material including bamboo and iron. Small loom used to weaving coarse cloth and plain silk fabric in the local, which should be identical with side loom in *Tiangongkaiwu*, namely: where weaving Hangxi, Luodi etc. silk fabric, plain silk fabric, and towels, hats and other fabrics, not with jacquard machine, only with a small loom. Local silk weaving machine with the small loom general home use only, and for the silk trade is the use of special silk looms, there are four different types of silk looms, including horizontal silk loom, big jacquard machine, small jacquard machine, and twill loom, 3000 for the small jacquard machine, 6000 for the big jacquard machine, 50 for the twill loom, 200 for the horizontal silk loom, producing mainly concentrated in Lu silk origin.

Fig. 4 Laying loom

3.2 Large loom

Large loom, also known horizontal loom in the local, appeared slightly later than the small loom. Interview, we met 91-year-old skilled weaver, Lv Ruilian, who mentioned her large loom is modeled by a local craftsman Henan loom manufacturing. According to speculate, the age of its formation should be the beginning of the 20th century, a lot of promotion and use is probably in the early days after liberation. *Chinese industrial annals-Volume of Shanxi* also records that Jincheng city, which belong to Ze Zhou region, whose cottage industry has weaving, paper making etc., "weaving began in the previous generation, the technology introduced by Henan province, and intense place was near Gaodu town"[7]. This also further confirmed the technical and weaving origins. The frame of large loom is horizontal, which is relatively with laying loom; horizontal body was about 2 meters long, 1.2 meters wide. At present the local production of coarse cloth used is this old type of loom. From the figure, mainly consists of frame, reed, track pieces (2), pedal (2), lever, sitting board, rolling axis, rolling-warp shaft. The Bouyei loom, which is in Luoping County, Yunnan province, is similar to mentioned above, despite some differences in appearance [8].

Fig. 5 Horizontal loom

4. Preliminary conclusions

From the Kaihua Temple mural of Gaoping city during the North Song Dynasty to *Zirenyizhi* written by Xue Jingshi in Wanrong county (the neighborhood of Ze Zhou region), combining with folk samples and oral historical material, it makes a basically presumption: traditional loom of Ze Zhou region experienced several development stages: standing loom, jacquard machine, small cloth laying loom during the Song Dynasty, through side loom of the Ming and Qing Dynasties, to small loom, big loom from the 1920s to 1980s. Standing loom in the Song and Ming Dynasties has not yet seen in present Ze Zhou region among people, but some people say that there is still retained in the middle region of Shanxi province[9], to some extent, which illustrates its promotion case in Shanxi Province. Large looms can be thought as the most popular model in the 20th century in Shanxi and Henan provinces, spread along the general model, but also are better in efficiency than small looms.

Existing traditional spinning wheel, spindle-whorl and loom in Ze Zhou region can be traced back to the Late Qing Dynasty period, and are inseparable with the local ancient agricultural civilization and its textile, as of today, which can also be seen in the rural areas faintly. The development of modern industry makes people no longer engaged in spinning and weaving all day, but spinning and weaving, as a culture, has rooted in women's lives. Although the modern mass production provides necessary human material products, but the local rural women are still using traditional woven textiles and tools to show their interpretation of nature and life, expressing their good wishes, and to future generations.

References

[1] 田秉志. 中国工艺美术史. 上海：东方出版中心，2009.289
[2] 李仁溥. 中国古代纺织史稿. 长沙：岳麓书社，1982.176
[3] 张正明，薛慧林. 明清晋商资料选编. 太原：山西人民出版社，1989.11
[4] 〔元〕薛景石著，巨欣注释. 梓人遗志图说. 济南：山东画报出版社，2006.59
[5] 赵承泽. 中国科学技术史（纺织卷）. 北京：科学出版社，2002.160
[6] 赵丰. 卧机的类型与传播. 浙江丝绸工学院学报，1996，13（5）
[7] 实业部国际贸易局. 全国实业调查报告·华北编（第二种）. 中国实业志-山西省（第一册），台北：宗青图书公司，1936.178
[8] 李云，李晓岑. 云南少数民族传统织机研究卧机的类型与传播. 广西民族大学学报（自然科学版），2008，14（1）
[9] 徐岩红. 宋元时期山西寺观壁画中的技术成就. 山西大学科学技术哲学中心2008届博士学位论文

Titles and Classifications of the Ancient Artisans in Dunhuang

Wang Jingyu

(Dunhuang Academy, Dunhuang, Gansu Province, China)

The library cave of Dunhuang preserves large quantity of social economic manuscripts and the inscriptions on the murals record a lot of artisan's titles, names as well as their working scenes from Tang, Five Dynasties and the Song Dynasty. According to the records from ancient documents, the artisans in ancient Dunhuang were managed by local authorities, temples and individuals.

Local government set up workshops to organize for some industries the production and manufacture. The ministry of workshop, which had already been established in the early Tang Dynasty, is the specialized agency to supervise handicraft industry. Artisans' ranks of various industries can be roughly classified into Duliao, Lushi, Boshi, Teacher, Artisan, Master, Yuansheng and students etc., the highest rank of which is Duliao. However, Artisans ranking Duliao were not covered in all industries. According to the latest research, Duliao was set in among such artisans as blacksmith, carpenter, sculptor, painter, paperhanger, masonry, as well as in some industries as bullion houses, bow store, saddle store and so on.

The ancient artisans of Dunhuang can be approximately classified into two types: the first are those whose work related to the social production and people's lives, that is, artisans working on making production tools and daily necessities for the people, such as blacksmith, carpenter, masonry, hat maker, shoemaker, silversmith, jade craftsman, locksmith, bow maker, etc; the second are those who were engaged in cultural and artistic activities like sculptors, painters, paperhangers etc. In addition, Duliao was also found in some households working on various industries, e.g. families specialized in making weapons, wine and extracting oil. The artisans' titles even in the same industry are different. According to the preliminary statistics, there are about 100 different titles to the various craftsmen in the ancient times of Dunhuang. Such detailed social divisions reflect the prosperity of the handicraft industry and the progress of the technologies, and they are important materials for us to study the commercial economy and the science and technology in medieval China.

敦煌文物中古代工匠的名称与分类[*]

王进玉

(中国敦煌研究院)

1. 敦煌古代工匠史料

根据笔者的初步研究，敦煌古代工匠史料主要有以下几个方面：

(1) 在有关敦煌石窟营造的碑、铭、记、赞文书中，几乎都有要对从事石窟营造的"良工"、"巧匠"们进行描述[1]。

(2) 在敦煌石窟各个时代的壁画中，有大量反映古代工匠们劳动的画面[2]，如房屋（塔、庙）建造、钉马掌、凿石磨、制陶、酿酒、打铁、纺线、织布、制皮、做靴等，这些都生动地表现了敦煌地方古代手工业劳动的具体情景。10世纪前期建成的莫高窟第72窟还绘有《修塑大佛图》和《临摹佛像图》，我们从中可以直接了解到敦煌古代的塑匠、画匠们从事敦煌石窟艺术创造活动的情景。另外，在公元10世纪的石窟壁画（藏经洞绢画）中，出现了较多工匠的供养像及题名，但这只是敦煌几十代千千万万工匠之中的极少数。就是这些各行各业的工匠和广大劳动人民创造了敦煌的古代文明[3]。

(3) 敦煌藏经洞保存的数百件社会经济遗书等资料中，记载了唐、五代、北宋时期的许多工匠名称、姓名及其生产活动。主要有入历，亦名得物历、领得历、得物抄录等。收入账目，有官府、寺院、个体工匠三种，以寺院的最多。记单项收入的，如油入历、布入历等，记多种收入的叫诸色入历。破用历，亦名用历、破历、破除历、使历、使用历、所用抄录等。支出账目，敦煌写本中有官府、寺院、个体工匠三种，以寺院的最多；记载支出日期、货物品名、数量、用途。便物历，借贷账目，亦名出便历、出便与人名目、出便与人抄录；又因便物不同而有便粟历、便面历、贷粟历、借贷油面物历等名。敦煌写本中有数十件，多出自寺院。各种簿籍等，如 S.542V⁰ (4)《戌年沙州诸寺丁持车牛役簿（附亥年至卯年注记）》[4]，为都司行政文书。原题"戌年六月十八日诸寺丁持车牛役簿"，存196行，后缺。系吐蕃统治时期戌年以来六年间（818~823）敦煌诸寺寺户及车牛在沙州僧团统领机构"都司"上役的登记簿，登录有龙兴等14寺191名寺户丁口的姓名、役作时间及内容。记载为寺院从事农业生产和其他劳动的分工就有20多种。其中第121行记载灵图寺寺户"葵曹八，纸匠"。官府、寺院、个体工匠的与此有关的其他遗书等等，此不赘举。

---

[*] 本文为国家社会科学基金西部项目"敦煌古代工匠研究"（06XZS0108）的部分成果之一。

(4) 史书中与此有关的资料。
(5) 敦煌等地考古发现文物中与此有关的资料。

2. 敦煌古代工匠及其管理方式

2.1 工匠的管理方式

从文献记载可知，敦煌古代工匠大体为官府、寺院的和个体工匠三种管理方式。

9~10世纪的瓜、沙归义军时代，手工业劳动者们摆脱了寺院的控制，变为各种行会作坊、专业户、官府匠人和零散工匠四类。

2.1.1 官府工匠

官府设立作坊组织某些行业的规模生产和制造，而作坊司是管理手工业的专门机构，敦煌地区在唐朝早期已有设置。

关于行会，已见于记载者有木行、画行、弓行、刺鞍行、皱文行、金银行等，专业户有弩户、酒户、梁户、砲户[5]、弩家等，当府匠人有纸匠、笔匠、皮匠及各类纺织行业的工匠等，这几类工匠都是在官府直接控制下从事本行业劳动，行会的头目由官府委任，并在归义军府衙内挂一"押衙"之类的虚衔；泥匠、石匠一类可能是不受官府控制的工匠，但他们要像普通农户那样承担所有赋税徭役。总的来说，敦煌瓜沙归义军时代的工匠组织，特别是作坊和行会组织，在社会变革方面有一定的进步意义[6]。

作坊及作坊司

关于敦煌古代的作坊、作坊司、作坊使的情况，在 S.11453H-L（参 S.11459C-H）《唐瀚海军典抄牒状文事目历》、P.2942《唐永泰年间河西巡抚使判集》、S.7384B《光启三年（887）二月及三月作坊使康文通牒并张淮深判（二通）》、P.4640V⁰《归义军己未至辛酉（899-901）布纸破用历》、P.3234V⁰（7）《癸卯年（934）正月一日已后直岁沙弥广进面破》、S.8666《戊寅年七月都头知作坊使邓守兴请判凭状并判》、S.8673《丁丑年八月都头知作坊使邓守兴请判凭状并判》、S.9455《丁丑年九月都头知作坊使邓守兴为造佛轴用毯杖柄请处分状并判》、P.3347《天福三年（938）十一月五日归义军节度使曹授张员进衙前正十将牒》、S.4700、S.4121、S.4643 和北图新 1450 号（周字 45 号）《甲午年五月十五日阴家婢子小娘子荣亲客目》、P.3942《荣亲客目》、P.3763V⁰《布缯褐麦粟入破历》、P.3440《丙申年三月十六日见纳贺天子物色历》、S.1153《诸杂人名一本》等近 20 件遗书中都有记载。

作坊司是归义军管理手工业的专门机构，在归义军前期已设置。长官为作坊使，通常又称为作坊，在敦煌文书中记载甚多。作坊使是归义军使衙中地位较高的官职，与酒司、草场司、柴场司、营田司等都属同级官职[7]。

2.1.2 寺院工匠

属于官府和寺院的工匠，他们的身份基本是世袭的，属于奴隶或农奴性质的被役使者，没有人身自由。他们所从事的技术劳动，实际上是一种"常役"。包括看料、博士等高级工匠在内的所有工匠，都受官府的控制。这类工匠可以根据需要互相派遣和役使，并由役使一方提供饮食和适当的报酬。也有一部分官府匠人，他们在从事手工业的同时，还耕种一部分官府所分的土地。

在古代敦煌，一部分僧侣也从事工匠的劳动。一部分官家、贵族子弟或已在军政部门为官者也从事工匠劳作。这些人亲自操作工匠活计的事实向我们说明，写经、绘画（包括石窟营造）当时是在一种神圣的信念支配下的艺术活动，是一种社会性的活动。至于博士与寺院的关系，一般来说，只是雇主与雇工的关系。S.4705《某寺诸色物破历》记载"博士手工价物七斗"是其明证。

2.1.3 个体工匠

第三类工匠是平民身份的手工业劳动者，他们有一定的土地、财产和庄园，也不受官府和寺院的管辖，属于自由民，但也常为官府和寺院有偿使役，即赚取雇价，并受到一定尊崇。另外，平民中取得"匠人"以上资格者，可免除部分徭役。这可能是官府或寺院给予的部分特权，不过也有可能是因为平时工匠们特别是高级工匠任务繁重所致。

3. 工匠的名称与分类

3.1 工匠的分类

据敦煌文献和敦煌石窟题记记载，敦煌古代称"匠"者共有20余种[8]。这些工匠，大都以历代形成的行业门类进行分类，有些也以具体单一产品分类，其中大部分是与社会生产及人们的生活直接相关的，即为人们提供劳动工具和衣、食、住、行等生产、日常生活密切相关用品的各行业工匠，有专门从事兵器等特殊产品的工匠，而且出现了大批从事绘画、彩塑艺术的独具敦煌地方特色的工匠等，如铁匠、木匠、石匠、泥匠、灰匠、染布匠、毡匠、桑匠、洗褋匠、褐袋匠、罗筋匠、帽子匠、皮匠、鞋匠、金银匠、玉匠、瓮匠、鞍匠、索匠、弓匠、箭匠、胡禄匠、塔匠、纸匠、笔匠、画匠、塑匠等。另外，还有一些专门从事各行业劳动的家、户等，如制作武器的弩家、榨油的梁户、酿酒的酒户等。从9～10世纪敦煌的工匠史料可以看出，当时已经有较细致的社会分工，手工业制作、加工技术十分发达。

俄藏敦煌文献 Дх.2822《蒙学字书》[9]，可能是敦煌莫高窟藏经洞之外所出西夏时期的写本，因其中各类字、字词中均有大量与西夏相关者。有学者据其他敦煌文书考证，将此文书定名为《杂集时要用字》[10]。但是，关于该文献的出处、时代及名称方面目前还存在着分歧[11]。全文共分20个类别，其中第七类即为工匠专类：

12－2　诸匠部第七
12－3　银匠　鞍匠　花匠　甲匠　石匠
12－4　桶匠　木匠　泥匠　索匠　纸匠
12－5　金匠　银匠　铁匠　针匠　漆匠
12－6　鞘秋　秋綮　伞盖　赤白　弓箭
12－7　销金　捻塑　砌垒　扎抓　铸泻
13－1　结瓦　生铁　针工　彩画　雕刻
13－2　剜刀　镞剪　结绾　旋匠　笔匠
13－3　结丝匠　□□匠

从上列关于工匠的记述中可以看出，《杂集时要用字》中"诸匠"的分类有的以工匠的工种来分，这些工匠和工种工艺在敦煌藏经洞文献中基本上都出现过，如银匠、鞍匠、石匠、木匠、泥匠、索匠、纸匠、金匠、铁匠、笔匠，以及赤白、弓箭、伞盖、彩画、铸泻、雕刻、砌垒等。有的则是以制作的工艺来分。这份文献所记载的西夏时代的工匠，与敦煌藏经洞文献所记唐五代宋初的工匠相比，变化不大。

3.2 工匠的级别

各个行业的工匠级别大体分为都料、录事、博士、师、匠、先生、院生及生等多种，其中，级别最高的是都料。

3.2.1 都料

关于工匠行业设都料的问题，一些学者已有涉及，有学者认为：从遗书记载看，都料又名都匠，是唐五代敦煌手工业阶层中非常活跃的人。他们既是工匠，又是都料[12]。

也有学者认为：都料是工匠中技术级别最高者，也是本行业工程的规划与指挥者。他们除了具备本行业工程的设计、规划和组织施工的才能之外，作为高级工匠，一般都具有过硬并超越其他工匠的本专业技术，经常参与施工造作[13]。从文书及壁画题记看，都料级工匠并不是每个行业都有，从笔者目前查阅的资料来看，敦煌文物中记载至少有11个种类或行业中设置都料。如木匠都料、铁匠都料、金银匠都料、纸匠都料、塑匠都料、泥匠都料、毡匠都料，画行都料、弓行都料、刺鞍行都料、皱文行都料等。虽然在敦煌遗书中目前没有发现"弓行都料"的记载，但是，在榆林窟第34窟五代东壁南侧供养人榜题中记载："社耆镇兵马使兼弓行都料赵安定一心供养。"另外一些行业如石匠、灰匠、皮匠、瓮匠、桑匠、染布匠、帽子匠、索匠等，都未见有都料的记载[14]。

3.2.2 博士

博士是唐代对具有专门学识的一种官职称谓。《新唐书·百官志》、《唐六典》、《旧唐书·职官志》等多种文献中皆有记载。从敦煌文书记载来看，敦煌工匠称博士最早出现在吐蕃统治时期。这一时期在多种类别的工匠中都设置了博士。显然与官职称谓不同。工匠中的博士是具备较高的专业技术，可以从事高难度技术劳动并可独立完成所承担的每一项工种的施工任务的工匠，在各行各业中都有。博士之名在开始使用时只是一般雇匠的俗称。9世纪以后的敦煌文献中，博士显然是区别于一般工匠的高级工匠。一般被称作匠者，当为能独立从事一般技术劳动者，在工匠队伍中占多数。匠和博士两级工匠，是敦煌工匠队伍中的基本力量。工匠的最低一级，是作为学徒的"生"。

3.3 工匠的报酬

工匠们平常在为官府或寺院所役使时，一般是由官府或寺院按定量供给饮食。但这并不是官府或寺院对役使工匠们的关心，而是为增加工匠们的劳动时间，最大限度地提高工匠们的劳动效率。据有关文献记载得知，平时供给工匠们的主食，各类工种、各个季节的标准大都是一样的。工匠们在这里属于服役的劳工待遇。

前引大量的收支账目（入破历）、借贷账目（便物历）表明，寺院所有收入的粟中，其开支主要有以下几项：用于牧子、园子、工匠等为寺院服役者的日常口粮；用于寺院劳作的僧人的食物；用于为寺院佛事或其他活动中造食女人的食物；用于寺院僧人死亡等的纳赠等；用于卧酒和沽酒。

研究表明，官府、寺院等为工匠供应酒也是敦煌古代的一大特色。通过对晚唐至北宋敦煌官府（归义军衙府）、寺院僧尼用酒账册的统计、分析、研究，这些账册大体分为五类：一为《付酒本历》，即官府、寺院为卧酒、沽酒支付麦粟的专账，实即酿酒、沽酒的专账；二为《酒破历》，即酒的消费破用专账；三为《算酒历》，即酒的收支结算账；四为《诸色斛斗破用历》，即官府、寺院的《出入流水日记账》；五为《算会牒》（"会"音kuài），即今所谓《决算账》，其中亦有不少关于酿酒用酒的记载。现敦煌研究院收藏的DY.001[15]、DY.369[16]《归义军衙府酒破历》，这是一件记载于宋乾德二年（964）的《归义军衙府酒破历》之残卷，为某种原因支出饮酒若干的账目。自某年4月9日开始，到6月20日，共记酒账100笔，P.2629《归义军酒破历》[17]则是与DY.001、DY.369残卷相衔接的另一部分酒账，它起自6月25日，迄于10月16日，以下缺。仅这件酒账中记载的工匠就有泻匠、皮匠、酿皮（匠）、灰匠、索匠、缚箔子僧、箭匠、工匠、索子匠、画匠、缝皮人、弓匠、皮文匠、木匠等10多种[18]。当时曹府酒账单上有支某某工匠酒多少的记载，在曹府"宴设司"的供应单上还有"大厅设画匠并塑匠用细供"，而其他人员中有用"下中次料"或"下次料"。可见对于画院匠师在生活上有一定的优待。

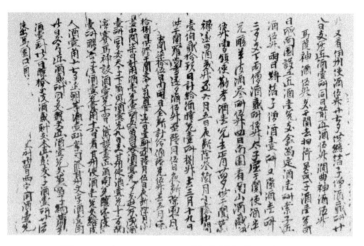

图 1 DY.001、DY.369《归义军衙府酒破历》

4. 分工细致的工匠名称

同一行业的工匠名称又有多种不同的叫法。据初步统计，敦煌古代各种工匠名称多达近百种。我们从铁匠、皮匠、画匠等几个行业为例，可以知道当时工匠的细致分工和名称。

4.1 铁匠

从敦煌汉代以来铁器的使用，石窟开凿中大量铁工具，矾类铁矿的开采、冶炼[19]，以及遗书中大量与铁及其金属生产、使用有关的史料记载可知，冶铁业及其铁器生产是唐代以来敦煌的一种特殊的手工业，分官营、私营两大类。敦煌需要的各种铸造农具、家具、用具、器皿、寺院供养具、建筑构件及普通兵器等也大都是当地生产的。

唐、五代、宋时期的敦煌遗书中，明确记载有当地的铁器制作作坊及其行业中铁匠的姓名。其中，熟铁工艺主要是锻造，工匠有铁匠、铁博士、铁匠都料等多种。敦煌遗书中把用生铁铸造釜、镬、钟、铧等铁器的工匠叫泻匠（又作写匠、泻博士、生铁匠等），所谓"泻",就是铸造的意思,《集韵》注:"写,范金也,即铸炼。"遗书中明确记载有名有姓的铁匠有索海全、史都料、史奴奴、陈丑子等,其余大都没有记载姓名。此外,还有锢鏴匠、锢鏴博士、锅子匠等[20]。榆林窟西夏第3窟的"观音经变"壁画中两幅形象相同的"锻铁图"与梯形木风箱,反映出当时的冶炼技术和先进的鼓风设备[21]。

4.2 皮匠

根据笔者的研究,唐代以来,敦煌文献和敦煌绘画题记中出现了专门从事皮革加工、皮革制品制作的工种、工匠,据初步调查、统计,共有:持韦皮、持羊皮、酿皮、缝皮人、缝皮裘人、除皮人、皮条匠、皮匠、鞋匠、靴匠、缝鞋靴匠、缝鞋靴录事、缝皮鞋博士、皱匠、皱文匠、皱文行录事、皱文行都料等近20种之多。敦煌畜牧业的发展促进了相适应而发展起来的皮革加工、制作等相关的手工业。而畜皮的管理和使用又为皮革加工、制作提供了足够的原材料[22]。

4.3 画匠

为了营造寺院和石窟,除僧尼之外,还需要有一批从事开窟、造像、绘壁画的专门人才。从隋代开始,石窟壁画中就出现了工匠的供养像及题名,如303窟的"画师平咄子";当时曹氏政权仿照中原设立了画院。莫高窟和榆林窟供养人画像题记中有:"沙州工匠都勾当画院使"、"节度押衙知画行都料"、"节度押衙口左厢都画匠作"、"节度押衙知画手"、"衙前正兵马使兼绘画手"、"⋯⋯画匠"、"塑匠"、"节度押衙知左右厢⋯⋯书手"、"雕板押衙"以及"社官知打窟"、"押衙知打窟"等,可知当时画院里包括了石匠、塑匠、画师和管理画院的"都勾当画院使"[23]。敦煌文献中也有大量为画匠、塑匠等工匠提供饮食的记录。

这种细致的社会分工反映了当时手工业的繁荣和技术的进步,是研究我国中古时期商品经济和科学技术的珍贵资料。

4.4 纸匠

造纸业是唐五代敦煌重要的手工业,从事这种手工业的工匠称作纸匠。有造纸作坊[24],敦煌遗书中涉及造纸、用纸的文献非常多[25]。敦煌石窟发现的古纸一直持续到西夏和蒙元时期[26]。

参 考 文 献

[1] 王进玉. 敦煌学和科技史. 兰州:甘肃教育出版社,2010. 200-211
[2] 王进玉. 敦煌壁画中的科学技术. 自然杂志,1988,11(11):862-868 转808
[3] 王进玉. 敦煌古代的手工业生产技术(论文摘要). 开发研究,1992,(6):62
[4] 中国社会科学院历史研究所,中国敦煌吐鲁番学会敦煌古文献编辑委员会,英国国家图书馆,伦敦大学亚非学院编. 英藏敦煌文献(汉文佛经以外部分). 2. 成都:四川人民出版社,1990. 28-34
[5] 姜伯勤. 唐五代敦煌寺户制度. 北京:中华书局. 1987. 246-268,239-246
[6] 马德. 敦煌工匠史料. 兰州:甘肃人民出版社,1997. 39
[7] 同[1]. 222-227
[8] 王进玉. 敦煌石窟全集·科学技术画卷. 香港:商务印书馆(香港)有限公司,2001. 225-226
[9] 〔俄〕孟列夫(Л. H. 缅希科夫)主编,俄罗斯科学院东方研究所圣彼得堡分所,俄罗斯科学院出版社东方文学部,上海古籍出版社编. 俄藏敦煌汉文写卷. 10. 上海:上海古籍出版社,1998. 58-67

[10] 马德. 《敦煌工匠史料》补遗与订误. 敦煌学会编. 敦煌学. 25. 台北：乐学书局有限公司, 2004. 293-301
[11] 同［1］. 228-230
[12] 郑炳林. 唐五代敦煌手工业研究. 敦煌学辑刊. 1996, (1)：20-38
[13] 同［6］. 9-11
[14] 同［1］. 228-230
[15] 甘肃藏敦煌文献编委会, 甘肃人民出版社, 甘肃省文物局编. 甘肃藏敦煌文献. 1, 2. 兰州：甘肃人民出版社, 1999. 1
[16] 同［15］. 166-167
[17] 上海古籍出版社、法国国家图书馆编. 法藏敦煌西域文献. 16. 上海：上海古籍出版社, 2001. 362-363
[18] 唐耕耦, 陆宏基编. 敦煌社会经济文献真迹释录. 3. 北京：全国图书馆文献缩微复制中心, 1990. 271-276
[19] 王进玉. 敦煌矾石的初步研究. 考古与文物, 1986, (4)：102-105
[20] 同［1］. 228-230
[21] 王进玉. 漫步敦煌艺术科技画廊. 北京：科学普及出版社, 1989. 116-117
[22] 同［1］. 351-362
[23] 王进玉. 敦煌石窟全集·科学技术画卷. 香港：商务印书馆, 2001. 235-237
[24] 同［1］. 264-272
[25] 同［1］. 272-296
[26] 王进玉. 中国少数民族科学技术史丛书·化学与化工卷. 南宁：广西科学技术出版社, 2003. 504-505, 531-532

Guo Songtao and the Western Telegram Civilization

Xia Weiqi

(Huainan Normal University)

Guo Songtao, the first Minister stationed abroad of the late Qing Dynasty, inspected widely the western telegram civilization after he arrived at England and France, and appealed to the Qing government to introduce this technology intensely while introduced it positively. Then Guo Songtao launched the debates with the conservative strength, and became an important public opinion manufacturer and the backbone of the new strength during the late Qing Dynasty' the telegram construction. The formation of the status of Guo Songtao reflected by a case the circle of thinkers' differentiation of the late Qing Dynasty after China experienced the impact of the formidable west tide, as well as numerous appearances during the vicissitude of the society and idea of the late Qing Dynasty when the modern strength challenged the tradition.

郭嵩焘与西方电报文明[*]

夏维奇

(淮南师范学院)

1. 引言

在晚清电报建设史上,清朝第一位驻外公使郭嵩焘是一位重要人物,却为一些研究者所忽视。郭嵩焘曾利用亲临西土之便,广泛考察西方电报文明,并在积极予以介绍的同时,强烈吁请清政府引进这项技术,进而同守旧力量展开论争,成为晚清电报建设的重要舆论制造者及趋新力量的中坚。郭嵩焘的这一身份的形成以一个案领域揭示出晚清中国社会在西潮冲击下思想界的分化,以及晚清社会和观念变迁过程中近代向传统挑战的众多面相。

---

[*] 本文系国家社会科学基金资助项目"晚清电报与社会变迁研究"(08CZS012)阶段性成果。

学术界于郭嵩焘洋务思想多有论涉，近年来对其之于万国公法会、图书馆等西方近代社会科学之认知与推介更有专论①，却不曾述及其于电报这一科学技术之识见与态度，更鲜见将此置于晚清社会及观念变迁的大背景下加以审视。这无论是就郭嵩焘研究，抑或是就晚清电报史研究而言，皆不能不说是一缺憾。鉴于此，笔者特作是文初步考察之，并期对进一步认知晚清社会有所助益。

2. 电报知识的推介

1875年8月28日，清政府任命郭嵩焘为出使英国大臣[1]。1876年12月2日郭嵩焘自沪启程，翌年1月21日抵伦敦，1879年1月31日离任回国。此恰值西方电报技术成熟、电报事业昌盛之期。伦敦街头"之飞线纵横，如蛛丝也"[2]，"电线通达各处，约二千条"[3]。而该市电报局"楼舍高大"，"各屋数百间"[4]，共设电报机一千余台[5]。这种情状不仅促进了郭嵩焘对电报文明的深入考察②，无疑也为其向国人推介电报知识提供极为有利的条件。

尚在赴英途中，郭嵩焘即敏锐察见苏伊士运河沿岸电报置设的一些景况[6]，抵英未及两月更亲访伦敦电报局③[7]，嗣后又参观刊伦电气厂[8]、银城电线制造厂[9]，了解电报相关企业产品及生产状况。通过上述一系列考察，郭嵩焘对电报文明的认知大增，由此甚感让国人了解之必要，遂就其所识择要介绍。概言之，主要包括以下两个方面：

其一，电报的技术层面知识。电报机型是电报技术研发水准的重要表征之一，故而它的不断更新可从一个侧面反映出电报技术的演进历程。郭嵩焘考察伦敦电报局后指出，"电报各异式，而总分为三等"：一是旧式电报机，"设二十六字母"，使用时直接"用指按之"；再是普通莫尔斯电报机，使用时"盘纸转而运之，以着点长短成文，而视其继续成句"，即通过点横构成的"莫尔斯电码"，查检相应文字；三是新式莫尔斯电报机，通过报音时长以代替点横之法而确定"莫尔斯电码"，再查检相应文字。郭嵩焘尤其注意到第三种电报机"辨声知字，运用尤灵，其机尤速"，从而赞之为"尤奇者"[10]。可见，郭嵩焘对电报机型及其操作特征的介绍十分精细。郭嵩焘此举不仅仅是让国人了解电报机型的基本种类及最新成果，主旨当是让国人通过了解此一情状而认清西方科学技术日新月异之发展大势。正因为如此，郭嵩焘进而指出："西洋取用电气，穷极心力搜索，出奇无穷。"[11]

电报一经发明，即不断拓展应用领域，传报各地天气情形即是其新开拓的重要领域之一。郭嵩焘对伦敦电报局天气电报的操作流程作出细致描述：由新闻电报室分向利物浦、

---

① 关于郭嵩焘的研究现状，可参考（美）汪荣祖的著作《走向世界的挫折——郭嵩焘与道咸同光时代》（长沙：岳麓书社，2000.329-330, 338-340）"参考书目"第二部分"论述郭嵩焘作品"及第六部分"英文史料与近著"中"一般论著"。近年来张建华（郭嵩焘与万国公法会，近代史研究，2003，1：280-295）、傅敏（郭嵩焘与西方图书馆文明的东传，图书馆，2004，4：14-16）以及孟泽（洋务先知——郭嵩焘，南京：凤凰出版社，2009）深化了对郭嵩焘的研究。

② 郭嵩焘出使本来即奉"朝命"："在西洋三年，考究利病，知无不言"（合肥李鸿章胪陈事实疏．载：郭嵩焘．玉池老人自叙．上海：上海古籍出版社，藏清光绪十九年养知书屋影印，2）。

③ 郭嵩焘称"波斯阿非司-得利喀纳福"，是Post Office-Telegraph的音译，意为邮电局。但据李圭、张德彝的描述知，伦敦电报局与邮政局实分开，左为邮政局，右为电报局（李圭．环游地球新录．长沙：岳麓书社，1985.282；钟叔河．英轺私记·随使英俄记．长沙：岳麓书社，1986.332）。

· 204 ·

曼彻斯特、伯明翰三处拍发电报，"问彼处天气阴晴"状况。郭嵩焘指出，"才问而三处回信齐至"：利物浦"回言天气甚好"；曼彻斯特"回言阴晴有雾"；伯明翰"回言天气好"[12]。由此，国人不仅可了解电报具有快速传递信息的这一常识性功能，以及电报应用的新领域，进而可认知电报潜在的巨大价值及广阔的应用前景。

其二，电报的社会层面知识。电报的研制在经过数代人接力式努力后最终得以完成。1844年5月24日，美国人莫尔斯（Samuel Finley Breege Morse）用其编订的"莫尔斯电码（Morse Code）"由华盛顿向巴尔的摩（Baltimore）发出人类历史上第一份长途电报。自此，电报正式宣告诞生。这一具有划时代意义的信息传播工具在嗣后二十余年间获得迅猛发展，并越洋过海，由欧美地区设至中国近邻。对于电报的这一发展沿革，郭嵩焘介绍说：

> 其电报起于一千八百二〇年，有安恩柏者初用指南针作之，是为嘉庆二十五年。一千八百三十六年，惠子登又创作吸铁电报，是为道光十六年。一千八百三十七年，摩西氏始创作点、画为号以记字母之电报机器，是为道光十七年。一千八百三十八年始设电报于伦敦西铁路旁，是为道光十八年。一千八百四十年设电报白赖克华尔，是为道光二十年。一千八百四十一年设电报苏格兰之葛赖斯哥，是为道光二十一年。一千八百五十一年始由海通电报于法国，是为咸丰元年。一千八百六十五年始通至印度孟买，是为同治四年。[13]

郭嵩焘在此详细胪列电报的研发及展延历程，虽有舛误①，但并不影响他欲使国人知晓：电报发展大有突飞猛进、锐不可当之势，故各国各地区对之不应抵制，而当积极引进，此方是正确态度。职是之故，郭嵩焘又说：电报等在西方，"其开创才数十年，乘中国之衰敝，七万里一瞬而至，然亦足见天地之气机，一发不可遏；中国士大夫自怙其私以求遏抑天地之机，未有能胜者也"[14]。

1878年2月23日郭嵩焘受命兼任驻法公使。是年4月26日日本驻英公使上野景范得知郭嵩焘将去巴黎，"过谈，并约同赴万国公法会及电报会"。郭嵩焘遂向上野"询知电报会在伦敦都城，以西历六月"，并在当日日记中写道：

> 电报会由国家主持（日本长崎之那噶萨奇，由极南以至极北皆有电报；而那噶萨奇电报，西洋主之，非公例也），商定各国互相交涉之电报，故先须画诺入会。万国公法会由各国读书有学识者为之，不待画约也。然电报会派员往视亦无不可行。[15]

郭嵩焘此记关涉电报最为重要的国际组织——万国电报公会的性质、功能、入会规则以及相关会议制度等问题。1865年3月1日欧洲一些国家派员在巴黎议商各国间电报接线及拍发收费等事宜，标志该组织成立。至清末1908年，万国电报公会共召开10届大会，为电报在世界范围内的推展作出了重要贡献。由于其为政府间国际组织，各国只有在批准《国际电报公约》而成为会员国之后，方能派出正式代表参会。所以郭嵩焘称："电报会由国家主持……商定各国互相交涉之电报，故先须画诺入会。"然非会员国虽不能派出正式代表，却可列席旁听。郭嵩焘特别指出这一规则，当有中国应派员列席，以了解世界电

---

① 例如引文中"安恩柏"当指丹麦人奥斯特（H. C. Oersted）。1820年奥斯特发现罗盘通电后指针发生偏转现象，俄国驻慕尼黑使馆馆员希林（Baron Pawel Lwowitsch Shilling）受此启发于1832年研制出利用电流计指针偏转以接收信息的针式电报机（Needle Telegraph）（Ken Beauchamp. History of Telegraph. The Institution of Electrical Engineers. London: United Kingdom, 2001. 29）。不过，奥斯特并未制造电报。

报发展态势之意①。

综上可知，郭嵩焘对电报文明的介绍既有电报机型、最新应用领域等技术层面知识，亦有电报沿革、国际组织等社会层面知识，这一切的背后有其深刻寓意：即世界科技正日新月异，各国因电报具有巨大价值而广泛展延，并协调推进，发展之势锐不可当，故中国应取的态度当是大力引进，万不可抵拒。

3. 电报建设的吁请

如果说，对电报知识的推介是郭嵩焘间接告知国人中国应建电报的话，那么，郭嵩焘并未止于此，还多次发出明确而强烈的吁请。早在出使前，郭嵩焘即认为"电信、铁路为必行"[16]，亲临西土、实地考察电报后更是强调指出："泰西遍国皆机器也，中国无能效之；其必宜效者二：一曰电报，一曰汽轮车。"进而又言："是二者之宜行也，无待再计决也。"[17]可见其吁请不仅恳切，且甚急迫。这一切有其深刻的认知基础。

其一，电报等业是西方国家富强之基。郭嵩焘指出，西方轮船航行海上始于嘉庆六年，火车起自嘉庆十八年，"至道光十八年始设电报于其国都"。可见这些科技实业开创并非久远，然短短数十年便获突飞猛进的发展[18]，各国因之迅速强大："用此以横行天下，战必胜，攻必取，诚有以致之，尽泰西十余国比合以尽其利者也。"[19]由此当知电报等业确是西方各国富强之基。既如此，在郭嵩焘看来，中国应当大力建设。故郭嵩焘致函李鸿章提出："窃以为方今治国之要，其应行者多端，而莫切于急图内治以立富强之基，如此二者（电报、铁路）可以立国千年而不敝，其为利之远且大者不具论也。"[20]

其二，电报等业便于朝廷行政理事。郭嵩焘认为，中国幅员万里，驿传谕旨奏章，"远者数十日乃达，声气常苦隔绝"。倘建有电报、铁路，情形将为之一变："万里犹庭户也，骤有水旱盗贼，朝发夕闻"，从而不必担心民人叛乱；另一方面，中国地方官吏常压制民怨，使得民情不能上达，朝廷因之与百姓隔膜甚深。倘建有电报、铁路，则如同人体血脉畅通，"政治美恶无能自掩"，从而不必担心贪官遏抑民情而专营私利[21]。基于此，郭嵩焘指出，电报等西方科技"可以便民，可以备乱，可以通远近之气，而又行之甚易、历久而必无弊"[22]，故中国应积极引入而不能再犹豫不决[23]。

其三，日本电报等业发展迅猛于中国不利。自19世纪60年代末明治维新之始，日本即引进电报技术[24]，并广派国人赴欧洲学习，电报等业由此日臻发达："至大小塾房、邮政局、电报局、开矿局、轮船公司，皆仿西法，而设官为经理，举国殆遍。而于电报、邮政两端，尤为加意，几堪与泰西比美。"[25]对此，郭嵩焘至英国后不久就注意到，"日本在英国学习技艺者二百余人，各海口皆有之，而在伦敦者九十人。……所立电报信局，亦在伦敦学习有成即设局办理。而学兵法者甚少。"进而指出："盖兵者末也，各种创制皆立国之本也。"[26]需说明的是，清政府较日本更早推行洋务新政，然一二十年来着力建设军事事业。故郭嵩焘之言的背后，是其对中国时下洋务路径的不以为然乃至批判，认为是本末

---

① 不过，对于此届伦敦大会（万国电报公会第五届大会），英国公使威妥玛曾函知总署（台湾中央研究院近代史研究所. 海防档·丁·电线. 台北：艺文印书馆, 1957. 247-248）。然郭嵩焘虽有上野之约，却未得总署任何指示，故未派员列席。

倒置、舍本逐末。《左传》有"邻之厚，君之薄也"之训，郭嵩焘据此警醒国人："日本为中国近邻，其势且相逼日甚。吾君大夫，其旰食乎！"[27]观上可知，日本自明治维新以来的积极进取精神与迅速发展态势，以及中国儒家的经世安邦理念使得郭嵩焘产生强烈的危机感，由此而呼吁中国引入电报等科技当是合乎逻辑的发展。

与此同时，郭嵩焘还就电报建设的一些宏观问题提出看法。首先，就策略言，中国宜先建电报，后修铁路。尽管在郭嵩焘看来，轮船、火车、电报等同为富强之业，因而皆甚重要，然若同时举办，需费太巨，在清朝政府财力有限的情形下，显然难以施行，故只能次第展开。既如此，相对而言，铁路工程更大，这不仅表现在需款更巨，且开山辟土，影响亦大，郭嵩焘因此提出："轮船、电报必宜通行，铁路暂必不能行。"郭嵩焘认为，行此方略，"无已，则小试之，徐徐推广之，庶无大失也"[28]。即通过由简入繁的渐进方式，不仅可基本解决经费问题，且在保守思想浓厚、对西方科技多持抵制态度的情形下，亦易让部分国人接受。

其次，就方式言，电报建设当由国家主之。此一方面源自郭嵩焘在西方所见：英法等欧洲国家的电报建设主要由政府经办；另一方面在郭嵩焘看来，电报架建国家最为受益，故理当由政府筹资兴办。况且，电报建设需资不菲，私力难支，即便从此角度言亦只能由国家任之。基于上述考虑，郭嵩焘遂称："电报者，通所治行省之气，有事则急先知之，可以国家之力任之者也。"[29]

综上可知，曾任封疆大吏、后又出使西洋的郭嵩焘以其对电报之巨大价值及中国之严峻形势的深刻认知，从政权的巩固乃至国家的安危出发，广泛指陈电报之于经世安邦之意义，较深入地申论了中国引入电报的必要性与紧迫性。在此基础上，郭嵩焘进而根据西方的电报建设经验及中国的当下实际情状，还就如何建设电报问题提出战略性看法，从而初步形成其电报建设的思想认知体系。

4. 余论

随着电报在西方各国的广泛架建并向东方的不断展延，自19世纪60年代初起，西人屡向清政府鼓吹电报价值，并请求在中国架设电报。对此，总理各国事务衙门先是感到，如允则将"失我险阻、害我田庐、妨碍我风水"，继又认为"占我民间生计"，从而坚决抵拒。[30]尽管总署的一些担忧不尽合理，如"妨碍风水"之类，但其抵拒的主旨却是力维主权、保国安民，故不可厚非，且有必要。

然需指出，其时清廷内外官员多视电报为"奇技淫巧"，故在坚拒西人请求的同时，亦不主张甚至严厉反对自己架建。如两广总督瑞麟、广东巡抚蒋益澧认为，政贵从民所欲，治贵因地制宜。电报等西洋器物，不过技艺之末，无关治道①[31]。三口通商大臣崇厚

---

① 与之相近的还有浙江巡抚马新贻、署直隶总督官文、醇郡王奕谖等论。马新贻认为富民强国，只能靠发展农桑，"一切求之在己之本务"，电报之类末务，于中国之大计难有裨益（浙江巡抚马新贻奏［同治五年十月二十一日］. 载：宝鋆等. 同治朝筹办夷务始末. 卷45. 台北：文海出版社，1971.47）。官文亦指出，电报铁路，不过是为往来迅速起见，只便于贸易，仍是奇技淫巧，因而不是中国所尚（署直隶总督官文奏［同治六年十二月二十二日］. 载：宝鋆等. 同治朝筹办夷务始末. 卷56. 台北：文海出版社，1971.11）。奕谖则进一步指出，电线等物"尽可一概不用，无损于国计民生，有裨于人心世道"（醇郡王等奏［同治八年正月初三日］. 载：宝鋆等. 同治朝筹办夷务始末. 卷64. 台北：文海出版社，1971.5）。

更是强调指出:"铜线铁路二事……于中国毫无所益,而徒贻害于无穷。"[32]这些言论当是封建士大夫昧于世界发展大势,力持传统治国理念之反映,显然,在进步的时代面前已经落伍,但却是斯时清廷上下对电报等科技的主流认识。

不过,此间亦有部分官宪初识电报价值。江苏巡抚李鸿章认为,"铜线费钱不多,递信极速","传播自远,应较驿递尤速"①。福州船政大臣沈葆桢更是指出中国应建电报:"秦筑长城,当时以为殃,后世赖之。铜线铁路,如其有成,亦中国将来之利也。且为工甚巨,目前亦颇便于穷民。"[33]可见,李鸿章、沈葆桢等已看到电报对于中国的现实及长远意义。此反映出的是,在西来的滚滚电报洪流冲击下晚清思想界的分化,趋新认知渐露端倪。然从总体上看,这种论识在19世纪60年代,可谓曲高和寡,势孤力弱。

至70年代初,随着西人将电报展至上海,沈葆桢、李鸿章、丁日昌等已感到中国自设电报以维主权与利益之紧迫。而沈葆桢、丁日昌更于日本以生番事件侵台后,相继在台湾等地试办电报,以加强海防[34]。然"内地若果议及,必致群起相攻",李鸿章对此颇有感触。自1874年海防之议起,李沥陈"电线铁路必应仿设",遭大理寺少卿王家璧、通政使于凌辰等痛诋②[35]。其时,工科给事中陈彝亦上折反对③。甚至连光绪帝师翁同龢及随郭嵩焘一同出洋的驻英副使刘锡鸿等都坚决抵制④。可见,即便是在70年代中后期郭嵩焘出使前后,反对中国引入电报之声仍众,提示此时守旧力量的强大及守旧氛围的浓厚。

对于上述情形,郭嵩焘有着深刻体认,称国人"一闻修造铁路、电报,痛心疾首,群起阻难,至有以见洋人机器为公愤者"[36]。正因为如此,驻英期间,郭嵩焘在广泛考察西方电报文明的同时,努力向国人推介,并以其强烈的忧患意识,全面申论中国架建电报的必要性与紧迫性。此外,郭嵩焘还就国人的反对态度以及一些错误认知,进行有力批驳与抨击。指出,国人抵制,"是甘心承人之害,以使朘吾之脂膏而挟全力自塞其利源"[37]。

---

① 因持此识,李鸿章遂在西人一再请设的情形下,提出"若至万不能禁时,惟有自置铜线以敌彼飞线之一法"。但李鸿章深知,斯时中国风气未开,故又提出将此设想"存而勿论"(台湾中央研究院近代史研究所. 海防档·丁·电线. 台北: 艺文印书馆, 1957.9),不久重申:"前曾设为自置铜线以敌彼飞线之议,原以备万不得已之时,存此一说,并未稍涉假借。"(台湾中央研究院近代史研究所. 海防档·丁·电线. 台北: 艺文印书馆, 1957.23)足见其保守势力之强,李鸿章对此深为了解与畏惧,然在其内心深处,却并未改变对电报之看法,只是觉知架建时机未熟,故两年后再度提出:"将来通商各口洋商私设电线,在所不免,但由此口至彼口,官不允行,总做不到。……或待承平数十年以后,然与其任洋人在内地开设铁路电线,又不若中国自行仿办,权自我操,彼亦无可置喙耳。"(湖广总督李鸿章奏 [同治六年十二月初六日]. 载: 宝鋆等. 同治朝筹办夷务始末. 卷55. 台北: 文海出版社, 1971.13-14)表明李鸿章对电报的态度之一贯性。

② 1875年4月1日王家璧奏称:"事事法西人,以逐彼奇技淫巧之小慧,而失我尊君亲上之民心也。"(中国史学会. 中国近代史资料丛刊·洋务运动. 第1册. 上海: 上海人民出版社, 1961.134)王家璧上折的同日,于凌辰亦奏称:"是古圣先贤所谓用夏变夷者,李鸿章、丁日昌直欲不用夷变夏不止!……洋人之所长在机器,中国之所贵在人心。……复不可购买洋器、洋船,为敌人所饵取。又不可仿照制造,暗销我中国有数之帑项,掷之汪洋也。"(中国史学会. 中国近代史资料丛刊·洋务运动. 第1册. 上海: 上海人民出版社, 1961.121-122)

③ 1875年10月18日陈彝奏称:"电线一事可以用于外洋、不可用于中国。"(中国史学会. 中国近代史资料丛刊·洋务运动. 第6册, 上海: 上海人民出版社, 1961.329)

④ 翁同龢在其《日记》"光绪二年二月初一日"则中记有:"适郭筠仙来,遂论洋务,其云滇事将来必至大费大辱者,是也; 其以电信、铁路为必行,及洋税加倍、厘金尽撤者,谬也。至援引古书,伸其妄辩,直是丧心狂徒矣。"(孔祥吉、村田雄二郎. 翁文恭公日记稿本与刊本之比较——兼论翁同龢对日记的删改. 历史研究, 2004, 3: 183)痛诋郭嵩焘提倡架设电报等事。刘锡鸿"谓电报铁路虽于不慊于心之夷鬼有用,于汉人全不相宜"(郭嵩焘. 使西纪程. 沈阳: 辽宁人民出版社, 1994.84-85)。

而针对时论有"洋人机器所至,有害地方风水"之说,郭嵩焘细致剖析后指出:"其说大谬。"[38]这一切从本质上讲,当是近代向传统的挑战,既有力地冲击了守旧的思想阵地,壮大并增强了支持者的声势,同时亦使得郭嵩焘成为晚清电报建设的重要舆论制造者①。

尤为关键的是,郭嵩焘对电报的一些重要见解与态度,驻英期间通过书信方式与晚清重臣李鸿章有过广泛交流,大为李鸿章赏识与推崇[39]。更有甚者,1877年5月23日郭嵩焘致函李鸿章,直促其"赶办铁路电报,以树富强之基"[40]。这一切在一定程度上推动了李鸿章后来的电报建设实践活动。在与郭嵩焘信函往复后不久,李鸿章即先是试造天津总督衙门至机械局电报、继又试设天津至大沽电报,1880年9月16日更是直接向清廷奏设中国第一条电报大干线——天津至上海电报线[41],由此全面开启中国通信的近代化历程。从这个意义上讲,郭嵩焘又是晚清电报建设的直接推动者,并为中国大规模自建电报活动的兴起提供部分的理论支持,故当是彼时趋新力量的中坚。郭嵩焘的这一身份的形成以一个案领域透视出晚清社会及其观念的变迁。

参 考 文 献

[1] 世续等. 清实录. 第52册. 北京:中华书局,1987. 253
[2] 李凤苞等. 使德日记及其它二种:英轺私记、澳大利亚洲新志. 北京:中华书局,1985. 17
[3] 李圭. 环游地球新录. 长沙:岳麓书社,1985. 279
[4] 钟叔河. 英轺私记·随使英俄记. 长沙:岳麓书社,1986. 332
[5] 郭嵩焘. 郭嵩焘日记. 第3册. 长沙:湖南人民出版社,1982. 157
[6] 郭嵩焘. 使西纪程. 沈阳:辽宁人民出版社,1994. 29
[7] 同[5]
[8] 同[5]. 307-308
[9] 同[4]. 487-488
[10] 同[5]
[11] 同[5]. 307-308
[12] 同[5]
[13] 同[5]. 183
[14] 郭嵩焘. 伦敦致李伯相. 载:中国史学会. 中国近代史资料丛刊·洋务运动. 第1册. 上海:上海人民出版社,1961. 304,304-305,305
[15] 同[5]. 489
[16] 孔祥吉,村田雄二郎. 翁文恭公日记稿本与刊本之比较——兼论翁同龢对日记的删改. 历史研究,2004,3:183
[17] 同[6]. 157
[18] 同[14]

---

① 郭嵩焘一直未参与晚清电报建设的实践活动,不过,在其出使前任福建按察使期间曾涉电报交涉事。1874年5月日本侵犯台湾,沈葆桢为此奏设闽台电报。然承造方丹麦大北电报公司(The Great Northern Telegraph Co.)在未与闽省议妥之际即先建成马尾段,乘势再请办福州厦门线,遭一些官民抵制。闽省遂与大北电报公司展开交涉,反复未果后,闽浙总督李鹤年荐郭嵩焘接办此事。郭嵩焘最终迫使大北电报公司作出让步,随即离闽(台湾中央研究院近代史研究所. 海防档·丁·电线. 台北:艺文印书馆,1957. 217;郭嵩焘. 玉池老人自叙. 上海:上海古籍出版社藏清光绪十九年养知书屋影印. 22)。

[19] 同［6］.158
[20] 同［14］.305
[21] 郭嵩焘. 郭嵩焘诗文集. 长沙：岳麓书社，1984.191
[22] 郭嵩焘. 与友人论访行西法. 载：中国史学会. 中国近代史资料丛刊·洋务运动. 第1册. 上海：上海人民出版社，1961.322
[23] 同［21］.555-556
[24] 曾鲲化. 祝中国交通界之前途. 交通官报，1910，1：8
[25] 同［3］.319
[26] 同［14］.304
[27] 同［5］.804
[28] 同［21］.243
[29] 同［6］.158
[30] 总理各国事务恭亲王等奏（同治六年九月初九日）. 载：宝鋆等. 同治朝筹办夷务始末. 卷50. 台北：文海出版社，1971.32-33
[31] 两广总督瑞麟、广东巡抚蒋益澧奏（同治五年六月十九日）. 载：宝鋆等. 同治朝筹办夷务始末. 卷42. 台北：文海出版社，1971.65
[32] 三口通商大臣崇厚奏（同治六年十一月二十六日）. 载：宝鋆等. 同治朝筹办夷务始末. 卷54. 台北：文海出版社，1971.18
[33] 总理船政前江西巡抚沈葆桢奏（同治六年十一月二十一日）. 载：宝鋆等. 同治朝筹办夷务始末. 卷53. 台北：文海出版社，1971.4-5
[34] 夏维奇. 清季台湾电报发展述论. 台湾研究，2005，3：57-62
[35] 李鸿章. 复郭筠仙星使. 载：中国史学会. 中国近代史资料丛刊·洋务运动. 第1册. 上海：上海人民出版社，1961.269
[36] 同［6］.135
[37] 同［14］.304-305
[38] 同［14］.305
[39] 同［35］
[40] 窦宗一. 李鸿章年（日）谱. 台北：文海出版社，1980.4876
[41] 朱寿朋. 光绪朝东华录. 第1册. 北京：中华书局，1958.966-967

On the Manufacturing Technology of Traditional Curved-Beam Plough in China

Feng Lisheng[1]　Hang Xing[2]

(1 Institute for History of Science and Technology & Ancient Texts, Tsinghua University;
2 Institute of Historical Metallurgy and Materials,
University of Science and Technology Beijing)

Plough has a long history in China. However, with the unbalanced development in different environmental conditions and applications in different ethnic groups, it was created in different forms. As one kind of traditional ploughs, the curved-beam plough spread more widely, and its form became dominant in China after the improvement by Lu Guimeng in the Tang Dynasty. Until the end of 20th century, it had been the most effective farming tools in rural areas north and south, and played a very important role in agricultural production. Since 2008, we have made field surveys in Hebei, Shanxi and Inner Mongolia to find out the production and use of curved-beam plough in rural areas, interviewed skilled craftsmen who mastered the manufacturing skills in making traditional ploughs, and asked Ning Dasheng, who had over 50 years of experience in making farming tools, to make a wooden plough according to traditional process. We recorded the process by videotaping, taking pictures, surveying and mapping, etc. In this paper, based on historical documents and field research, we introduced the construction, production process and key technologies of the traditional curved-beam plough.

中国传统曲辕犁制作工艺初探

冯立昇[1]　黄　兴[2]

(1 清华大学科学技术史暨古文献研究所;
2 北京科技大学冶金与材料史研究所)

在传统耕犁中,曲辕犁流传更为广泛,它经唐代陆龟蒙的改进后,成为中国传统犁具最主要的形式。直到20世纪后期,它一直是南北各地广大农村最得力的耕作农具,在我国农业生产中扮演着极其重要的角色。

唐陆龟蒙在总结民间制作耕犁的基础写成的《耒耜经》一书,详细记述了江东地区普遍使用的耕犁的部件、尺寸和功用。江东犁由铁质的犁镜与犁壁和木材制造的其他九个部件犁底、压镜、策额、犁箭、犁辕、犁梢、犁评、犁建、犁槃组成。其中犁辕为动力牵引件；犁梢为耕作操纵件；犁箭、犁评、犁建是耕深调节件；策额、压镜为犁壁固定件；犁底是用作犁镜固定及保持耕作平稳件；犁镜（即犁铧，也称犁头）为耕作件，犁盘则是动力与犁的连接件。唐代这种曲辕犁经千年的发展，在保持其基本功能不变的情况下，全国各地根据当地的实际情况，在构件上删繁就简，制作了形式多样的、带有地方特色的曲辕犁。如有的省去了犁箭、犁评、犁建，将调节耕深的功能，放在了犁辕与犁梢的连接处，有的使犁梢与犁底一体化，整架犁只有犁辕、犁梢、犁底一体与犁铧。图1是笔者在江苏吴江县的甪直水乡农具博物馆拍摄的耕犁照片，这种犁在陆龟蒙的家乡曾经是广泛使用的曲辕犁类型，但目前当地已不再生产这种传统的耕犁。二三十年前，江南地区（江苏、上海等地）还在大制造和使用这种犁，图2为其构造示意图[1]，其中千斤板即为犁箭（或犁柱），也是承受牵引作用的主要部件，其上部与犁辕结合处为活动配合，犁辕上开有榫眼呈梯形状的孔，千斤板上端贯穿犁辕，并在固定犁辕位置处开有一高一低两孔，用木楔锁住，以固定犁辕调节后的位置。这种犁的犁辕不仅有上下的弯势，而且还有左右的弯势，因此对选材要求很高，这样制作出来的犁，在耕田时土块才向上翻起，不会堵塞在犁辕和犁壁之间，翻土比较容易。

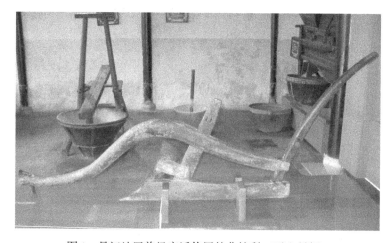

图1 吴江地区曾经广泛使用的曲辕犁（冯立昇摄）

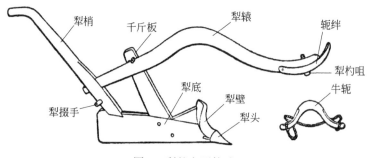

图2 犁的主要构造

目前在河北、山西、内蒙古一些地方的农村中仍有一些木匠掌握制作传统犁的技艺，2008年以来，我们对河北、山西、内蒙古一些地方的农村中制作和使用的旱地曲辕犁进行了实地调查，采访了掌握传统犁的制作技艺的乡村工匠。

由于耕犁的部件包括木制和铁制两大类，因此制作犁需要由木匠和铁匠协作完成。其中犁铧、犁壁、犁辕、犁底、犁梢及犁箭都是犁的主要部件，犁铧、犁壁和铁犁辕由铁匠制作，其余部件及装配要由木匠完成。犁的制作技术要求很高，专门的木匠才能胜任。

在河北张家口市宣化县深井镇乡丁家坊村，我们采访了有50多年制作农具经验的木匠宁大胜老人。他现年75岁，从十几岁开始师从一位本家的叔父学习木工手艺。从20世纪50年代初至今，他先后制作、维修过犁、耧、马车、独轮车、风箱等多种农具和工具。他目前每年农闲时仍要给村民制作和维修一定数量的农具。

2009年7月，我们又请宁大胜老人按照传统的工艺制作了一张木犁。材料已经提前准备好，他向我们强调了选材的重要性，如犁辕一定要选用合适的弯曲树干制作（图3），否则或达不到使用要求，或易断裂。

图3 宁大胜老人向笔者之一展示制作犁辕的木料并讲述制作方法（黄兴摄）

图4所示老人做犁工具都是木匠工具，包括推刨（大、小两个）、斧子、小锤子、凿子（大、中、小刃口各一个）、墨斗、墨笔、锯子、方尺、木经尺、锛子和一条尼龙绳子。他用了一天的时间基本完成了犁的木质结构部分的制作。

犁的制作过程可分为粗坯加工、铺犁、精加工、组装四个阶段。

（1）粗坯加工

就是把犁弯（犁辕）、犁底、犁脊（犁箭）等构件制作成型。

第一步，劈犁弯（犁辕）

制作犁弯的材料选用弯曲的榆木或杏木树干，树干随材就料，长度1.5米至1.7米，直径15厘米左右，没有特别严格的要求。犁其他各部分的尺寸根据犁弯确定。

犁弯制作过程如下：首先用脚踏住树干，用锛子劈去树皮和部分木头，把树干的横截面加工成矩形，上下较粗，左右较细（图5(a)）。再用刨将各个边、棱加工光滑（图5(b)）。

· 213 ·

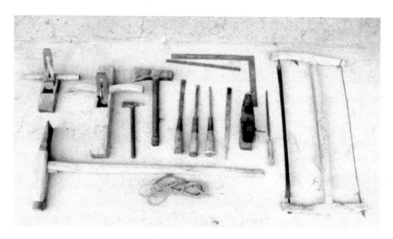

图 4　制作犁的工具

图 5　加工犁弯

第二步，劈犁底

犁底用杏木制成，质地较硬，而且耐磨。制作犁底的杏木材料长约 57 厘米。先用锛子和锯子把杏木加工成 57 厘米 × 20 厘米 × 20 厘米的长方体。

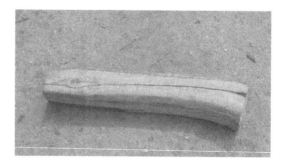

图 6　犁底

第三步，加工犁尾（犁梢）、犁脊（犁箭）

犁尾（犁梢）和犁脊（犁箭）都是连接犁底的构件。犁尾还能供人手扶，控制耕地方向和深浅。平地使用的犁，犁尾长一些，减少弯腰；坡地使用的犁，犁尾短一些，以免翘得太高。在这一步，先将犁尾和犁脊用榆木锯成长条状木料。

第四步，制作犁托头

犁托头是一块矩形木料，安装在犁弯的前端（图7）。用来支撑犁弯，协助控制犁铧入土的倾角。在犁托头和犁弯结合部，楔入一块矩形楔子（图8）。

图7 犁托头

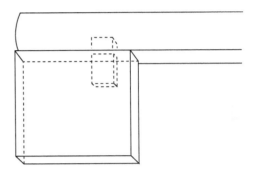
图8 犁托头结构示意

（2）铺犁

铺犁就是将上述部件按照各自位置摆放，确定犁整体框架，并借助方尺、木经尺等，在需要凿眼的犁底、犁弯、犁尾等部位画线（图9）。这一步很重要，宁大胜老人讲，各部分位置、角度摆放合适，犁看起来就谐调、美观，使用时也能节省畜力、结实耐用。我们估量了各部分之间的角度，犁底与犁弯后部夹角5°，犁脊大约后倾10°，犁尾约后倾30°。但宁大胜老人完全凭经验和感觉摆放，没有确定的夹角度数。

图9 铺犁

(3) 精加工

第一步，加工犁弯

按照铺犁时画的线，将犁弯的尾部凿成图 10 的形状。方法是先在将要去掉的部分上锯出几列槽痕，再用凿子加工成最后形状。

如图 11，在犁弯的中部偏后，与犁脊相接的位置要凿一通透孔。犁弯前端与犁托头相接触的地方凿约 1 寸深的孔，不凿通。

图 10　犁弯后端形状

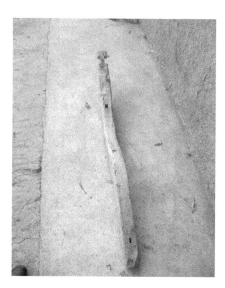

图 11　犁弯的下底面

第二步，加工犁脊

用锛子将犁脊的上部和下端的左右两个侧面加工得窄一些，分别与犁弯和犁底上的孔形成过盈配合。

第三步，加工犁底

加工犁底工作量大、材质硬，要求也比较精细，所以最费时间。由于时间和老人体力的关系，没有继续加工上面的犁底坯子，而是使用了一个旧犁底（图 12）。

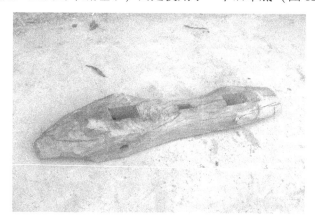

图 12　旧犁底

犁底头部削尖，前部大，后部小，略呈流线型；与犁脊、犁尾结合部较粗大一些，并凿贯通孔，其余地方用斧子砍去，以减轻犁底重量。犁底前部与犁脊结合处横凿一贯通孔，插入销子，以固定犁脊、绑定犁铧。中部凿一孔安装卧牛，协助绑定犁铧。

第四步，加工犁尾

如图13，在犁尾与犁弯结合部凿一贯通孔14厘米×2.5厘米，下端锯窄，与犁底上的孔形成过盈配合。在犁尾下部人手高度处凿圆形槽，安装犁把手，供人手持。从犁尾下端，沿中线锯通，一直锯到贯通孔。在贯通孔上沿，用尼龙绳扎紧。在贯通孔附近的木料上撒些水，增加木料韧性，以备安装犁尾。

（4）组装

首先是把犁尾与犁弯组装在一起。方法是先用一较大楔子将犁尾下端的锯缝撑开。再把犁弯后端架在板凳上，把撑开的犁尾下端插到犁弯后端，取下楔子，慢慢敲打往下犁尾，使贯通孔套住犁弯后端的颈部，将犁弯锁住，从而保证耕地时犁弯不会从犁尾中被拉出。

接下来把犁脊上端插入犁弯中部的孔内，下端插入犁底前部的孔内。然后把犁尾下端插入犁底后部的孔中。再用斧头把各个构件敲紧（如图14）。

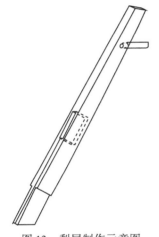
图13　犁尾制作示意图

图14　捆绑后的犁尾

在犁脊与犁底、犁脊与犁弯、犁尾与犁底、犁弯与犁尾等各个结合部都打入楔子，进行加固。犁弯与犁尾结合部的孔比较长，上下都可以打入楔子，改变楔子数目可以调节犁弯与地面的角度，影响犁铧入地的倾角，从而控制耕地的深度。

犁钩和犁铧由铁匠打制，雇主买来交由木匠安装。犁钩的下半部打制成钉子形状，从犁弯前端向下钉透，一直钉入犁托头内，将两者固定；犁钩的上半部向后弯曲，抵住犁弯，耕地时在此绑上绳索。犁铧安装在犁底前端，用铁丝与犁脊绑定。当地使用的犁铧为平头犁铧，耕地时向两侧翻土。

过去市场上出售犁铧、犁钩，但现在已经买不到了。至此，犁的构架已经基本完成。接下来还要对各个部件再进行精细加工，使之轻便、美观、光滑，更加适用。

在内蒙古凉城县的调查中，我们了解到凉城县六苏木秃小子、三苏木鞭墙村宋焕真、查汉营乡大洼大队后五号村阎喜连以及丰镇市元山乡榆树沟村张秀龙都能制作木犁、耧车等农具,他们祖辈或父辈都是从事木匠的手艺人。目前传统犁已逐渐被机耕犁所取代,使用的人越来越少。昔日使用的传统犁虽然大都被淘汰,但当时犁镜、犁头要由铁匠生产,有些已经专业化批量生产,因此尚有不少备件。

铁辕犁的加工制作,关键部件犁辕、犁镜(犁壁)和犁头都以铁为材料,因此制作工艺以金属工艺为主。山西阳城地区拥有犁镜、犁铧铸造的成套技术,犁炉炼铁和铁范的犁镜铸造一直延续到20世纪90年代,技艺十分精湛,至今仍有多个传承人和作坊存世,并有犁炉、铁范、犁镜等大量实物,堪称活化石。

参 考 文 献

[1] 上海市嘉定家具厂《农村木工》编写组. 农村木工. 上海:上海科学技术出版社,1979.86

Study on Indigenous Sugar-making Technology in Naman Tun of Daxin County

Liu Anding

(STS Research Center of Guangxi University for Nationalities)

Daxin County of Guangxi province has a long history of sugarcane planting and sugar-making. When all the sugarcane are sold to sugar factories in other places, especially mechanical equipments are widely used in sugar-making process, the traditional sugar-making methods are still used in Naman. By means of fieldwork, indigenous sugar-making technology was investigated in Naman, the result shows that some villagers in order to improve the production efficiency are more likely to use mechanical equipments instead of traditional tools in some sugar-making process. Moreover, villagers' attitudes on indigenous sugar-making technology are changed. From this transition, the author holds that more attention should be paid to the problem of conflict and adaptation between traditional technology and modern one when we discuss the inheritance of traditional technology.

大新县那满屯土法制糖工艺初探

刘安定

(广西民族大学科技史研究中心)

广西大新县种蔗与制糖的历史悠久，时至今日，其他地方的甘蔗已经全部卖给糖厂，或完全采用机械制糖，该县的黎明村那满屯却仍保留着土法制糖的传统。本文采用田野调查的方法，对那满屯土法制糖工艺进行调查，发现村民为提高生产效率，在一些工序中已采用机械设备替代了传统的制糖工具，村民对这项工艺的认可度也发生了变化，透过这样的变迁，作者认为在讨论传统工艺的传承时更应关注传统技术与现代工艺的冲突与适应的问题。

广西种蔗制糖的历史悠久，南宋《方舆胜览》载："藤州①土人，沿江种甘蔗。冬初压汁作糖，以净器贮之，蘸以竹枝，皆结霜。"再如，《本草纲木》引唐孟诜《食疗本草》上的一段："……竹蔗，以蜀及岭南者为胜。"等等，类似的有关广西地区种蔗制糖的文献较多，足可说明广西在很古老的时候就有种蔗制糖的活动。大新县位于广西的西南部，常年光照充足，雨量充沛，非常适宜甘蔗种植。据记载，土法制糖一直是该县一项传统的手工加工业②，但是自改革开放以后受工业化和城市化进程的影响，土榨点已逐渐被机械化制糖工厂所替代。那满屯是该县为数不多的土榨点之一，本文将以此点为例对现存的土榨工艺作一个简介。

1. 那满屯土法制糖工艺

现存的那满屯土榨工艺无论是工具设备还是工序都有了一定的变化，主要特点是在部分工序上采用了小型机械设备，工序相对原始的土榨过程有所简化。

1.1 原料

那满屯制糖的原料主要是当地种植的甘蔗。甘蔗的品种很多，但就功能来分主要可分为两大类，一类是果蔗，蔗茎比较粗，蔗皮颜色也较深，水分足、甜味浓，多用于食用；另一种竹蔗，根茎细，水分和甜分都不及果蔗，且茎肉质硬，但对种植条件要求不高，所以较适用于榨糖。一般每年的腊月或春节期间开始种植，次年十月份收割。收割回来的甘蔗务必要经过除叶除尘，尤其是表面的包叶、泥沙等附着杂物一定要用刀剔除，否则会影响后来的糖汁的洁净度和纯度。甘蔗的梢尖也要被砍除，因为这部分含糖分少，还含有一些淀粉、多元酚、色素、酶类等杂质，如果不除去都会影响糖分的品质[1]。

1.2 工具

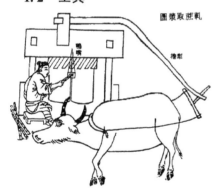

图1 古代糖车（引自《天工开物》，图中两轴（辊）齿向应相反）
资料来源：《天工开物》

1.2.1 榨蔗设备（糖车与榨蔗机）

据大新县志记载，早期的榨蔗设备（糖车），有两个木质的圆柱体，即木辊，大新当地大多用龙眼木作木辊；也有用石头做的石辊，一大一小两个，当地人称大的为车公，小的为车母。车公辊轴上连有一根弓形弯木，弯木转动后带动主辊也就是公辊转动，然后又带动母辊转动。榨糖时将甘蔗送入两个辊中间的缝隙，两边的辊通过轮上的辊齿的咬合向内转动挤压，蔗汁就顺着辊流下。另外还要使用到一种辅助性附件③帮助将头次压榨不够充分的甘蔗再次送入压榨口进行二次压榨。这与《天工开物》上的记载是吻合的，如图1所示。但是随着

---

① 藤州，即广西藤县，糖霜即冰糖。
② 1987年版的《大新县志》记载："土榨糖是本县一项传统的主要手工加工业，历史上凡是种甘蔗的村屯都用土法榨制。这种土榨（工具）是用坚硬的龙眼木做成的（也有用石头做）……"
③ 江浙一带称这种附件为"狮口"。

现代机械化进程的推进,传统榨蔗机已经逐渐退出人们的视线,这几年在那满已基本见不到这样的工具,所能看到的糖车只不过是一块菜板了①。现在那满屯使用的基本都是小型的机械榨蔗机。

1.2.2 孔明灶

"孔明灶"又叫"连环灶",民间传说的"孔明灶"是三国时期蜀国军师诸葛亮带兵打仗时所发明的,"孔明灶的形状是个长方体,上面可以安放四个大铁釜,后面有大烟囱,灶底为前低后高的斜坡。这种灶的优点是可以节省燃料"[2]。那满屯的村主任介绍说他们这种灶是学习20世纪60年代潮汕地区的,那时潮汕地区使用多锅同煮,需要很多劳动力,后来为节省人力,那满屯只采用五六口锅,如图2所示。这五六口锅共一个灶膛,一个烟囱,生火的时候,火苗会从灶口处向烟囱的方向飘去,靠近灶口的温度会比较高,越往烟囱的方向火势越小。在靠近灶口处还砌有一个地下灶堂,上面用几块木板条封盖,留有缝隙,当灶口处堆积的燃料较多时,挪开木板条时就可以让更多的氧气进入灶内促进燃烧,等火熄灭后,可以将木板掀开清除灶灰。

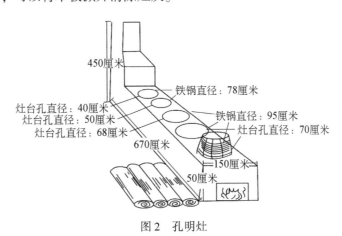

图2 孔明灶

1.2.3 晒台

晒台是用来将糖浆凝固成片的工具,宽一尺(约33厘米)左右,长约5米,主要有三层,最底一层通常是木棉板,也有少数人家用的是水泥预制板——水泥预制板上的糖浆冷却过快,制得的糖片易碎裂,木棉板质就不存在这个问题,它不易变形,但是由于某些历史原因造成了人为的砍伐(20世纪70年代末越南人因发表"有木棉树的地方是越南领土"的不当言论导致中国公民大量砍伐木棉树)[3],木棉树已经越来越少,现在,村民已经使用人工制造的"五合板"代替木棉板;中间一层是箱纸,箱纸有很好的隔热功能,防止过快冷却的糖片过脆易碎。最上面一层则为竹席,它有一定的隔水功能,糖汁倒上后不会渗漏到下面的箱纸层浸湿箱纸。凉席上面还要有木质压条,呈"口"状放置,为了加固通常还要用两条横条上下夹住长压条的两端。现在压条也有采用铝合金的,铝条不易变形也易于清洗。

---

① 2010年1月份,笔者在那满屯调研时,村主任介绍说以前村里还有一台木辊的糖车,但是村民认为闲置可惜就将木辊锯了当菜板使用。

1.3 工艺步骤

1.3.1 压榨

电力带动的小型榨蔗机，一般安置在村民家门口，村民开始将事先堆放在门口的甘蔗拉出3~5根伸进压榨口，扶住另一端蔗杆，直到蔗汁源源不断地流向放置在机器出汁口下的桶里，在另一端被压榨过的甘蔗变成蔗渣被吐出，蔗渣可以作为燃料供熬煮蔗汁用。每一批甘蔗经过一次压榨即可压榨完全，而在过去没有机械榨蔗机时，采用人力、畜力或者水力压榨，一批甘蔗往往要经过三次以上的压榨才能将蔗汁榨净。

1.3.2 熬煮

将蔗汁倒入灶上的五口锅内，灶火烧旺直到锅内的糖汁被完全加热至沸腾。约30分钟后，用过滤网将浮于液面的杂质捞出，（这些杂质被捞出后放置在一口专门用来盛放杂质的大缸内，可以用来酿酒）一边加热一边去杂质。由于灶口的温度较高，靠近灶口的三口锅水分蒸发较快，熬煮当中要将另外两口靠近烟囱的锅内的蔗汁调过来。为防止糖液溢出，将无底的竹筐罩在锅上（筐高约80厘米，直径略小于最小的一口锅的直径），当糖液沸腾时，液体沿着筐壁流下，很多杂质也同时附着在筐壁上。靠近灶口的锅内的糖汁浓度不断增大，在这过程中要加入适量猪油，问及加猪油的作用，那满屯人的说法是加入猪油可以减少泡沫并使成品糖口感更好。在潮州也有"加油土"的说法，"油土"可以是花生油、茶油或是菜油的沉淀物，潮州人俗称"未盖糖寮，先买油土"（没盖糖坊就要先买油土），可见加"油土"是多么重要[4]。另外，那满屯人说加油土还可以防止糖汁黏上锅底损坏铁锅和影响糖片色泽。当浓郁的糖香不断飘出，糖液呈现深黄色并逐渐加深变暗时，有经验的煮糖师傅会用一把长柄勺不断搅拌糖浆并根据糖浆的黏稠度判断起锅的时机。据说在那满屯，以前都是靠老人用嘴咬凝固的糖块，根据口感来判断起锅的时机的，而现在仅凭肉眼就可以判断了。当糖浆起锅的时机到来时，锅内的糖浆已是土糖的半成品，人们将其盛入备用锅等待运往晒台。第一锅糖浆从开始蒸煮到出锅总共历时1小时10分钟左右，而这之后每锅糖浆出锅的时间只要50分钟左右。

1.3.3 晒糖成品

将糖浆搬运到堂屋后，并不立即倒上晒台，而是将整锅放置在一个轮胎上（尖底铁锅放在轮胎中间非常平稳），用一支长柄小铁铲不停搅拌加快糖浆冷却，同时也防止糖浆结块黏住锅底。大约10分钟以后，糖浆变得更黏稠，这时就需要将糖浆倒上晒台。由两人抬起锅稍稍倾斜，从晒台一端开始边移动边将糖浆慢慢浇入晒台，由于晒台的槽子较浅，所以倒入糖浆时绝不能过急过快，否则糖浆会溢出槽外，糖浆浇入晒台后，还要用小平铲辅助着将糖浆的浆面挑拨平整，糖浆完全冷却后凝固在晒台上，呈浅咖啡色，用直角尺来量好切片的位置，将整条晒台的糖片切成大小均匀的片糖。切好的片糖被整齐地装入纸箱内等待出售。

2. 那满屯土法制糖工艺传承现状

那满屯几乎家家制糖，全村的90户制糖人家中绝大多数都集中在这个屯。虽然土榨糖在新中国成立前就有，但是从20世纪80年代后才真正大规模兴起。如今的那满屯制糖现状主要可以总结为这几点：第一，制糖点仍以户为单位，采用作坊式生产形式代代相传，父业子承，兄弟合作。父辈经验足指导操作，子孙辈体力充沛执行一些具等工作。第二，制糖主体以男性为主。由于工序中的几个关键步骤对体力有要求，工艺的实施主要以男性为主。虽然偶有女性参与辅助，如添加燃料、调拨蔗汁，但是相对可参与的工序非常有限。第三，工艺认同的人数在减少。在制糖的队伍中，25周岁以下的年轻者较少，他们认为这种土榨糖的方法效率低下，又受季节性限制，所以宁愿在外打工也不愿参与。

3. 总结与反思

总之，从以上简介中可以看出，变迁贯穿于整个工艺的发展。在设备上，孔明灶的改造、现代机械榨蔗机取代糖车等，无疑都提高了土糖的生产效率，但是这些变化后的工艺，却可能让后人无从知晓前人所做的工作，所以尽可能记录那些即将消失的事物是必要的，至少不会让那满屯的后人误以为他们的前人所用过的糖车只是一块菜板。在制糖的工序上，那满人屯尽量减少或根本不使用添加剂，这种做法提高了产品的绿色等级却同时缩短了它的保质期，这也体现出传统工艺与现代技术间日益加深的冲突。同时，新一代的年轻人对于土糖的价值观念也日趋淡薄，如何唤起他们的传统价值观都将是这项工艺所要面临的问题。

参 考 文 献

[1] 陈初尧，袁幼菊. 四川土法制糖工艺. 北京：轻工业出版社，1958.17

[2] 刘绎如. 潮洲制糖的方法及糖的加工. 化学通报，1958，(6)：383

[3] 边陲木棉. 越南为啥说：有木棉树的地方是越南领土. http://www.beihai365.com/bbs/viewthread.php? tid = 198460 [2005-10-09]

[4] 高纲勉，蔡正成. 潮州制糖的补充. 化学通报，1958，(9)

British Iron and Steel Technology's Transfer in Early Modern East Asia: The Case of Qingxi Iron Works, China and Kamaishi Iron Works, Japan

Fang Yibing

(The Institute for the History of Natural Science, CAS)

Kamaishi（釜石）National Iron Works, founded in 1874, was the first modern iron works in Japan, also Qingxi（青溪）Iron Works in Guizhou province founded in 1885 was the first modern iron works in China. It is worth noting that both the two enterprises chose Britain as their technology supplying nation, which means that it was Britain transferred the earliest modern iron and steel technology into East Asia. But the processes of technology transfer at Kamaishi and Qingxi can't be considered to be successful because of the failure of the two enterprises. In order to find out the characters of the early modern western technology transfer in East Asia, and make some theory construction in such field, it is valuable to discuss the details of the technology transfer process happened in Kaimaishi and Qingxi and do some comparative study between these two enterprises.

1. The construction of Qingxi Iron Works and its failure

It was in 1885 that the design and construction of Qingxi Iron Works started. From plan making, location selecting, machines purchasing to getting into operation, it lasted 5 years. The leader of such pioneering project was Pan Wei（潘霨，1815-1894）, the Governor of Guizhou province（贵州巡抚）. Pan Wei was a doctor with high medical skill before he became an official. He fortunately caught a chance to cure the wife of YI Xin（奕䜣）, and began to get into official career. In 1885, Pan Wei was appointed as the Governor of Guizhou province. In order to solve the financial difficulties, he decided to set up a western style mining and metallurgy company to provide raw materials to the modern companies of Nanyang and Beiyang（南北洋）because of the rich mineral resources in Guizhou province, especially the iron mines of fairly good quality.

In spring of 1886, the Guizhou Bureau of Mines（贵州矿务总局）was set up. To find out the quality of the mineral resources of Guizhou, Pan Wei sent 4 kinds of mineral, including coal, iron ore, saltpeter and sulphur, to the factories of Nanyang and Beiyang. [1] According to the reports from Nanyang and Beiyang, only the iron ore suit for the modern production. [2] So

Pan Wei decided to build up a modern iron works at Guizhou province.

Construction of a modern iron works was entirely a pioneering project in the late of 19th Century's China. It was almost impossible to find a professional expert in the field of modern iron and steel making. There were mainly 4 persons Pan Wei could rely on to make the technical decisions, such as the choice of the factory's site, the import of machines.

The first one was Pan Lu （潘露, 1827-1890）, the Younger brother of Pan Wei. Pan Lu was an experienced manager of Yangwu companies. He was involved in the start-up of Guangdong Bureau of Military Equipment （广东军装机器局） and Guangdong Bureau of Gunpowder （广东火药局） from 1873 to 1875, and worked as an manager in these companies till 1883.[3] In 1883, Pan Lu was appointed as the general manager （总办） of the Kiangnan Arsenal （江南制造局） by Zuo Zhongtang （左宗棠, 1812-1885）, the Governor of Jiangxi and Jiangsu （两江总督）, who appreciated his ability in western weapon manufacturing very much.[4] The rich experience in western arsenals and the family relative to Pan Wei make Pan Lu to be the most suitable person when Pan Wei selected the general manager of the Guizhou Bureau of Mines and the iron works. It was in the beginning of 1887 that Pan Lu formally took the post of the general manager of the Guizhou Bureau of Mines. However, in order to choose the site of the iron works, he had already traveled to Guizhou to carry on an investigation before, and chose a town locate by the side of River Wuyang （舞阳河） called Qingxi as the site of the factory.

In addition to the location, the machine's purchase is another most important matter. From a historical point of view, the initial purchase of machine sometimes may determine the track of technology development.

In the late of 1886, Pan Wei decided to send persons to Britain to study and purchase equipment of the iron works when he knew that the equipment can't be purchased in Shanghai. According to the historical data we can find currently, at least three persons were sent to Britain: Pan Zhijun （潘志俊, 1857-1919）, who was the son of Pan Wei; Xu Qinruan （徐庆沅, 1854-?）, an manager of the iron works; Qi Zhuyi （祁祖彝, 1863-?）, an interpreter.

Pan Zhijun was the second son of Pan Wei, he passed the provincial-level examination in 1876. From then on he got into his official career. Being appointed as an third-class counselor by Ambassador Liu Ruifen （刘瑞芬, 1827-1892）, Pan Zhijun arrived in London in the spring of 1886 together with Liu Ruifen and other colleagues of the Chinese embassy in London.[5] According to a letter written by Li Hongzhang （李鸿章, 1823-1901） in 1889, it is certain that to study and purchase machines for Qingxi Iron Works was one of his most important mission in Britain.[6]

The other two persons who were sent to UK were Xu Qinruan and Qi Zhuyi. Xu Qinruan was appointed as a manager of the Qingxi Iron Works by Pan Wei, under the command of Pan Lu in around 1886. From then on, He worked for Qingxi Iron Works till 1896, when left

Guizhou to join the Hanyang Iron Works (汉阳铁厂) as an iron and steel engineer. Qi Zhuyi was a member of the third batch of Chinese educational mission students (留美幼童) in 1874. He has been in United State for 10 years, studied western manufacture engineering. He joined the Kiangnan Arsenal when he returned to China, and was appointed as an interpreter by Pan Lu in around 1886 to go to UK together with Xu Qinruan. The only historical data which mentions these two young man's experience in UK is a memoir of Eugene Ruppert, the chief engineer of Hanyang Iron Works from 1904 to 1912. In this article, Ruppert has written down what he heard from his colleague Xu Qinruan:

> The two young men learned a wide range of knowledge about the construction, management and operation of an iron works in a short period of time. Afterwards, they brought a lot of advanced equipments back to China. Under the leadership of Pan Wei and Pan Lu, they devoted what they learned to make the pioneering iron works get into operation, entirely without the help of foreign experts. (von Eug. Ruppert. Chinesischer Bergban und Eisenindustrie)

According to the memoir of Eugene Ruppert, the Equipments of Qingxi Iron Works were purchased from the Tees-side Company of Middlesbrough in north England.

From the details described above, a structure of the personnel relations in the process of the technology import of the Qingxi Iron Works can be drawn out as below, which shows some characters of the initial iron and steel technology import in early modern China.

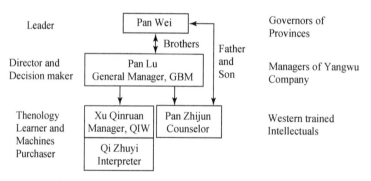

Fig. 1 The structure of the personnel relation in the process of the technology import of Qingxi Iron Works

Firstly, it is a network of relationship composed of three levels of personnel: the governors who engaged in the Yangwu Movement (洋务派督抚); the officials in the Bureau of Yangwu enterprise; and the intellectuals in Yangwu enterprise. It is worth noting that none in such personnel network really mastered the iron and steel technology.

Secondly, it is a decision-making system totally composed by Chinese, which is quite different from the Kiangnan Arsenal and the Fuzhou Shipyard (福州船政局).

Thirdly, on the side of China, the process of technology information receiving was of very accidental. Lack of the systematic professional knowledge made the decision-maker could not select equipments depend on the applicability of technology.

Under such situation, a 25 tons blast furnace, two 1ton bessemer convertors, fourteen puddling-furnaces, and a rolling mill were imported to Qingxi. The construction of the iron works completed in 1890. The furnace got into operation on 17th of July, 1890. However, the fuel used in the melting process was obviously unsuitable for the blast furnace. According to Ruppert's memoir, it was a blend of bad coke and anthracite. Such poor fuel caused the accidents of the furnace happened continually. The melting process only lasted for around 50 days and stopped in the middle of August. Pan Lu died on the 31st of August, 1890. The furnace never reopened from then on. The initial technology transfer was end in failure of the Qingxi Iron Works.

Fig. 2　The 25tons blast furnace of Qingxi Iron Works

It worth noting that the failure of Qingxi Iron Works didn't end the British iron and steel technology transfer into China. The Hanyang Iron Works imported the equipments from the same company in UK a few years later.

2. The construction of the Kamaishi National Iron Works

Although the Kamaishi National Iron Works was the first modern iron works in Japan, ten western-style blast furnaces had already been built at Kamaishi by Oshima Takato (大岛高任, 1826-1901), the father of Japanese modern iron and steel industry before 1868. However, these furnaces were quite small as their product ability per day was only 1-2 tons. The tradition waterwheels were used as the power equipment of the furnace. Strictly speaking, these ten furnaces were not the real modern blast furnace.

After the Meiji Restoration, the Japanese government pursued the modernization of military and industry firmly. However, Japan didn't have a modern structure of society which suit for industrialization yet. The new government then played the key role in the industrialization process through construction of the national enterprises. The Kamaishi Iron Works was one typical case of the national enterprise at that time.

The decision to build an iron works at Kamaishi was based on recommendations in a foreign adviser's report. J. G. H. Godfrey (1841-?), hired as the Chief geologist and mining engineer for Meiji government from 1871 to 1877, went to the Tohoku region of north Japan in the summer of 1872 to inspect the iron mines in the region where Kamaishi is located, he noted that the supply and quality of the iron ore was good and that working these mines should be profitable. His estimate was confirmed by the further survey carried by Japanese government. [7] During his tour, Godfrey scrutinized every detail of iron production in the region. He observed that there were a number of problems with production including the expense and difficulty of transporting raw materials through the mountainous area. He recommended the construction of larger furnace closer to the sea in his report. [8]

Based on Godfrey's report, the Meiji government decided to requisition the Kamaishi deposit in 1873. Then a proposal for an iron works at Kamaishi was submitted by Yamao (山尾庸三, 1837-1917) and Ito Hirobumi (伊藤博文, 1840-1909) to the Council of State and Ministry of Finance on 15 February 1874. [8] The plans include a railroad from the mines to Kamaishi harbor and three modern blast furnaces deemed capable of producing 12 000 tons of pig iron per year. There would also be a refinery at Nagasaki (长崎), with 12 furnaces. On 21 May 1874, the Meiji government established the Kamaishi branch office of the Ministry of Public Works Mining Department. Oshima Takato and another technologist, Koma Rinosuke, were appointed as Kamaishi's managers. A ground-breaking ceremony for the new iron works was held on 10 August 1874, and the actual construction began in January of the following year. [9]

The process of deciding the blast furnace's location was accompany with dispute between two technologists, Oshima Takato and Louis Bianchi, the Meiji government's foreign adviser for mining and metallurgy at the iron works. Bianchi was a German who studied at the Freiberg Mining Academy from 1856 to 1859. He remained in good standing at the academy before he was hired by the Meiji government in 1874. During late spring and early summer of 1874, Bianchi surveyed the Kamaishi region with Oshima Takato in an effort to find the best location of the iron works. It is said that the two men could not come to an agreement and fought bitterly.

Bianchi proposed a site at a location named Suzuko (铃子). Selection criteria were based on the site's proximity to transportation, access to water, proximity to ore and fuel, available land, and cost. For Bianchi's opinion, Suzuko was an ideal site for large modern iron works. Oshima proposed the site Otadagoe, which was surrounded by mountains on three sides: north, west, and east. In Oshima's opinion, there are no rainstorms throughout the four seasons, the cold will not stop production day and night. Moreover, Oshima suggested building 5 small blast furnaces similar to those he built at Kamaishi before. It is obvious that Oshima's proposal violated every convention of nineteenth century iron works. It was "traditional" and "backwards" and unacceptable from the Meiji government's perspective. The Meiji government accepted Bianchi's proposal finally. In October 1875, Oshima was sent to the Ikuno and Kosaka silver mines and the Sado gold mine where he would serve out his days in the Ministry of Public Works as Executive Mining Director.

The equipments of the iron works were all purchased from England based on Bianchi's recommendation. Actually, Bianchi was not the person responsible for designing Kamaishi, his job was to oversee construction. The design of the furnaces and related equipment, based on the public works ministry's demands, was the project of David Forbes, a British consulting mining engineer who never actually visited Kamaishi.

When Bianchi's contract expired in March 1877, he left his position because that he did not like the working conditions and the British staffs with whom he worked. Furnace construction was the job of William H. B. Casley, a British iron works manager from Stockton-on-Tees, who arrived at Kamaishi with the furnace components in early 1876. A Japanese manager, Yamada Jun'an (山田淳庵) who had studied in the London Mining College, was also responsible for daily operation.[10]

Based on the details presented above, a structure of personnel relations in the process of technology import of the Kamaishi iron works can be drawn out as below (Fig. 3), which shows some different characters from Qingxi Iron Works. Most importantly, the decision-making system was a network composed of both Japanese and foreigners. Foreign experts were involved in every level of the decision-making in the technology import's process. Secondly, a careful survey had been carried on before the project started. Thirdly, the Japanese involved in the process were not just the officials who had little knowledge of iron and steel. On the contrary, many of them are had studied related knowledge in Europe. Even senior official such as Yamao Yozo toured with the foreign expert survey the mine. All the characters shows the eagerness of Meiji government to learn western technology and realize industrialization.

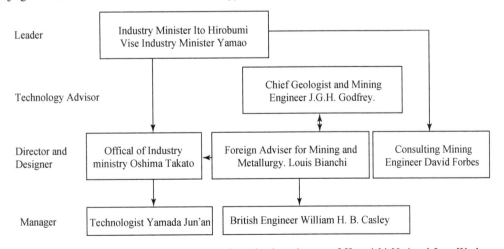

Fig. 3 The structure of personnel relations in technology import of Kamaishi National Iron Works

Forbes designed two blast furnaces for Kamaishi Iron Works, each of which was configured to produce approximately 75 to 80 tons of charcoal pig iron per week. According to Forbes, "the contract for the iron and brickwork fittings is placed in the hands of Messrs. Head, Wrightson and Co., of the Teesdale Iron Works, Stockton-on-Tees, whose patent hydraulic arrangement for lowering the bell will be employed as well as Whitwell's hot-air stoves, Lurmann's closed breast, and all the most modern improvements"[11]. It is obvious that the new iron works at Kamaishi was to be large and modern.

Fig. 4 The blast furnace of Kamaishi Iron Works

The construction of the iron works completed in 1880, one furnace was blown-in on 10 September 1880. Three days later, the iron began to flow. On 9 December 1880, less than three months after the beginning production, a fire destroyed the Kogawa (小川) charcoal facility, the majority of Kamaishi's fuel reserves. The shortage of charcoal forced managers to shut down operations on 15 December.

Kamaishi's furnaces were restarted in March 1882. During the month prior to resuming operations, Kamaishi's managers had brought 10 000 tons of coal to the iron works and ordered the construction of 48 coke ovens. It was said that at least one furnace made the transition to coke. When the switch was made to coke, little changed within the furnace, and output steadily decreased because the furnace was chilling. At first the furnace started forming clinkers, partially smelted agglomerations of iron ore, fuel, slag, and flux. Later the charge fused into one solid mass. Shortly thereafter, the facility was declared a failure. Within a few months, the Meiji government abandoned its first modern iron works.

It was around 1885 when Kamaishi Iron Works was sold to a business man named Tanaka Chobei（田中长兵卫）. The Kamaishi Tanaka Iron Works was then founded in July 1887, and developed gradually throughout the 1890s with the technical assistance of Professor Noro Kageyoshi（野吕景义，1854-1923）of the Imperial University of Tokyo and his disciple, Komura Koroku（香村小録）.

3. Conclusion

To compare the two first entrepreneurial ventures in iron industry, some related questions can be discussed.

Firstly, sometimes the selection of technology supplier is an important factor may affect the further development of technology, especially at the initial stage of the technology transfer. Why did both Japan and China select Britain as their first supplier of iron and steel technology? It is nearly impossible to find an undoubted answer just based on the historical data. However, it is certain that the strong desire of "civilization building" made Meiji government want to build a large and modern iron works to support the new growth of rail and sea transport, military and other industry. Therefore, it is reasonable for Japanese government to select Britain, the largest steel producer in the world, as its preferred technology supplier. On Chinese side, to exploit the local mine resource and stop the money flowing out was the main purpose of the first entrepreneurial venture of iron works. The decision makers didn't have enough knowledge of such complicated technology. I prefer that it is more like an occasional decision for Chinese to select Britain as the technology supplier.

Secondly, unlike other industry, many of the design and much of the smelting process in nineteen-century iron production relied on trial and error. Once in-blast, a furnace still required constant adjustments to compensate for variations in the fuel and ore. The experience of furnace operation is of much useful, sometimes it may become a key factor in a new iron works. This is what Japan and China lacked in nineteenth Century. Therefore, we can not say that the failure of Qingxi Iron Works and Kamaishi Iron Works was just a result of an improper technology selection.

Thirdly, further development of the two countries presented different scene. China continued to import the British equipment for the Hanyang Iron Works shortly after the close of the Qingxi Iron Works, presented a strong technological dependency at the early stage of the technology transfer. On the contrary, the early failure make Japan gave up Britain soon. When Yawata Iron Works was built, German equipments became the main content of its plants. Thereafter, Japan went on a way to technology independent gradually in a half century development of iron and steel industry.

参 考 文 献

[1] 光绪十二年三月十九日贵州巡抚潘霨奏. 载：中国史学会编. 洋务运动（七）. 上海：上海人民出

版社,2000.173-174
[2] 复潘伟帅. 载:中国史学会编. 洋务运动(七). 上海:上海人民出版社,2000.191-192
[3] 中国近代兵器工业编审委员会. 中国近代兵器工业:清末至民国的兵器工业. 北京:国防工业出版社,1998
[4] 左宗棠奏任潘露陈鸣志片. 载:《中国近代兵器工业档案史料》编委会. 中国近代兵器工业档案史料1. 北京:兵器工业出版社,1993.1227
[5] 刘瑞芬. 刘中丞(芝田)奏稿. 近代中国史料丛刊第61辑. 台北:文海出版社,1971.151,157,257
[6] 光绪十五年十一月初八日. 直隶李鸿章致盛宣怀电. 载:孙毓棠编. 中国近代工业史资料第一辑. 北京:科学出版社,1957.684
[7] Yonekura S. The Japanese Iron and Steel Industry, 1850-1990. New York: St. Martin's Press, 1994. 22
[8] Wittner D G. Technology and the Culture of Progress in Meiji Japan. Oxon: Routledge, 2008. 77-78
[9] Wittner D G. Technology and the Culture of Progress in Meiji Japan. Oxon: Routledge, 2008. 79
[10] Wittner D G. Technology and the Culture of Progress in Meiji Japan. Oxon: Routledge, 2008. 87
[11] Forbes D. Report on the Progress of the Iron and Steel Industries in Foreign Countries. Journal of the Iron and Steel Institute (part 1). London: the Iron and Steel Institute, 1875. 299

How the Government Deal with the Drought from 989 AD to 992 AD in the Northern Song Dynasty

Dong Yuyu

(Department for History and Philosophy of Sciences, Shanghai Jiao Tong University)

It is a serious disaster that the drought happened from 989AD to 992AD in the Northern Song Dynasty. First the paper gives an historical overview of the drought. Second, the paper focuses on how the drought impact on society. Third, by analyzing famine's rescue, the payment of taxes reduced, the epidemic prevention, the refugee to comfort, the water conservation to start construction, warehousing aspects and so on, the paper shows how the government deal with the drought.

1. Introduction

In Taizong's reign of the North ern Song, one drought took place in many places and lasted a long time. This topic was often mentioned by scholars when they talked about the most serious droughts in recent 1000 years. [1] However, how large was the spatial and temporal distribution of the drought, what did it bring to the society, and what has the government adopted to response this disaster? Questions like these are still lack of systematic research. This paper is on the basis of previous research and tries to give the answer about this serious drought between 989AD to 992AD.

2. The spatial and temporal distribution of the severe drought

Serious disasters have different appellations in Ancient China. In the Song Dynasty, people used "drought", "small drought", "big drought" and so on, to describe the situation of the drought. Severe droughts between 989AD to 992AD in the Northern Song Dynasty are qualitatively named as extreme drought, depending on its duration, the scope and extent of damage.

From the historical data, it is easy to see a wider range of the drought affected, especially most parts of north and a few parts of northwest regions of today's China. About the time distribution, we can see the drought last for a long time in spring, summer and winter. In terms of years, Shanxi, Shandong, and part of Henan province, the drought disaster last in successive years. Considering on the description "very dry", "drought" and other words in historical books, we can image the drought was very serious.

3. The harm to the society

3.1 Famine

Lasted long and occurred in suck a big range, the drought destroyed the agriculture badly. From the post-disaster government survey data[2], the droughts caused a large area of the cut or crops and a massive famine.

3.2 Locusts

From the *Five Elements of the Song Dynasty* (《宋史五行志》), the extreme drought also triggered a series of locusts.

3.3 Knave

Shanxi province is one of the most serious regions in this drought. because this drought was so serious that many people had to leave their hometown and to be beggars. [3]

3.4 Disease

Pandemic often happened after calamity. In this extra-large disasters, capital area was one of the worst hit areas. Disease out broke in the capital on a large scale. [4]

4. Social response to extreme droughts

Whenever the disasters occur, every aspect of society will be mobilized to deal with crises. Face the extreme drought in "absolute" and "disaster" traditional theory of scourge, represent the " dominant ideas into the nature of the emperor's ruling will stop the action, to respond to god, and the recognition that their subjects as soon as possible to eliminate to. On the other hand, remittent would have adopted some specific treatment measures to actively respond to the disaster.

4.1 The concrete measures to stop disasters

So-called stoping disasters refers to stop disaster after disaster, in "absolute" and "disaster" traditional theory of scourge, "under the thought of god" rule by the emperor's era, take some measures to respond to the scourge, hope to seek god's forgiveness, ameliorate disasters.

4.1.1 Admit the guilt

Two years after the droughts, according to the records of medium, the Emperor Taizong did say sorry to his people all over the country.

4.1.2 Praying for rain

From the historical emperor Taizong respectively[5], it is easy to see rain sent people to move mountains, hoping to heaven, abundant showers, alleviate the drought.

4.1.3 Seek respects

After the disaster, the Emperor often make use of the opportunity to review their own action and seek the views of people on the political affairs of the gains and losses, in order to turn over a new leaf. There will often be opportunity to characterize current affairs ministered out. [6]

In addition, the emperor also took measures to reduce disaster such as Mi diet, withdrawal of music, taking into account measures such as prisoners, even though a large extent, these measures only were the emperor a concrete manifestation of the religious teaching form, but also highlighting the King care of the disaster and livelihood concerns, helping alleviate the social crisis brought about by the disaster, the people in the spirit of solidarity can play a positive effect on disaster.

4.2.1 Famine relief

In the Song Dynasty after the disaster occurred, according to the severity of the disaster and the disaster of different objects to different relief measures. Some local officials of the Southern Song Dynasty have been summed up: "civil and military officials made: There are three court Famine: First of relief Tiao, second is relief loans, third is relief. Although both famine, and its method of different, appropriate relief Tiao marketplace, Rural appropriate relief loans, but relief of the poor can not self-preservation Zheyi. If the diffuse thereof, there must be the feasible, the official Stuart fee less than the public benefit." 989- 992 of the extraordinary drought of the government response also basically reflects the relief Tiao, relief loans, relief three basic measures.

4.2.2 Tax relief

In the Song Dynasty after the disaster occurred, the Government will first verify the specific disaster situation, and then take concrete under the disaster relief measures, and gradually formed a disaster appeal, the prosecution put, copy bar, relief to the tight program. In four areas, the prosecution put the key process in the past and it contains two tasks, first check the disaster injury, and second, to determine the extent of exemptions under the famine farm rents of the scores, it said "put tax". Can be seen from the historical data[7], the government has done a very thorough inspection and detailed mining work, according to the drought damage to the actual situation in the seedlings were free to determine, is in sixth place, is in the fifth cut, is in third place, and concrete relief scores, the government targeted for relief.

4.2.3 Prevention and treatment for preventing diseases

After a large disaster, it is often followed large-scale epidemic, 989-992, after successive years of drought it have also sustained serious outbreaks occurred, including the Capital area. The face of this situation, the court order just one hand edited "Holy Benevolent Prescriptions", enactment of the country, selected a good doctor the other hand, an important junction in the capital value 10 patients. [8]

4.2.4 Water for irrigating

In all of these measures, building water conservancy projects is of course important government response measures. The original irrigation facilities for some dilapidated, raging in the drought can not fully play its function. So some people took the opportunity to petition the court to re-Shuji.[9]

4.2.5 The Protection Offered by Refugee

After a little stability in the disaster, the government will introduce measures to pacify displaced persons return to their homes. Drought, the Government has repeatedly displaced persons under the enlistment.[10]

4.2.6 Build Warehouse

Chunhua three years (992 AD) in June, the court building in the capital often open.[11] This is the beginning of the Song Dynasty often build open by Shinshu promotion, was established at the local wide.

5. Conclusion

From 989AD to 992AD, years of severe drought in the Northern Song Dynasty made the community a huge crisis. In this crisis, the government took a positive response. Famine relief, tax relief, prevention and treatment for preventing diseases, displaced persons in comfort, the irrigating water, storage of settings and other measures have achieved good results. The government gradually got through the famine. Government accumulated a number of well-established experience, such as making warehouse build, setting up disaster management system, etc. We may learn from the history of this story, for improving disaster response management system today. （教育部人文社科基金项目基金09YJC770054支持）

References

[1] 李维京等. 中国干旱的气候特征及其成因的初步研究. 干旱气象, 2003, (4)
[2] 〔清〕徐松食货. 载：宋会要辑稿. 62
[3] 宋诗钞. 卷1
[4] 宋史. 67. 志二十. 五行五
[5] 〔元〕脱脱. 宋史. 卷5. 本纪五. 太宗二
[6] 〔宋〕李焘. 续资治通鉴长编. 卷30
[7] 〔清〕徐松. 宋会要辑稿. 食货62
[8] 〔宋〕李焘. 续资治通鉴长编. 卷33
[9] 宋史. 卷97. 志第四十七. 河渠四
[10] 宋大诏令集，卷185
[11] 〔宋〕李焘. 续资治通鉴长编. 卷33

Technologized Science:
Representational Theories *vs*. Epistemological Engines[*]

Byron Kaldis

I

This singular character of the techno-scientific phenomenon has been approached from a number of different angles. One crucial line of thought that I believe best captures this special character that science exhibits in our times is the approach which charts the novelty or singularity involved as emerging at the point when a purely mathematized science has gradually given over into a "*technologized*" one.

Thus the main point of contention for those who claim the singularity of a technoscientific fusion of science and technology is to show that the major shift has been from mathematics to technology. Surely, no one claims that mathematics is no more at the center of science or, obversely, that techniques were not present even before the official birth of modern mathematical science in the 17th century. Also, a number of other lines of thought that emphasize different core components for singling out the novel character of current science are not without their merits or completely unrelated to the present one.

However, what is in addition an extra valuable asset of the current approach that privileges advanced technology as core component is that it enables us to run together in a fruitful manner both a philosophical analysis of the phenomenon as well as a meta-historical one regarding the way science has developed. The way our thinking about it—whether philosophically or sociologically or historically—has equally developed accordingly, and this should be a factor in our analysis. What I mean by this is that the present way of looking at things that I choose to adopt—namely, construing current science as essentially *technologized* science—enables us to draw lines of convergence uniting how the history of the philosophy of science has proceeded, on the one hand, with the rise of the parallel and converging field of the history and philosophy of technology. In that way it enables us further, by means of that convergence, to address

---

[*] Though the term comes from. Ihde's seminal work on how technologically mediated practical action or embodied mechanisms can and have influenced directly the way theories about human knowledge are formed or how the mind Works in representing the outside world, my use of the term is different, though not unconnected: I denote by its technoscience's pillar, namely advanced converging technologies/artifacts that are expressly used as engines, or mechanisms that produce scientific knowledge. [1]

ethical worries of a special kind arising from modern technoscience.

Here is one way this convergence of analyses or paths enabling an added value to our programme of analysis can be understood as a fruitful methodological strategy. Both the original orthodox philosophy of science at its inception—heavily imbibing in logical positivism—as well as its somewhat later phase which, despite its distance from hard core logical empiricism, was in many ways a clear continuation of it, let us call it "new empiricism" (vide H. Feigl's philosophy of science) of the mid 50's-both of them, it can be safely said, regarded science as predominantly a logical machine producing theories or hypotheses. [1] The general tenor of this school of thought was that science amounts essentially to a theory-production. Science amounts to a well grounded logical construction that has in addition to its internal systematic coherence an external interpretation: a model and its interpretative domain. It is not hard to see that behind this idea, or even better engraved on its forefront, was the basic set-theoretic mathematical idea (present in the intellectual environment at the end of the 19th century) of an uninterpreted model—only here the empirical world acted as a necessary domain of application. Emphasis in all this is placed on the logical presuppositions that must be met. It is therefore a clear case where the ideal of systematization holds sway imparting its influence on all modes of learning, science and mathematics being the primary modes of authentic knowledge that need to be true to this ideal even more than anything else.

When philosophy of science begins to stir clear from such a vision of science as theory-production it is precisely when a major turn takes place away from mathematization and towards the historical or sociological interpretation of science as a certain kind of practice and of scientific practice, in its turn, as basically a social phenomenon carried out by an organized scientific community. The interpretative movement is from construing science as a body of statements exhibiting logical structure[2] to viewing it as a body of researches and their practice exhibiting a scientific community's structure: i. e. from a mathematical-cum-logical structure to a social structure with its own historical development inherent to it and affecting research programmes comprising it. Research programmes can be admitted to have a history i. e. they can develop, or go through phases. Notice that as the 20th century moved on Lakatos' principal idea of degenerating or not research programmes had gained ground. Even talking in terms of research programmes was a far cry from the erstwhile view of science as ahistorical. However to see more deeply into this turn and differentiate this line of exegesis from the one I chose to underline, that of science becoming technoscience—I propose to show the integral role that the foundational idea of *representation* plays in both the orthodox positivist view and its antithesis, i. e. that of the sociological turn. It may thus be seen that the latter turn towards an analysis of science in

---

[1] It should be noted however that M Schlick, the patriarch of the Vienna Logical Positivism, has some very interesting and positive things to say about metaphysics, alas neglected by standard histories.

[2] Recent philosophy of science has been alerted to the cognitive and/or methodological significance of non-literal language, in particular metaphor, for science from which it had been hitherto excommunicated. [2]

historical or sociological terms may not in the end be so revolutionary from an epistemological point of view: the core idea of representationalism is still present in both, where as in the alternative technoscientific view the novel approach has clearly kept its distance from that. It proved rather more authentically novel in that it shows current science to be itself more authentically novel when representationalism is dissociated from it.

Here is R. Merton's—notice: a sociologist's—repetition of the classic view of science as a theory-production mechanism: in his 1973 book *The Sociology of Science: Theoretical and Empirical Investigations* he singles out the following as the quartet of epistemic norms defining science proper: universalism-communal validation-disinterestedness-and organized skepticism. I mention Merton at this early point of my paper because his epistemological characterization matches his normative judgement and thus, curiously, he turns out to be an appropriate representative of a meta-methodological line of thought I am adopting here—though different from mine. We shall therefore have to come back to him below. Even before trying to spell out each of these different kinds of epistemic norms it is I believe evident that they all form together the ideal of scientific authenticity encapsulated in representational objectivity or universalism: "representational" because the aim is to represent in theory (or in the mind) an externally existing physical world, "objectivity" because ex hypothesi the representational ideal incorporates the medieval "adequatio rei et intellectus" thus demanding objective or some bona fide authentic mirroring of the external world by intellect's internal vehicles of representation, and "universalism" because the theoretical postulates of scientific theory-production that are present as a kind of pillar on which all kinds of theories are erected do not, and should not, be allowed to have any historical dimension or admit of social contingency. [1]

This is in a way a Platonic view that both logical empiricism and the Mertonian definition—malgré eux—promote. The Platonic view inherent in this construal of science as ahistorical and authentically representational—despite its explicitly un-platonic official pronouncements—nevertheless encapsulates the idea that the mind that produces the scientific theories is akin (vide the *Republic*), in its logical architectonic to the logical structure of the theories through which the equivalent logical structure of the external physical world is representationally reproduced. The language of mathematics is of course pivotal here. With the exception of Locke (perhaps the most original of empiricists) who warned us against conflating real substances with nominal essences of our our own making or representation, a warning not heeded by his modern logical offspring, the essentially rationalist optimism of a representational authenticity was shared by this view of science as an a-historical description of eternal verities (essences or natural kinds). It is crucial here to note that the representational ideal in this version includes already

---

[1] It should be noted in passing here, though the moral concerns come later in this paper, that universalism as a social ideal to the effect that science must be democratic in its application—if not in its direction as well, being under the watchful eye of the general public, thus damning expertise—is for some something accomplished by modern information technology thanks to all its everyday gadgets being in everybody's hands, more or less, unlike old heavy industrial technology the province of the very few only.

the notion that the representation must be authentic in itself, i. e. that there cannot be, or count as, a representation at all, something that falsely portrays the external physical world. Representation already means "correct representation"—and is thus truth functional in line with the tacit mathematical ideal behind all this. It should be emphasized that this is a crucial way of conceiving such an intentional quality of the mind as present only in the case of authentic knowledge—and this can be none other than science. Mathematics (geometry) is the model again. ①

An interesting case in point that subsumes the above latter part of the epistemic idea given in Merton's formulation is the fact that Astronomy started as a clearly mathematical discipline portrayed by a consciously mathematical model (classically in the case of Ptolemy) up until ontological questions, regarding the constitution and the background mechanism responsible for the way the planets move, began to be asked in the modern times. Would this be a case of de-mathematization and technologization of a science? Even then, nevertheless, the questions were still put in terms of clearly defined philosophical terms regarding causation that went back to antiquity, i. e. in terms independently of any concern with actual ontological make-up of the erstwhile firmament. Or so it seems. The debate on how to explain astronomical phenomena (for instance, as opposed to explaining movement in space) was carried out on grounds clearly required by the logical structure of notions borrowed from Aristotelian language such as "demonstratio quia" vs. "demonstratio propter quid" —in other words demonstrating a planetary phenomenon by giving an (or any, at least geometrically sound) explanation (model) that accounted for it, i. e. accounting for observed effects by offering possible models (more than one) of its possible cause—as against demonstrating from identified necessary causes to effects. The rise of modern astronomy is explained by some historians as the move from the former type of demonstration to the latter, or from a posteriori explanation to a priori explanation. What thus revolutionized this science was moving from, I would say, something resembling a mathematical uninterpreted model to an ontological quest. The movement from one type of demonstration to the other, intermingled with, or spurred by, the theological dogmas about how to understand a geometer God, in mutual conflict at the time, brought to the fore how explanation ought to proceed. But this move or radical change was guided by strict adherence to philosophical principles, in this case the Aristotelian distinction between a priori and a posteriori demonstration plus the medieval foundational idea of sciences arranged in a certain hierarchy, a higher one lending its principles to the lower: so mathematics could bequeath its own principles to a lesser science such as astronomy.

In all this I discern another case of the representational ideal. The dispute between those who were satisfied with a Ptolemaic strategy of explaining planetary phenomena by means of

---

① It should not surprise us that the search for such an authentic intentional (representational) object that would issue in real essences after a phenomenological epoche was also Husserl's ideal of thus founding a veritable science.

more than one possible models (quia) and those who pressed on for veritable (actual) cause→effect explanation (propter quid) may be seen to invite as interpretation the antithesis between the instrumentalist view of science vs. the opposite, realist, one. One however could question critically such an absolute separation of demonstrations (and hence this interpretative move). What is the exactly difference between demonstratio quia and propter quid? Ptolemaic epicycles do not explain in terms of the existence of this or that cause, ontologically speaking, but they are supposed to be simply offering geometer's models. We could maintain however—without reverting to instrumentalism—that a given Ptolemaic a posteriori explanation that the phenomenon e. g. of planetary retrogression (an optical illusion) is produced by the causes operative at the ontological background (the causal backdrop) which the Ptolemaic models of possible cause postulated does not completely sever ontological links—i. e. the postulated movements of the planets must have been regarded as real (at least "regarded"), hence the optical illusion; i. e. for the optical illusion to be the case, the postulated model must at least regard some segments of the causal background from which it draws its models as real. I point to this correction of the usual story[3], I briefly attempted here, as a correction to a simplistic espousal of instrumentalism as a model of what that kind of demonstration must always be following in order to strengthen further my more general interpretation that the idea of representationalism must be seen as covering the whole field of classical scientific methodology philosophically speaking.

Such an across-the-board faith in representationalism as the central feature any bona fide scientific theory must exhibit necessarily, present both in instrumentalism and realism spuriously contrasted according to my position, yields the further view of an ahistorical, universal, objective theory-producing science. Technology was then but an added extra from the outside. However, the epistemological disputes about the telescope that Galileo's contemporary opponents began should make us pause and reconsider this simplistic view of technology being an outsider even then. The challenge to technology voiced against him by putting forth very subtle arguments from optics in order to question the authenticity of the results collected from his novel instrument—now reappearing in the epistemology of Nanotechnoscience[4] though no one has noticed the kinship—are according to my position still disputes within representationalism. Even when we throw off the simplistic view of technology as an extra optional or science's "sidekick", something we should certainly do, even in those early modern times, it still remains the case that the way the disputes were carried out was always in terms of the epistemic ideal of mirroring nature that a theory-producing must incorporate. It is of course to be expected that even in such a slightly more sophisticated view of technology posing at least epistemological worries regarding what it delivers (as in the case of the telescope), discussions about ethical challenges that could possibly be associated with it or are found to actually stem from it are correspondingly less sophisticated. It is only when we view science as seamlessly fused with advanced modern technology, as technoscience, that the moral discourse becomes interesting.

Going back to the Mertonian formulation we find in it clear pronouncements on how to understand the epistemic norms identified as essential requirements: the first one, objectivity (corresponding to universalism) precludes particularism stemming from personal opinion or social considerations; the second norm refers to the fact that authentic science is a common inheritance, or better understood, the truths contained therein cannot be subject to obstacles barring their dissemination—here we should notice that this criterion can assume both an epistemic and a moral formulation: either that the findings of science have a universal validity depending on no human factors, i. e. they are invariably representing the physical world as it is in itself for all human beings (unlike the social world the principles of which permit variation without though precluding validity—as best expressed by Montesquieu), while in its ethical version the criterion of common possession appears to have an equally forcible normative dimension whereby it is proclaimed that science is the property of all people—something crucial for technology and its uses worldwide (also something crucial as far as the morality of patenting is concerned); the third epistemic norm "disinterestedness" in Mertonian terminology can be glossed over as the third basic aspect of scientific endeavour whereby intersubjectivity in the practices of validation in the pursuit of truth is the ultimate standard: here objective representationalism takes on the form of a set of self-contained legitimation processes whereby the norms to be followed so as to guarantee such an objective representationalism must be internal to science itself—here we have on the one hand a confirmation of my own interpretation above, namely that the ideal of representationalism bears the demand for objectivity inherently or "always already" in itself, and is thus self-legitimizing, while on the other hand it raises rather controversial issues regarding the danger of reflective or self-validating norms imposed by science itself; in the latter case the history of science has shown a remarkable ability in pushing this possibly question-begging nature of scientific confirmation under the guise of a philosophical questioning that assumes the role of an outside judge (Popper for instance was one of those who clearly discerned the problems in such self-validating procedure; the Aristotelian types of demonstration discussed above in connection to early modern astronomy were introduced as also such an example of philosophy lending to scientific methodology); finally the fourth epistemic norm, suspension of judgment or "organized scepticism" is connected to the third one whereby intersubjective validation rests on practices of controlled or healthy scepticim—i. e. rule-following and structured self-criticism—that allows for mandatory criticism both at the institutional and at the methodological (individual) level: empirical and logical criteria come to the fore as indispensable for any process where the pursuit of truth is a process that must be checked at all times (hence at no time there must be a guarantee that truth is arrived at). Again, here the indispensable logical props central to theory-production, the official view of what science is, according to this representational ideal are evident.

In all the above we encounter the ideal of science as conforming to the requirements that its essential nature, i. e. theory-production, must meet, i. e. that of universality, objectivity, intersubjective validation, and constant revisability (as long as the representational ideal is not justified). Behind this cluster of criteria conforming to the overarching ideal of science as theory-production there always lies the analogy with mathematics and proof-theoretic ideas. So in the end, on the official view, science as theory-production conforms to a quasi mathematically structured representational theory of the physical world.

What the major revolutionary turn away from this model of science to more or less its exact opposite did was basically to replace the mathematical structure and what it inspired in terms of what we have seen science above by that of history, i. e. science is now seen as a historical exercise, as a practice. Seen as such, is not theory any more but praxis. In that sense it has a history, i. e. it evolves, not so much in the sense of the previous model i. e. as approaching asymptotically towards truth, but as having its basic concepts in constant revision. Sometimes—according to one of the more revolutionary or nonrepresentational schools—its concepts are even revolutionary altered so as to have discontinuous jumps to incommensurable paradigms. In its original version the model of a sociologized science though viewed principally as a practice it was still a case of theory-producing. In some more recent versions there is no theory-production but a constant engagement by means of a praxis. In both cases the results are not eternal, objective, universal or whatever (though the ideal of constant revisability and scepticism are still operative) but something that is fallible. Not because it is a faulty or missed representation but because it is inherently embedded in historical, cultural and social practices and corresponding values of contingency and temporal relativity that mark it essentially (or "contaminate" it depending on how you look at it). Not a-historical, timeless, a-cultural (i. e. neither Chinese, nor European), not universal truth-producing but something embedded in human practices where even verifiability is something that needs to be discussed (and as we know from Quine by now, the analytic/synthetic distinction is equally suspect).

However my line of thought follows a different route—bypassing as it were the historical or sociologizing turn. I regard this alternative to the former official view—despite its acclaimed revolutionary target—as in the end a version of representationalism. And the recent "wars" waged by and against postmodernisms of all kinds on relativism and antirealism or irrealism are equally fought, to my mind, on the grounds of the representationalist ideal. I regard on the contrary as more true to the current trend characterizing science its fusion with technology thus bypassing the above considerations and underlining the contrast instead as, not one between ahistorical objectivity vs. historical paradigms, but rather as one between mathematized science vs. technologized science or technoscience. What, to my mind, is revolutionary in this interpretation is that the radical incorporation of technology into science allows the emergence of a distributed kind of cognitive system (similar to that in the philosophy of mind) whereby the advanced technological tools assume for themselves part of the cognitive process: they become tools of knowledge. Glossing over Heidegger's dictum—but going back to Bacon really, something that Heidegger missed, we may say that: contra to the by now disgraced naïve view of technology as simply applied science or merely tools embodying pre-existing science, what we have always had is the setting up of (modern) science in such a way so that the way it questions (or probes, Bacon would say) nature is carried out in a way that nature itself gets viewed technologically viewed. Nature is seen as technological.

II

A grand statement favouring representational theorizing was that of Gassendi's: "if we know anything, we know it by mathematics."[①] Is this replaced by modern Technoscience? -a term we owe to Lyotard that captures this total merging of the two fields wrongly conceived as separated i. e. that of science vs. instrumentation? Techno-science has replaced both pure and applied science (without any intermediary in-between) by an epistemic activity that privileges instrumentation and technical engineering. The latter are not simply tools. They are the privileged routes towards disclosing the properties of whatever constitutes ultimate reality. They are epistemological engines. So nowadays we, as technoscientific versions of monoscientific overconfidence, may well assert, in equally unconditional terms as Gassendi, that "if we know anything, we know it by techno-science". Technological powers are not only detected behind merely beneficial instrumental advances; rather, they are principally seen as powers of ultimate ontological clarification, and, therefore, are as such presupposed by the success of mere instrumentation. There is this peculiarity about them: they are instruments of knowledge which, however, coincide with knowledge for, and by means of, instruments. And this further coincides with the power to transform nature.

Understanding Techno-science is not unrelated to pressing ethical and social issues that stem from it. The starting point is that no one should be so naïve as to hold on to erstwhile simplistic views forgetting that technology cannot be self-critical. Technological secularization's dominant mode of thinking contains an ingrained inability to equip us with critical values, both of what itself does as well as what of happens in other spheres of human endeavour. This point must be linked with the critical observation that technology and science have succeeded in imposing their own principles on what were supposed to be technologically uncontaminated spheres of social or intellectual life; spheres where reasoning along technical lines of maximization of efficiency has always been considered as quite inappropriate and distorting in its effects. This is particularly so when technology is seen as effacing all types of distinct authenticity expected to define the specific, separate, character of these other spheres such as, education, the production of art, professional ethics, and so on. In certain areas, as in biomedical technology (organ transplants, so-called transhumanist extensions of bodies by means of nano-machines, and the like.) new

---

① As a further confirmation of my general interpretative strategy we may note that, historically, the initiators of modern philosophy of science (Duhem, Mach, Poincaré, et al.) as practicing scientists and mathematicians could not fail to be self-influenced by their profession.

advances in Techno-science are expected to create wholly new attitudes and unforeseen ethical responses to relevant questions otherwise never posited before.

The crucial question here is, though, whether Techno-scientific epistemological engines, taken as the dominant mode of thinking, tend to shape their own ends and purposes. Technologized science self-propels itself into a position of generating important evaluative human ends. This runs counter to a standard view of any technique involved in human affairs, expected to be merely a vehicle of application and a neutral instrument for the advancement of desirable human ends themselves posited in other spheres and given to technique only to implement them. Such a self-determining mode of thinking and acting is considered deplorable for it blocks an authentic way of being human—if such thing exists.

There is I claim a shift of techno-scientific knowledge towards the centre of social life displacing of straightforward economic conditioning from a determining position. It's not simply economics and nothing else anymore. The implication is that Techno-scientific knowledge becomes somehow "autonomous". Does then technologized science cease to be propelled by economic forces? Such economic forces become now endogenous to technoscience-driven knowledge and computer information-based industries. It is a general aspect of all knowledge that it generates its own economic potential and values—unleashing economic powers to be appropriately channelled and exploited. The same pattern seems to be present here too. But in this case the important difference lies in the fact that this type of novel form of knowledge qua epistemological engine generates the only economic potential available or at best the most dominant one. This is something quite evident in the case of the special nature and place of advanced information and communication technologies performing the role of indispensable substrate or prerequisite of all other types of advanced Techno-science. At the same time politically, not only economically, there is the indisputable fact of a special, separate institutionalization of the science-establishment which attests to the further fact that Techno-science has by no means been domesticated by the state.

So epistemological vs. political conclusions are to be drawn. As a preliminary result, there is an overall sway of Techno-science both over economic as well as state-related or political forces. This double predominance could not have happened otherwise. For this specially technologized science could not have predominated over the former without also doing so with respect to the latter type of forces; it could not domesticate economic factors leaving political structures untouched. But accepting all this—however tentative—necessitates a corresponding re-examination of the question of values, which such drastic technoscientific developments engender. We must accept the possibility that setting, discerning and delineating such a special set of values cannot be of an instrumental kind nor can it be the case that such appraisal and evaluation could be left lying outside the Techno-science's own domain. If Techno-science has indeed become a state-independent social institution, has spread so much and in such a non-instrumental fashion so as to underlie as an indispensable basis the erstwhile most central state-related elements of social structuring, it cannot then be the case that Techno-science can allow anything outside itself to provide the ground of its value; for there is nothing outside it. Its social predominance together with its logical indispensability attest to the fact that all values related to Techno-science cannot be but its own products. Hence the ensuing circularity of any kind of hackneyed explanations in terms of placing the value of science upon the use made of it by society, having the latter based in its turn on society's own values themselves the product of (at least) Techno-science. If Techno-science is powerful enough to posit its own questions, and given its centrality and logical indispensability, it can indeed provide post-industrial societies with a new ethos (elements of which, e. g. , are the subversion of the idea of delayed satisfaction, or the speed of technical responses as ultimate and unique solutions to socio-political, international, problems—information technology setting the triumphant tone underlying biotechnological research). Far from being a passive recipient of already existing social values with a mandate to have them transformed and embodied into technical appliances, technologized science is itself a new ethos.

Some have voiced the concern that society in the grip of advanced technologies may be devoid of any normative ideals or transcendent meanings by means of which people can make sense of their lives at the normative level. Let us call this possible predicament: nihilistic technologized self-absorption. If such were the predicament of modern sophisticated Techno-science, namely to be absorbed into the instrumentality of social needs and political aims or conform to economic imperatives, then obviously society is left with no external-to-itself axiological reference point (religion apart) from which its members could possibly draw meaning. I believe to the contrary that once we grant due recognition to the full extent of the special nature and power of Techno-science as a self-positing axiological system relatively independent of the social, political and economic structure, then we can envision Techno-science as such an outer reference point.

In a prescient analysis (drawing on Merton's epistemic norms which I used above for different purposes) R Sassower[5] pinpoints as pivotal the role that rationality plays in the eventual breakdown of the presumed self-policing that the scientific community is supposed to exercise-given these epistemic norms on which it was erected and from which such an ethical role is expected to follow. For instance objectivity and (in my sense) representationalism in science may well lead to "depersonalize scientific activities... entail [ing] unforeseen consequences" or (he borrows Z. Bauman's formulation) "emancipating the desiderata of rationality from interference of ethical norms or moral inhibitions. And this is crucial for technoscience if it turns out, given my analysis above in this section, to actually entail such dangers as a "separation between science and ethics" [5]:7-8. If public control of advanced technoscience is relegated to the experts (giving rise to how I pictured technoscience's usurping of both the political and the economic domains) then the obvious deficit in democratic control can be seen to reflect the unlimited faith in the rationality of representational ideal I started with. There is then an epistemic cause behind the current ethical concern and its political consequence (of technoscience's unaccountability). But it is even more so when we think of it in terms of the alternative paradigm, that of epistemological engines. If the authority of representationalist view of science as rational theory-producing was limited to that of an expertise that would allow political control over it on grounds of moral concern from outside science itself, the totalizing effect of technoscience captured in the model of epistemological engine by giving rise to the possible lack of any outside or "transcendent realm of values" (a possibility I discussed above), then the political control operative in the other paradigm may not be forthcoming here at all. I do not concur in holding responsible epistemic representationalism's rationality and its value-free ideal ("science should be free to investigate everything"). Rather I see the danger in the new paradigm of an epistemological engine: for it is the latter that lacks any components comparable to one of those we saw above in the form of one of the quartet of epistemic norms, "organized scepticism". Seen as "engine of knowledge" and not as a "theory" technoscience cannot even avail itself of a modicum of Enlightenment-like self-critical attitude. Curiously the old paradigm was, to my mind, less prone to ethical silence (though its untamed pursuit of "innocent" rationality was responsible for the horrific crimes to humanity in the name of science). Should you construe science as theory-producing and in addition come up with the well-known problems of building a widely accepted logic of confirmation (or just relying on Bayesian probability for strengthening inferences to the best explanation as science's best strategy) then inevitably science limits its own predictability on logical grounds. And this may be good for its being controlled by morality. But if no such logical grounds are any longer the case, but rather technoscientific epistemological engines become the norm of the new phenomenon of *converging technology*[6] where the fusion of nano-bio-information-cognitive-artificial intelligence technology defines knowledge, then such total convergence (vide e. g. machine consciousness[7]) and its total applicability to all parts of human (and even post-human) life, deflects any outside possible centre of ethical control.

References

[1] Ihde D. Epistemological Engines. Nature, 2000, (406)

[2] Kaldis B. Metaphor and Science. In: Kairos. A Journal for the Philosophy of Science (Universidad Lisboa). 2011

[3] Barker P. The Role of Religion in the Lutheran Response to Copernicus. In: Osler M J ed. Rethinking the Scientific Revolution. Cambridge: Cambridge University Press, 2000. 59-88

[4] Kaldis B. Epistemology of Nanotechnology. In: Encyclopedia of Nanotechnology and Society SAGE. 2011

[5] Sassower R. Technoscientific Angst: Ethics and Responsibility. Minneapolis: University of Minnesota Press, 1997

[6] Kaldis B. Converging Technologies. In: Encyclopedia of Nanotechnology and Society SAGE. 2011

[7] Kaldis B. Just Machines or Automatic Angels? Machine Consciousness and Moral Status. In: Lin P et al. eds. Robot Ethics. Cambridge: The MIT Press, 2011

A Shift in Interests to Science and Technology in the 11th China

Su Zhan

(The Institute for the History of Natural Science, CAS)

The Song Dynasty (10th to 13th century) is an important Dynasty of China which has been known with a lot of scientific and technological achievements. This paper investigated the shift in people's interests to science and technology in the Song Dynasty during 1001 to 1120 by a statistic to the *Dictionary of Chinese Names*. According the statistic, people's interests to science and technology were still declining since 1040s. And it was just the time when the famous Chinese philosophical school—well-known as *Songxue*—rose. Writer will analyse how and why the decline occurred in the Song Dynasty. Furthermore, writer tries to study the social tendency in the 11th century in the Song Dynasty and the academic trend of *Songxue* to discuss the relation between the former and the opinion to science and technology of people.

宋人科技兴趣的计量研究

苏 湛

(中国科学院自然科学史研究所)

宋朝是中国历史上的一个重要王朝，其在中国科技史上的重要地位尤其引人瞩目。本文选取北宋咸平四年（1001）至宣和二年（1120）之间的120年历史作为研究对象，以《中华人名大辞典》为主要资料来源，对活跃于这一时期的北宋社会精英群体进行计量研究，以分析这120年中北宋精英阶层科技兴趣的变化情况。本文将指出，北宋精英阶层对科技的关注自11世纪20年代（约宋仁宗天圣年间）以后就一直在持续下降，并尝试对可能造成这一现象的原因进行了讨论。

北宋是中国科技史上的一个重要时代。李约瑟曾指出："每当人们在中国的文献中查考任何一种具体的科技史料时，往往会发现它的主焦点就在宋代。不管在应用科学方面或在纯粹科学方面都是如此。"[1]另一方面，中国传统科技在北宋以后的迅速衰落同样引人注目，以至于对这种衰落及其原因的探究甚至形成了被称为"李约瑟问题"的专门问题域[2]。

美国科学社会学家罗伯特·金·默顿在研究17世纪英国科学的异军突起时曾指出，17世纪英国科学的加速发展与当时英格兰社会精英们的职业兴趣转移之间存在着关联。当时对科学产生兴趣的人数明显增加，而诗歌、宗教等领域所吸引的人数则明显减少[3]。如果这一发现具有普遍意义，那么同样有理由相信，在宋代科技的兴衰与宋人职业兴趣的变化之间，也可能存在类似的关联。本文将模仿默顿的研究方法，对这一猜测进行验证。

1. 资料来源

本文选用的资料来源是商务印书馆1921年出版的《中华人名大辞典》[4]（以下简称《大辞典》）。该书可以视作英国《国民传记辞典》在中国的对应物。它是中国第一部专门的人物传记词典，并且至今仍是关于中国历史人物的最全面和最优秀的传记词典之一。同时，为避免讹误，笔者还参照《宋史》[5]《隆平集》[6]、《续资治通鉴长编》[7]（以下简称《长编》）、各种墓志、行状、答和诗文等原始文献对所涉及的记录进行了逐条考证，同时参考了《宋人传记资料索引》[8]等今人研究成果，最大限度地确保史实的准确性。

2. 数据处理

2.1 基本处理

《大辞典》的正文和补遗共收录中国历史人物45 000余位，我们从中筛选出在1001年至1120年之间步入职业领域的2920人①，建立数据库，以每10年为单位划分为12组，并根据当时的分类习惯，将入选者的职业兴趣划分为政治、军事、经术、史地、文学、艺术、宗教、方技8个领域，统计出每10年中涉足过各领域的人数。对于涉足过不止一个兴趣领域的人物，都遵循默顿的处理方式，将其同时记入所涉足过的每一个领域。另外将其中参与过各种科学技术相关活动的人物单列为一表，并根据各项活动在现代科学技术体系中的位置进行分组。同样，对于不止参与过一种科学技术活动的人物，都将其同时记入所涉足过的每一个领域。此外，对于一些不足以被称为"职业兴趣"，但在很大程度上反映着宋代社会风气变迁的事迹，如积聚财富（或者按照古人的说法称为"治生"）、教育、各种美德，以及一些自称或被认为的超自然事件，也分别进行了统计。（表1、表2）

---

① 在《十七世纪英格兰的科学、技术与社会》中，默顿以每个人"初始兴趣发生的大致时间"作为时间划分的依据。但对于本文而言，这种资料几乎是不可获得的，相反，本文涉及的大部分人物步入自己职业领域的初始时间（以下简称"入职时间"）都是可知的。

表1　1001~1120年间北宋社会精英职业兴趣领域的转移（一般领域）

时间	政治		军事**		经术		史地	
	人数	比例*/%	人数	比例/%	人数	比例/%	人数	比例/%
1001~1010	73	59.35	39	31.71	24	19.51	20	16.26
1011~1020	96	62.34	63	40.91	36	23.38	17	11.04
1021~1030	142	65.44	73	33.64	57	26.27	43	19.82
1031~1040	142	65.74	87	40.28	50	23.15	27	12.50
1041~1050	134	52.76	64	25.20	62	24.41	22	8.66
1051~1060	130	55.32	53	22.55	69	29.36	24	10.21
1061~1070	145	53.51	49	18.08	81	29.89	34	12.55
1071~1080	130	50.19	48	18.53	66	25.48	26	10.04
1081~1090	130	49.43	46	17.49	66	25.10	19	7.22
1091~1100	169	55.05	66	21.50	62	20.20	31	10.10
1101~1110	172	53.92	91	28.53	63	19.75	36	11.29
1111~1120	176	58.28	107	35.43	72	23.84	38	12.58

时间	文学		艺术		方技***		宗教	
	人数	比例/%	人数	比例/%	人数	比例/%	人数	比例/%
1001~1010	38	30.89	24	19.51	9	7.32	15	12.20
1011~1020	50	32.47	27	17.53	13	8.44	12	7.79
1021~1030	79	36.41	50	23.04	20	9.22	25	11.52
1031~1040	66	30.56	42	19.44	17	7.87	21	9.72
1041~1050	83	32.68	45	17.72	15	5.91	22	8.66
1051~1060	87	37.02	50	21.28	21	8.94	24	10.21
1061~1070	95	35.06	67	24.72	18	6.64	36	13.28
1071~1080	83	32.05	51	19.69	19	7.34	24	9.27
1081~1090	79	30.04	60	22.81	13	4.94	23	8.75
1091~1100	87	28.34	74	24.10	13	4.23	27	8.79
1101~1110	97	30.41	65	20.38	12	3.76	34	10.66
1111~1120	90	29.80	56	18.54	10	3.31	18	5.96

* 表中的"比例"指各时段中进入此领域的人数与《大辞典》中属于这个时段的总人数之比。

** 既包括职业军人，也包括管理过军务的文官，以及在军事理论上有建树的人物。

*** 包括天文步推、医药炼丹等，参见《宋史·方技传》。

表2 1001~1120年间北宋社会精英职业兴趣领域的转移（科技和其他）

时间	科技（全部参与者）		各学科参与者人数						
	人数	比例/%	天文	地理	生物	物理	农学	技术	工程
1001~1010	36	29.27	4	9	1	5	4	5	18
1011~1020	54	35.06	4	4	1	7	9	11	34
1021~1030	77	35.48	7	13	4	14	13	15	51
1031~1040	75	34.72	8	8	1	7	13	14	45
1041~1050	67	26.38	8	10	4	2	16	13	42
1051~1060	65	27.66	5	7	7	7	14	10	40
1061~1070	56	20.66	5	4	2	5	4	11	30
1071~1080	53	20.46	3	4	3	6	10	8	26
1081~1090	39	14.83	1	8	2	1	5	10	11
1091~1100	41	13.36	6	7	3	4	6	10	19
1101~1110	49	15.36	2	6	3	1	7	14	27
1111~1120	49	16.23	4	8	3	2	9	10	21
总计	661		57	88	34	61	110	131	364

时间	各学科参与者人数（续）					特殊事迹人数			
	化学	数学	医学	气象	其他	治生	教育	美德*	超自然
1001~1010	2	3	5	1	1	4	7	7	3
1011~1020	—	4	9	4	9	3	9	6	4
1021~1030	1	3	8	4	3	2	15	8	4
1031~1040	2	5	12	2	8	—	16	5	5
1041~1050	1	4	9	2	2	3	12	13	9
1051~1060	3	4	13	—	3	5	10	12	3
1061~1070	3	4	8	—	1	1	15	17	4
1071~1080	1	4	13	2	2	4	8	15	1
1081~1090	3	3	10	—	1	1	13	13	3
1091~1100	2	5	4	1	3	—	9	15	3
1101~1110	1	3	9	—	3	3		19	5
1111~1120	1	3	4	—	4	4	6	13	—
总计	20	45	104	16	40				

*对一般政治人物或文人的程式化的溢美之词不计算在内。只计算那些单纯因美德事迹（如孝行、诚信）而入选《大辞典》的人物。

2.2 分组依据

本文划分职业兴趣领域的主要依据是《宋史》的列传分类、传统"四部"分类法（以宋仁宗《崇文总目》[9]和清《四库全书总目》[10]为参照），和南宋郑樵的"十二类"分类法[11]。如某人的作品在"四部"或"十二类"分科法中被列入经部，或其人按照当时的标准在此领域中有所贡献,则将其作为经术领域中的一员。

相对于一般职业兴趣领域的划分,科学技术活动的认定和分类相对复杂。甚至在讨论古代问题时,是否能够使用"科学技术"这一指称,都存在不同意见。然而无法否认的是,在古代确实存在着一些与现代科学技术有关的东西,归纳起来大致有以下四种情况:

(1) 一些学说,其观点与现代科学观点具有某些一致性(或曰"正确")。比如被认为包含了近代天文学中无限宇宙理论的"宣夜说"天文理论。

(2) 对某种自然现象的正确记录和描述。比如中国古代著名的超新星和日全食记录。

(3) 与现代科学活动有着相同旨趣,即以了解自然、理解自然、解释自然为目的而进行的活动。如沈括针对雁荡山、磁针等自然界对象进行的研究。

(4) 与现代科学有相同研究对象或使用某些相同的手段、工具的活动。比如与现代天文学有着千丝万缕联系的古代占星学。

以上四种知识或活动,就是本文在使用"科学技术"一词时所描述的概念范畴。使用这个词并不意味着笔者不理解它与现代"科学技术"之间的区别,而是为了强调这种古代"科学技术"在人类知识积累过程中的位置。诚然,现代科学诞生于欧洲,但欧洲科学也并非从天而降,它同样建立在古希腊、古罗马以及中世纪学者们遗留下来的以上四类知识的积累之上。在一定意义上,这四类知识的积累程度,以及相关活动的繁荣程度(至少是其中某类或某几类),直接影响了欧洲近代科学的产生。而中国人对这四类活动的态度有何不同,这正是本文所要讨论的。同样,在对不同的科技活动进行细分时,本文也以各种活动与现代科学知识之间的继承关系为准,把这些活动划分到现代学科体系的分类中去①。

2.3 对误差的讨论和修正

如默顿曾经指出的,资料来源中的倾向性可能影响计量结果的准确性[3]。为了增强结果的说服力,必须尽可能的对各种倾向性进行规避。

我们面对的倾向性主要来自职业、年代两方面。其中职业倾向性是默顿曾经讨论过的。但是鉴于一人"身兼数职"的情况在我们的研究中远比在默顿的研究中普遍,在本文中,职业倾向性可能有一个新的疑问:我们的计量结果最终反映的会不会仅仅是某一类人,比如官员,价值取向的改变呢。

这当然很有可能,但笔者认为这并不妨碍统计结果对于北宋社会精英阶层职业兴趣变化情况的指示意义——无论构成这一阶层的主体是什么,他们都是他们所处社会的主流价值观的最直接的代表。从这个角度说,即便考虑到职业倾向来的不利影响,统计结果也仍然是有意义的。

年代倾向性的问题在默顿的研究中没有提及,但在本文中却不容忽视。事实上入选《大辞典》的人数随着年代推移一直在迅速增加。这可能部分与人口的自然增长有关,但更可能与历史学的发展和史料档案的自然损耗有关——越靠近现代,历史学越发达、史料散佚越少,留下历史记录的人物就越多。针对这一问题,本文采用百分比修正法,计算每10年内进入某领域的人数在同时代入选者中的百分比②,这就显示出了该职业领域在社会

---

① 这种权宜之计并非首创,而是借鉴了包括李约瑟《中国科学技术史》在内的很多著作的实际做法。
② 需要强调的是,这个百分比数与默顿论文中的百分比含义是完全不同的。

中相对社会地位的变化情况。

除了以上两种倾向性,还有一个问题来自针对科学技术活动的统计:参与过工程活动的人物在全体科技活动参与者中所占的比例实在太高了,远远超出其他科技活动数倍。这必然导致这样的质疑:本文对宋人科技兴趣增减情况的描述,是否仅仅反映了宋人对工程建设的热情?

因此本文进行了如下验证:将那些仅仅因为参与过工程建设项目而被记录在案的人物从科技活动的名单中剔除,对剩余的数据重新统计(图1)。经过验证,可以看到,修正后科技兴趣指标的起伏情况不但与修正前完全一致,而且有趣的是,它与工程兴趣指标的起伏情况也高度一致。这说明,我们测到的宋人科技兴趣的变化情况,并非是由工程活动这一个领域单方面决定的。但他们对科技活动的热情,与对工程活动的热情,在时间分布上确实具有高度的一致性。

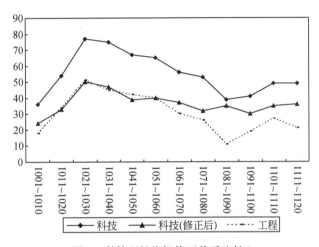

图1 科技兴趣指标修正前后比较*

*为说明工程兴趣指标与修正前后的科技兴趣指标间的关系,我们将1001~1120年间参与过工程建设工作的人数变化用虚线在图中标出

3. 结论与分析

3.1 基本结论:宋人科技兴趣的下降

统计结果显示了一个出人意料的结论:宋人对科技活动所表现出的兴趣自从在11世纪20年代短暂地达到最高值以后,就开始了迅速且持续的下降过程。而且在大部时间里,这种下降已不仅仅是比例上的,甚至在绝对数量上也出现了明显的下降——特别值得注意的是,这是在统计样本的总数迅速上升的情况下发生的。只是到了12世纪的头20年,这一情况才有所好转。通过分析相关科技活动参与者的事迹,可以看出,这20年中科技活动参与者数量和比例的回升显然与宋徽宗大兴土木和提高医学、算学地位的政策,以及建炎兵兴后百废待兴的社会环境所提供的发展机会有关。但即便在这种双重作用下,这20年间的科技人才产出率(无论从数量上还是从比例上)也仍然没有恢复到1071~1080年以前的水平。

3.2 结论的可靠性

作为一项基于不完全归纳的研究方法，如本文所使用的这种统计方法可能带来的谬误是显而易见的。特别是在经历了千年的朝代更迭，很多必不可少的信息已不可逆转地丢失了的情况下。

可能影响本文结论可靠性的因素首先来自资料来源本身。正如前人指出的"在历史上留存下来的资料中，主要是关于有身份、有地位的人们（即'精英'们）的记录，在社会系统中，身份和地位越低的人，文献记录就越不完整"[12]。因此，正如本文一开始就指出的，《大辞典》收录的人物，仅仅是中国各个时代社会精英阶层的代表，而并不是对整个社会各个阶层的全景式展现。这也就解释了为什么基于《大辞典》统计出的北宋精英阶层的科技兴趣自11世纪20年代起就一直在降低，而《中国科学技术史·年表卷》记载的北宋科技成果数量在70年代才到达高峰[13]。这一事实最好地证明了大量无名的科技工作者（尤其是技术工人）的存在。

其次，生活在一千年前的编史家们对记录对象的选择也在很大程度上影响着我们的统计结果。默顿曾经假设，对于固定的编纂者，我们"没有理由猜测"他们"在不同职业领域的相对重要性方面的态度上会有任何可察觉的变化"[3]。然而问题在于，《大辞典》所收录的人物并不是由同一批固定的编纂者所决定的，而是根据这些编纂者所能找到的史料所决定的，而这些史料则来自前后相隔一个世纪的好几代信奉不同学术观点和价值风尚的历史学家的记录。因此，我们的统计结果所反映的就很可能并不是北宋社会精英的职业兴趣变化，而仅仅是北宋历史学家们的兴趣变化。

这两种质疑都在很大程度上是正确的，但即便如此，作为北宋社会主流价值观变化的一个指标，我们的统计结果仍然是具有参考意义的。正如我们所知，一个社会的主流价值观乃是由这个社会中精英阶层的价值观所引领的。而在中国古代社会中，历史学家这一群体本身作为社会精英阶层的成员，更是社会主流价值观的最重要的代表。因此，即使统计中所显示出的变化真的主要是由历史学家导致的，这种变化所折射出的社会价值观方面的改变也是真实的。

更何况，精英阶层的兴趣转移对整个社会的影响绝对不会仅限于心理层面。在一定程度上，社会精英阶层本身必然是一个社会中最聪明、最优秀人物的聚集之所。因此社会精英阶层的兴趣转移很大程度上意味着这个社会中最优质的智力资源的转移。对于任何一个领域来说，如统计所示的这样剧烈且持续的智力流失，对其可持续发展能力所造成的打击都将是沉重的。事实上，这一推论同样可以得到某些统计数据的支持①。

当然，足以对本文的结论构成质疑的理由还不止于此。比如我们在时间上和职业领域上的分组方式，以及与史料散佚等因素有关的偶然性问题。但需要指出的是，我们并不是根据任何微弱的变化来作出结论的，而是一种非常令人印象深刻的显著变化。很难假设这样一种显著的变化是偶然发生的，而没有其他可理解原因。至于改变时间分组方式是否会

---

① 根据金观涛等人的研究，将中国古代所有科技成果按重要性加权评分后进行计量，发现这一计分值在11世纪上半叶达到整个中国古代史上的最高值2000分，而11世纪后半叶的计分值只有前者的一半左右[2]。尽管从《中国科学技术史·年表卷》看，这后50年的科技成果在数量上要更多些。显然，尽管宋人的创新能力在1020年以后没有立刻发生下降，甚至还略有提高，但作出重大创新的能力却下降了。

导致计量结果的改变这样的质疑,我想统计中所显示出来的变化的持续性,本身已经提供了一个很好的回应。

3.3 宋人科技兴趣下降的可能原因

默顿在《十七世纪英格兰的科学、技术与社会》一书中,将17世纪英国科学技术的发展与宗教、经济和军事三大社会因素联系在一起。不过默顿命题中的后两个显然无法为11世纪在北宋发生的刚好相反的情况提供合理的解释。从经济角度说,北宋不但没有出现过大的经济萧条(即使在靖康前后),而且还是中国历史上为数不多的商品经济可以相对自由的发展的时代。在军事上,尽管北宋国内一直相对和平,但在西北战场上,宋军与西夏间的相互攻伐自宋仁宗以来几乎从未停止。而宋军武器的科技含量之高,北宋朝廷在武器研发上投入的力量之大,也足以给人留下深刻的印象。事实上,《中国科学技术史·年表卷》中所记载的宋人在11世纪中期以后所取得的成就中,有相当一部分都要归功于军事活动和民间工商业活动。

那么,默顿所提出的宗教或意识形态变革影响科技兴趣和科技发展的理论是否能够解释宋人科技兴趣的衰落呢?

11世纪40年代是回答这一问题的一个关键点。假设我们画一条横线穿过标识宋人科技兴趣的百分比曲线的顶端,当我们慢慢地把这条线向下移,我们会发现,当移到某个位置的时候,横线与曲线之间刚好只有一个交点。而这个唯一的交叉点首次出现的地方,恰恰是在11世纪40年代的位置上。它把整个变化曲线一分为二,在40年代以前,是科技参与者比例一直高于水平线的时代,而在40年代以后,是科技参与者比例一直低于水平线的时代。而且这一现象不仅仅存在于科技活动领域,在关于教育活动参与者比例的统计中也存在同样的现象。而在军事领域,到1110年为止,所有统计数据也符合这一情况(只是在1111~1120年之间,军事兴趣百分比数才勉强回升到水平线之上,而这显然与"靖康之难"造成的筛选效应有关)。唯一与它们刚好相反的是关于美德的记载。在1040年以前,除了1001~1010年间的相关记载比例较高,其他3个10年中仅仅因为美德而被载入史籍的人物比例没有超过4%,而在1040年以后,则从未低过这一数字。

那么在11世纪40年代发生了什么呢?在这10年间确实发生了宋史上最重要的事件之一——庆历新政。而不容忽视的是,这同时也是历来被公认的"宋学"正式形成的时间点[14]。

关于宋学精神的特征及其与前代儒学之异趣,前辈学者多有论述,归纳起来,主要包括怀疑主义、经世精神、性理之学和崇尚秩序等项[14,15]。乍看之下似乎令人有些费解,因为很难看出宋学的这些精神气质对科学技术会有什么破坏性。怀疑主义历来是知识发展的动力;经世精神则似与新教的功利主义有相通之处;宋学对心性、义理的强调也似乎符合清教伦理崇尚清苦、勤奋的价值取向。事实上有学者恰恰根据宋学的这些精神,断言宋学对科技所起的作用是积极的。[16]

然而我们的调查却提供了对宋学不利的证词。除了宋人科技兴趣的下降与宋学兴起这两个事件在时间上的一致性,对北宋各大儒学学派的调查同样支持了这一结论。在北宋中后期最著名的四个相互竞争的儒学派别——司马光朔学、王安石新学、三苏蜀学和二程洛学中,参与过科技活动的人物的比例分别是27.27%、23.53%、30.00%、17.86%,洛学最低，新学其次。而后期宋学正是以这两个学派为主要基础发展起来的[14]。

那么导致宋学精神与新教伦理在科技态度上分歧的根本原因是什么呢？这种差别很可能植根在二者更深层次的教义里。比如，科学史家霍伊卡曾经指出，基督教教义为手工劳动赋予的神圣性在现代科学的兴起中起到了重要作用[17]。由于手工劳动被认同为一种"赞颂上帝"的可取手段，因此对于新教徒而言，相关的各种活动，包括科学实验、技术发明等，就都成了值得付出刻苦和努力的活动，得到认同。

而儒学对手工劳动的态度恰恰相反。著名的"樊迟学稼"的故事充分说明了儒家的这一态度：

> 樊迟请学稼……子曰："小人哉，樊须也。上好礼，则民莫敢不敬；上好义，则民莫敢不服；上好信，则民莫敢不用情。……焉用稼？"[18]

单就这段论述本身而言，孔子敏锐地指出了国家强盛的根本不在于技术因素，而在于统治者对待人民的态度，直指人心，发人深省。然而当这种教条渗透到价值观中，它所表现出的对劳动的鄙视就开始显示出副作用了。

经世是目的，但直接去从事具体的工作却是不可取的，因为这种方法是有局限性的，且不能治本。通过"向身上做功夫"，解决道德这个根本问题，这才是唯一值得努力的。这就是宋学家们的逻辑。有人将这种倾向的一系列外部表现总结为"唐宋变革中的道德至上倾向"[19]。这也解释了1040年以后，有关美德的记载增多的原因。

因此，古代的中国于17世纪的英国存在着完全不同的情况。在英国，清教的勤奋刻苦精神可以成为将身心奉献于科学的动力；而中国的情况恰恰相反，科学惨淡的萌芽只有在贵族与文人的闲情逸致中才能幸存，而那些正统的学者们，他们的刻苦只是让科学离他们越来越远。

参 考 文 献

[1] 〔英〕李约瑟．中国科学技术史．第一卷．《中国科学技术史》翻译小组译．北京：科学出版社，1975．287
[2] 金观涛，樊洪业，刘青峰．文化背景与科学技术结构的演变．载：刘钝，王扬宗．中国科学与科学革命：李约瑟难题及其相关问题研究论著选．沈阳：辽宁教育出版社，2002．326-393
[3] 〔美〕默顿 R K．十七世纪英格兰的科学、技术与社会．范岱年，吴忠，蒋效东译．北京：商务印书馆，2000．41-63
[4] 臧励和等．中国人名大辞典．上海：商务印书馆，1921
[5] 〔元〕脱脱等．宋史．北京：中华书局，1977
[6] 〔宋〕曾巩．隆平集．文渊阁四库全书
[7] 〔宋〕李焘．续资治通鉴长编．文渊阁四库全书
[8] 昌彼德，王德毅，程元敏等．宋人传记资料索引．台北：鼎文书局，1986
[9] 〔宋〕王尧臣等．崇文总目．文渊阁四库全书
[10] 〔清〕纪昀等．四库全书总目．文渊阁四库全书
[11] 左玉河．从四部之学到七科之学——学术分科与近代中国知识系统之创建．上海：上海书店出版社，2004．59-62
[12] 刘兵．关于科学史研究中的集体传记方法．自然辩证法通讯，1996，18（3）：49-54
[13] 苏湛．11世纪中国的科学、技术与社会．北京：北京师范大学哲学与社会学学院，2009．41-42

[14] 漆侠. 宋学的发展和演变. 石家庄：河北人民出版社，2002.7-16，291-299
[15] 宋晞. 论宋代学术之精神. 载：张其凡，范立舟. 宋代历史文化研究（续编）. 北京：人民出版社，2003.109-119
[16] 乐爱国. 宋代的儒学与科学. 北京：中国科学技术出版社，2007
[17] 〔荷〕霍伊卡. 宗教与现代科学的兴起. 第二版. 邱仲辉，钱福庭，许列民译. 成都：四川人民出版社，1999.100-102，111-117
[18] 〔春秋〕孔子. 论语章句集注. 北京：中国书店，1985
[19] 严耀中. 唐宋变革中的道德至上倾向. 江汉论坛，2006，(3)：104-106

Traditional Chinese Science Among Vietnamese Minorities: Preliminary Results

Alexei VOLKOV

(Tsing Hua University, Taiwan)

1. Introduction

General works by colonial French scholars, such as P. Huard and M. Durand[1,2], provided a very general description of the history of conventional sciences in Vietnam while avoiding considering as scientific the topics related to the traditional science (in a broader sense), that is, including such "unorthodox" disciplines as astrology, divination, traditional and "magical" medicine①, geomancy (风水), etc., often dubbed as "pseudo-sciences". These disciplines, however, are of considerable interest for the studies on the history of science and on the history of interaction between science and religion (in particular, Daoism). The history of the "pseudo-sciences" in Vietnam, however, has never been systematically studied, even though these disciplines occupy a prominent place in the extant texts on traditional sciences.

The first results of my work in the framework of a research project on the history of traditional science in Vietnam in 2006 revealed the existence of a vast literature devoted to these disciplines; I identified approximately 370 treatises devoted to traditional medicine and over 80 treatises on geomancy written by Vietnamese authors.② Moreover, I was able to locate more than 100 manuscripts devoted to astrology, divination, and other "pseudo-sciences" collected in the areas populated by minorities nationalities, in particular, the Nung (Chinese Nong 农), Dao (Chinese Yao 傜) and Tay (Chinese Tai 傣) still using classical Chinese language for (re)production of the religious and (pseudo-) scientific texts. These documents appear to be invaluable source of information for studies on the history of science and religion.③ However, as far as I know, there was no research conducted on these materials by Vietnamese or

---

① I follow M. Strickmann in using the term and the notion of "magical medicine"[3].

② This research project titled "Traditional sciences in pre-modern Vietnam: A preliminary investigation of the extant sources" was supported by the Science Council of Taiwan in 2006-2007.

③ The documents were collected, digitized and made available online (http://vnmss.twgg.org/) in the framework of the project "Traditional Chinese science among Vietnamese minorities" supported by the Science Council of Taiwan in 2007-2011.

international scholars, due to the following factors: ①the materials were preserved in private collections (often in the collections of the practitioners) and not available to the researchers; ②the disciplines of "pseudo-science" were neglected by positivistically oriented historians of science; ③the available texts produced by minority nationalities in Vietnam have not been given relevant attention by Vietnamese scholars; ④ a historical study of the "pseudo-sciences" among the minorities requires a theoretical framework which would adequately represent the studied disciplines; however, one can argue that such a framework has never been established for the disciplines under investigation.

The study of the history of the "pseudo-science" in the "Chinese cultural zone" as a whole (and, therefore, in Vietnam) is important for future research on the history of Chinese, East-Asian, and South-east Asian science for the following reasons:

(1) The history of the conventional pre-modern sciences cannot be fully understood without a thorough investigation of the "pseudo-sciences": both co-existed for long time and in certain cases overlapped considerably or even constituted one and the same discipline. For instance, the functioning of the Imperial Astronomical Bureau in pre-modern China performed systematic observations of celestial phenomena in order to prognosticate future political events;① both medical and alchemical practices comprised collection, investigation, classification, experimentation with, and application of, organic and non-organic drugs;② the numerological considerations based upon the divinatory formulas of the Book of Changes (《易经》) formed the theoretical framework of a large spectrum of disciplines, from metrology and calendar to esoteric practices of inner alchemy (内丹), traditional medicine, and geomancy. ③

(2) The history of scientific traditions (including "pseudo-sciences") in the zone of Chinese culture cannot be fully understood without a thorough investigation of the processes that took place not only in China but also in neighboring countries influenced by Chinese culture; moreover, certain countries (in particular, Vietnam) preserved parts of Chinese cultural heritage which was irretrievably lost in China and its study can be highly pertaining to the work on the history of science in China; conversely, an adequate understanding of the local scientific traditions in the countries-recipients of the traditional Chinese sciences (viz., Vietnam) is impossible without solid expertise in the history of Chinese science.

---

① This topic was discussed by numerous authors; for one of the best documented accounts see the unpublished dissertation of T. Dean[4].

② There exist a large number of publications concerning the early history of Chinese ("external") alchemy; I will mention here only the groundbreaking paper of M. Strickmann discussing the interrelationships between external alchemy and Daoist activities in the first half of the first millennium AD[5].

③ For a discussion of the definition of the term "science" in the context of historical investigation see publications of N. Sivin[6] and A. Volkov[7].

2. Historical background and current situation

2.1 Historical and cultural background

Northern part of the present day Vietnam formally became a province of the Han Empire in the 2nd century BC; however, one can suggest that various types of intellectual exchange between this region and other parts of the present-day China had existed even before this time. The contacts were considerably reinforced when a large number of Chinese *literati* took refuge in the southern province during the military chaos by the end of the Eastern Han Dynasty. When Vietnam ceased to be a Chinese province in the 10th century AD, the newly born Vietnamese state implemented the bureaucratic system similar to that of the Song Dynasty (960-1279). More than 2000 years of Chinese cultural and political influence played a major role in shaping Vietnamese culture and written language, political and social institutions, arts and sciences.

During the period called in modern Vietnam "Chinese domination", that is, the period of time when (Northern) Vietnam remained a province of Chinese empire, local government used Chinese language for official documentation and state examinations. The earliest documents (namely, inscriptions on stone steles of the first millennium AD) show no, or very little of, "local" characters. After the separation between Vietnam and China in the 10th century, an increasing number of local characters appear in stele inscriptions and written documents. The local script designed on the basis of Chinese characters and used to record Vietnamese literary language is known as Nom (喃); there is no consensus as to when exactly the first Nom characters were created. However, the above-mentioned inscriptions suggest that the first Nom characters already existed by the early second millennium AD. Under several Vietnamese administrations some attempts were made to use Nom as the language of official documentation and of scholarship instead of the classical Chinese (Han), yet Vietnamese government apparently had no consistent language policy in this respect, and, historically, the classical Chinese was used more often. Below I will use the term "Han-Nom books" referring to the entire corpus of old Vietnamese books written either in classical Chinese or in Vietnamese using Nom characters, or in a mixture of both.

The French colonial government since the conquest of Vietnam in 1859 systematically eradicated the traditional written culture partly naively understood, and partly intentionally positioned, as a sign of a cultural and political dependence of Vietnam from China. The use of phonetic writing system using Latin letters with diacritical signs designed by Catholic missionaries in the 17th century (and much later dubbed as *Quoc ngu* (《国语》), "national language") was suggested as the solution for problems related to the difficulties of French civil servants in learning Vietnamese and at the same time as a means to eradicate the Vietnamese dependence on Chinese scholarly tradition and to replace it by French educational system.① Later, the local anti-colonial movements, paradoxically, also championed the alphabetic script *Quoc ngu* perceived as an important tool for modernisation and progress of the country. When the instruction of French and in French was prohibited in 1940s, the *Quoc ngu* became the only efficient instrument for solving the problem of illiteracy of the underprivileged layers of the population. As a result, the literary heritage of more than ten centuries of independent development was lost in a few decades, and nowadays only a few individuals are capable of reading texts written in Han-Nom. Moreover, during the wars that took place in Vietnam in the 20th century, many libraries, including the Imperial Library in Hue, were destroyed and the books and archives were stolen, destroyed, or lost.

It is important to note that a number of minority nationalities until now use the classical Chinese or (slightly modified) Nom for writing purposes. There exist a large number of documents written in Chinese script mainly originating from the Nung, Dao (pronounced *zau* or *yau*, Chinese Yao) and Tay minorities groups; these documents attracted the attention of scholars worldwide only recently. In the following sections I will briefly introduce the current situation of minorities in Vietnam and, in particular, the Dao (Yao) literature.

2.2 Minorities in Vietnam and their cultural heritage

Vietnam is a country with one of the largest number of minority nationalities living on its territory. According to a recent National General Census, there are 53 minorities groups with the total population amounting to approximately 15% of the population of the whole country (ca. 77 million), that is, ca. 11.5 million.② The minorities live in various regions, in particular, in highland, remote and rural areas, while perpetuating traditional life-style, including various forms of economic activities (mainly cultivation of land and handicrafts) and of religious practices; they speak languages belonging to two linguistic families: the Austro-Asiatic family and the Malayo-Polynesian family, each of which consists of various language groups (such as Mon-Khmer, Tai-Kadai, Sino-Tibetan, Austronesian, Viet-Muong, etc). There are two main upland minority areas in Vietnam: one is in the Northwest of the country (dominated by various Tai groups, along with Hmong, Yao, and various Mon-Khmer speaking peoples most of which, except for the Mon-Khmer, came from China), and the Central Highlands (after 1975 also called Tây Nguyên, "Western Plateau"). In this paper I deal exclusively with the minorities populating the North of the country.

The changes in cultural and economical life (ongoing urbanization of Vietnam, economical reforms, implementation of schools in rural areas and introduction of compulsory

---

① For the ideological premises of the French colonial policy, especially in the field of education, see, for example, the articles by Lévi[8] and de Barthélemy[9]; on the early history of French education in Indochina, see the works of Roucoules[10] and Richomme[11], for its modern (and somewhat simplified) description see the publications of Vu Tam Ich[12] and A. Woodside[13], and for an analysis, the book by Trinh[14].

② These data are borrowed from the books by Nguyen Van Huy et al. [15], and by Neejfes et al. [16].

education in Vietnamese language) provoke fast erosion of traditional culture. The permanent loss of the language and cultural traditions by the younger generations under the increasing influence of the mass-media, urban life-style and of new types of industry often acutely conflicting with the traditional cultural values and representations endangers the cultural identity of the minorities and undermines the established structure of the traditional society. One of the outcomes of the ongoing illiteracy eradication campaign among the minorities is that the written languages of ethnic groups based on Chinese script are going to disappear in the nearest future. The remaining books most likely will be lost or disposed of, or, in the best of cases, sold to tourists. Also, as a recent research shows, the growing tourist industry constitutes a particular menace to the traditional way of life of the minorities, mainly those living in the Northern Vietnam. ① It can be concluded that the traditional types of knowledge related to the traditional religious, medical, and (pseudo-) scientific practices under such circumstances will be extinct very soon.

Recently certain steps in this direction were made by Vietnamese researchers: about 3000 manuscript books were acquired by Han-Nom Institute (Hanoi). Another large collection is hosted in the library of Son La province, it contains ca. 2700 books of Dao and Tay minorities. ② However, these books certainly do not represent the totality of the documents that circulate among the minorities. A cursory inspection of the items sold in the tourist shops of Hanoi reveals a large amount of books and scrolls put on sale for foreign tourists who buy them as objects of art or souvenirs without understanding their true value as cultural artefacts. These objects are usually purchased for small money by Hanoi dealers from the minorities located not far from the capital and then re-sold to tourists for much higher price, thus disappearing forever from the scope of the international scholarly community.

2.3 The Yao outside Vietnam

In the late 1990s a major effort to preserve the books and paintings of the Yao settled in China, Thailand, and Laos was made by a team of German researchers from München University (Germany) who catalogued a large collection of Yao manuscripts (ca. 2600 manuscripts) currently preserved in Bayern State Library in München; the results of their work was presented in a paper by L. Obi and S. Muller[20] and in the catalogue of the manuscripts edited by T. Höllmann and M. Friedrich[21]; see also the catalogue of exhibition of the religious texts and paintings of Yao compiled earlier by the same authors[22]. Another recent major

---

① See the book of Pham and Lam[17], esp. pp. 220-233. On pp. 238-239 Pham and Lam describe the disappearance of certain traditions in Sapa area caused by the changes in the traditional culture provoked by the boom of tourism industry. This conclusion corroborates the results obtained earlier by western experts, see the reports of M. DiGregorio et al. [18], and of M. Grindley[19].

② A private communication of Chu Tuyet Lan (Han Nom Institute, Hanoi, 2004).

publication devoted to Yao is the richly illustrated book by J. Pourret①[23], the materials collected by the author originated mainly from Thailand, China, and Laos. In his annotated bibliography B. ter Haar mentions other collections of Yao texts open to public located in Heidelberg University (Sinologisches Seminar), in the Volkenkundig Museum (Leiden), at Leiden University, at Oxford and Copenhagen[24]. Ter Haar also mentions the collection of Shiratori Yoshirō at Sophia University (Tōkyō), yet he does not know whether the latter collection is accessible to the researchers. ②

B. ter Haar's analysis of the state of the modern scholarship on Yao culture[24] can be summarized as follows: "the study of Yao religion is extremely limited in size"; there is no clear identification of minorities (some non-Yao communities have been identified as Yao, and vice-versa); the religion of Yao was identified as Daoism (道教), however, other minorities also may have undergone Daoist influence (and therefore Yao should be studied together with other minorities groups); source publications lack any dealing with the religious or social context; there is no complete survey of the written and oral texts of the minorities; the majority of publications authored by mainland Chinese scholars are but short and superficial ethnographic descriptions. Ter Haar is mainly interested in the religious culture of Yao, while his bibliography and analysis show that the majority of publications on minorities were authored by anthropologists and linguists, and not historians of religion. Ter Haar's interest in religious life of the Yao is fully justified: it was discovered in 1930s[26] (and re-discovered in 1970s) that Yao's religion is one of the forms of Daoism. In his article of 1982 M. Strickmann even suggested that Yao's religion originates from the Song Dynasty Daoism[27], but some doubts were later voiced by B. ter Haar[24].

If ter Haar regrets that the Yao written texts have not been studied thoroughly enough as sources on the history of religion, even more regretful is the complete lack of studies of the scientific dimensions of these texts. As far as I know, no systematic research has been done on the traditional science presented in the minorities' manuscripts, neither the task of exploring this aspect has ever been explicitly stated. However, the treatises of Vietnamese minorities which I located in the framework of my above-mentioned research projects "Traditional sciences in pre-modern Vietnam: a preliminary investigation of the extant sources" and "Traditional Chinese science among Vietnamese minorities" suggest that a considerable number of texts were related to sciences (esp., calendar), medicine, and pseudo-sciences (astrology, divination, magical medicine, talismans).

---

① According to Barend ter Haar, "This is quite an extraordinary book that introduces all aspects of Yao culture through a broad selection of photographs of artifacts, textual materials, and people (including historical pictures). Surpasses all existing collections of this kind."[24]

② Several books of this collection were published by Shiratori Yoshirō[25].

3. Science and Daoism

The above-mentioned suggestion that Yao religion represented a form of Daoism, even if not providing specific details as far as the precise time and location of the "conversion" of the Yao into Daoism are concerned, remains a plausible working hypothesis that can be tested and probably refined in future. My cursory investigation of the books in Han-Nom used by the minorities for religious and magical purposes from the collections I came across during my missions in Vietnam made rather clear that the minorities practiced various types of esoteric practices, astrology, and magical medicine combined with various versions of Daoism. This observation may be crucial for the present study, since certain networks of (religious) medieval Chinese Daoism, as I argued in earlier publications[7,28], played the part of the institutional framework for various types of scientific activities (in a broader sense), beginning from alchemy and ending with astronomy and calendrical computations. To support my hypothesis, I used the work of the prominent Daoist master of the Complete Perfection (全真) school, Zhao Youqin (赵友钦) (1271-1335) and his disciples.

My study of the case of Zhao suggested an approach alternative to that of J. Needham and N. Sivin to the problem of the relationship between science and religion in traditional China. This is why the study of the data found in Yao documents, even if fragmentary and distorted, can provide important evidence concerning the implementation and functioning of scientific knowledge (even if "proto-" or "pseudo-scientific") in the context of religious Daoism, and therefore reinforce the conclusions made on the basis of the study of the case of Zhao and his school. Moreover, if Strickmann's hypothesis is correct and the Daoism of Yao indeed goes back to the Song Dynasty, one may hope to discover at least some fragments of the contemporaneous scientific knowledge lost in China that could shed new light on the relationship between science and Daoism. On the contrary, if the investigation of the texts demonstrates that the relevant astrological, medical, geomantical and other "scientific" elements in them are no more than recent borrowings from respective disciplines, this finding will constitute the decisive evidence in support for more recent dates for the production of the Yao manuscripts and therefore will confirm ter Haar's doubts concerning Strickmann hypothesis.

4. Conclusions

The transmission of a (pseudo-) scientific tradition is a rather complex process including a number of activities such as adaptation (or invention) of the technical terms and of principal ideas of the transmitted disciplines, translation of treatises, transmission of non-verbal expertise, implementation of the institutions (or networks) performing relevant activities, providing instruction, etc. Such activities have never been studied in the case of Vietnam: on the one hand, the source texts on the history of pseudo-sciences in Vietnam, certainly one of the major recipients of the Chinese tradition, have never been located, collected, described, and published; on the other hand, no attempt has been made to put under scrutiny the available "pseudo-scientific" treatises, despite the fact that some of them are still extant and being practiced either by traditional doctors (in the case of traditional medicine) or by experts in divination, *fengshui*, and other disciplines.

One can conjecture that a large part of the Vietnamese scientific traditions, from the "official" disciplines①, such as the astronomy or mathematics, to fortune-telling and magical healing②, were the result of a long process of transmission and local adaptation of the traditions coming from China, and this is why the ongoing study of the collected manuscripts is also relevant to a further investigation of the history of traditional Chinese science. Even though the latter is arguably found among the best documented and studied pre-modern "traditions of knowledge", the transmission of Chinese science to neighboring countries (in particular, to Vietnam) has never been studied systematically③. Moreover, even in the case of China certain traditions (such as geomancy or divination) were not explored sufficiently well, and even more so were such complicated questions as the relationships between science and religion, especially Buddhism and Daoism. One can also conjecture that a given scientific tradition imported by neighboring countries from China may well have remained intact during several centuries even after it had partly or entirely disappeared in the country of origin, and therefore the study of these local traditions may shed a new light on the history of Chinese science.

References

[1] Pierre H, Maurice D. Connaissance du Viet-Nam. Paris: Imprémerie Nationale and Hanoi: Ecole Française d'Extrême-Orient, 1954

[2] Pierre H, Maurice D. La science au Viêt-nam. Bulletin de la Société des études indochinoises, Saigon N s t, 1963, 38 (3-4): 533-555

[3] Michel S. Chinese Magical medicine. Stanford: Stanford UP, 2002

[4] Deane T E. The Chinese Imperial astronomical bureau: form and function of the Ming Dynasty Qintianjian from 1365 to 1627. Unpublished PhD Dissertation. Ann Arbor (MI): UMI Dissertation Information Center, 1989. esp 18-50, 236-241

[5] Michel S. On the Alchemy of T'ao Hung-ching. In: Welch H, Seidel A eds. Facets of Taoism, New Haven & London: Yale UP, 1979. 123-192

[6] Nathan S. Introduction. In: Sivin N. Medicine, Philosophy and Religion in Ancient China. Researches and Reflections. Aldershot/Brookfield: Variorum, 1995. i ~ xvii, esp xiii-xiv

---

① On J. Needham's and N. Sivin's definition of "official" (or "orthodox") and "unorthodox" sciences, see, for example, my paper of 2004. 523-534[28].

② In some cases (e. g., geomancy), a discipline could be practiced at the highest level of the society as well as among its lowest layers.

③ Even less is known about the transmission of knowledge to China from outside, except for a few relatively well explored cases such as the introduction of the Western science to China in the 17th-18th century. However, it is not impossible that such "flows" from "cultural periphery" to the "center" may also have existed in the case of Vietnam.

[7] Alexei V. Science and Daoism: an introduction. Taiwanese Journal for Philosophy and History of Science, 1997, (8): 1-58, esp 9

[8] Sylvain L. L'enseignement en Indochine. Académie des Sciences Coloniales, comptes-rendus des séances/ communications, tome IV (1924-1925), 345-352. Paris: Société d'éditions géographiques, maritimes et coloniales, esp 351

[9] de Barthélemy. Quel est le Rôle que doit rechercher et remplir la Colonisation française en Indochine? Académie des Sciences Coloniales, comptes-rendus des séances/ communications, tome IV 1924-1925. 289-297, esp 289-292

[10] Roucoules E. Etude sur l'instruction publique en Cochinchine. Bulletin de la Société des Etudes Indochinoises, 1889, (2): 25-44

[11] Richomme M. De l'instruction publique en Indo-Chine. Thèse pour le doctorat. Paris: Emile Larose, 1905

[12] Vu Tam Ich. A historical survey of educational developments in Vietnam. A special issue of the Bulletin of the Bureau of School service, by College of education, University of Kentucky, Lexington, 1959, 32 (2)

[13] Woodside A B. Vietnam and the Chinese model: A Comparative Study of Vietnamese and Chinese Government in the First Half of the Nineteenth Century. (Harvard East Monographs, 140). Harvard: HUP, 1988

[14] Trinh Van T. L'école française en Indochine. Paris: Karthala, 1995

[15] Nguyen Van H et al. The cultural mosaic of ethnic groups in Vietnam. Hanoi: Education Publishing House, 2001

[16] Neejfes et al. Promoting Ethnic Minority Development, Hanoi: UNDP Poverty Task Force, 2002

[17] Pham Thi Mong Hoa, Lam Thi Mai Lan. Tourism among Ethnic Minority (sic) in Sapa. Hanoi: Nha Xuat Ban Van Hoa Dan Toc, 2000

[18] DiGregorio, Michael, Pham Thi Quynh Phuong, Minako Y. The Growth and Impact of Tourism in Sa Pa. Hanoi: Center for Natural Resources and Environmental Studies, and The East-West Center, 1996

[19] Grindley M E. Preliminary Study of Tourism in and Around Sapa, Lao Cai Province. Frontier-Vietnam Forest Research Programme. Lao Cai, 1997

[20] Lucia O, Shing M. Religiöse Schriften der Yao. Überblick über den Bestand der Yao-Handschriften in der Bayerischen Staatsbibliothek. Nachrichten der Gesellschaft für Natur-und Völkerkunde Ostasiens, 1996, 67 (1-2): 39-86

[21] Höllmann T O, Friedrich Michael, eds. Handschriften der Yao (Teil I): Bestände der Bayerischen Staatsbibliothek München, 2004

[22] Höllmann T O, Friedrich Michael, eds. Botschaften an die Götter. Religiöse Handschriften der Yao. Südchina, Vietnam, Laos, Thailand, Myanmar (The catalogue of the exhibition of Yao texts and paintings published in the series "Asiatische Studien"). Wiesbaden: Harrassowitz, 1999

[23] Pourret J G. The Yao: the Mien and Mun Yao in China, Vietnam, Laos and Thailand. Chicago: Art Media Resources, 2002

[24] Barend T H. Yao religious culture: bibliography by Barend ter Haar (last revision 28.1.2005). Retreived from http: //website. leidenuniv. nl/ ~ haarbjter/ yaotext. htm#Hollmann [2006-12-25]

[25] Yoshirō S. Yōjin monjo. Tōkyō: Kōdansha, 1975

[26] Franklin F R. Yao Society: A Study of a Group of Primitives in China. Lingnan Science Journal, 1939, 18 (3): 343-382, 18 (4): 397-455 (not seen; reviewed in American Anthropologist, Vol 42, No

4)
[27] Michel S. The Tao among the Yao: Taoism and the Sinification of South China. In: Rekishi ni okeru minshū to bunka-Sakai Tadao sensei koki jukuga kinen ronshū (历史における民众と文化: 酒井忠夫先生古稀祝贺记念论集). Tōkyō: Kokusho kankōkai. 1982. 23-30
[28] Alexei V. Astronomical Data in a Daoist Treatise: Chen Zhixu's commentary on the Scripture of Salvation (Duren jing). In: Lagerwey J ed. Religion and Chinese Society. Vol 2. Hong Kong: the Chinese University of Hong Kong Press, 2004. 519-552

First Steps of Russian-Chinese Scientific Cooperation Contacts: Pyotr Kozlov's Visit to Beijing in 1925

Tatyana Yusupova

(Institute for the History of Science and Technology,
St. Petersburg Branch, Russian Academy of Sciences)

This paper deals with the episode from the first quarter of the 20th century, when Russian-Chinese scientific links had just been established. This episode is the visit of a famous Russian explorer of Central Asia Pyotr Kozlov (1863-1935) to Beijing in May, 1925.

It should be noted that Russian expeditions to the Central Asia in the second half of 19-the beginning of 20th century had significant value for studying of this region by Europeans. P. Kozlov's books have became one of the important ways to broadcast image of another culture to Russian and European audience. The natural and archaeological collections brought by P. Kozlov's expeditions several helped generations of Russians to get acquainted with the nature, culture and history of the Central Asia in general and China specifically.

An eminent Russian traveler Pyotr Kozlov[1] belongs to the brilliant generation of enthusiastic explorers of Central Asia on the turn of 20 century. Kozlov was the last representative of a certain type of Russian travellers—he was a geographer, ethnographer and naturalist at the same time. From 1883 to 1926 he took part in 6 prominent large-scale expeditions to Mongolia, East and West China, and to East Tibet; three of above mentioned expeditions he led himself.

P. Kozlov's professional success was defined by his outstanding personality: his talent as a researcher-geographer, ample knowledge in different branches of regional studies, unique energy, persistence and devotion to his work. Besides the area of his research coincided with the region of state geopolitical interests of Russia. P. Kozlov took this fact into account when filing for financial support.

As an explorer and naturalist P. Kozlov made himself known already after the Mongolian-Kamian expedition (1899-1901) and after Mongolian-Sychuanian expedition (1907-1909)[2] he became worldwide famous. In that travel during the excavations of a "dead" city Khara-Khoto in Southern Gobi P. Kozlov made some extraordinary archaeological discoveries. In one of the sacral buildings— "suburgan", that afterwards was called "celebrated" —P. Kozlov had luck to discover the richest collection of books and manuscripts in Tangut, Chinese and

Tibetan, hundreds of sculptures and icons, sacred attributes from Buddhist temples. They are stored now in the State Hermitage in St. Petersburg, except manuscripts, which are stored at the Institute of Oriental Manuskripts, RAS[3]. Materials from the "celebrated" suburgan allowed the scholars to reconstruct the history of forgotten Tangut state Xi Xia, that had been flourishing for almost 250 years (982-1227) on the territory of modern northern China.

All the results of P. Kozlov's explorations were highly estimated by international scientific community and made him one of the most respected European researchers of Central Asia. P. Kozlov was elected an honorary member of Russian, Hungarian and Dutch Geographic Societies; was awarded with the gold medals of London and Italian Geographic Societies, as well as with Chihachev's prize of French Academy of Sciences.

There was one more extraordinary archaeological discovery which belonged to P. Kozlov. In 1923-1924 Kozlov's Mongolian-Tibetan expedition worked in Northen Mongolia where it conducted excavations in Noyon Uul mountains, in Xiougnu (Khunnu) barrows, dated back to the period of the Han Dynasty (206 B C-220 A D) In ancient barrows the expedition found lots of perfectly preserved funeral accessories, among them remarkably rich silk clothes and carpets decorated with images of mythic animals, several golden, silver and bronze articles, ceramics. This excavation attracted a great attention of scientific society all over the world.

However the final goal of this P. Kozlov's expedition was to continue Khara-Khoto excavations. That way he had to obtain a permission on carrying on works on Chinese territories. For this reason in May of 1925 P. Kozlov visited Beijing.

This trip is fully depicted in P. Kozlov's diary, published in 2003[4]. In this paper we will add some details to his description that, in my opinion, can broaden and enrich semantic horizons of his memoirs and show his visit not only as a fact of famous traveler's biography, but also as an independent event in the history of science, illustrating the first steps to cross-acquaintance of Soviet and Chinese scientists.

The visit of P. Kozlov in Beijing coincided with possibly the most favorable periods both in social situation in China, and in Soviet-Chinese partnership in general. Only one year before, in May 31, 1924, the first official agreement between the USSR and Republic of China had been signed. It constituted ordinary diplomatic and consular relationship between the participants. By the way, in Beijing P. Kozlov stayed at hotel "Wagons-Lits" (present Guoji hotel) in so-called embassy quarter, where diplomatic missions of Great Britain, the USA, France, Germany, Japan, the USSR and other countries were situated.

At that time several Russian orientalists lived in Beijing. First of all A. Ivanov, a specialist in Chinese and Japanese culture. He was one of the first who started analyzing the materials brought by Kozlov from Khara-Khoto and got a title of "a pioneer of Tangut studies". Also A. Ivanov had close connection with scholars of Beijing University. Another person was B. Vasiliev, a researcher from Asian Museum of RAS. He worked both as a secretary in consulate general and as a scholar in Chinese culture. His task as formulated by Russian Academy of

Sciences was "to initiate relations with new Chinese scientists, to identify the most aspects of modern Chinese humanities and to collect the necessary literature for Chinese collection in Asian Museum".

Close connections linked Russian orientalists with baron A. Stahl von Goldstein, a researcher at Chinese-Indian Institute of Harvard University; he also got acquainted with Kozlov and his Khara-Khoto collections. Vast and various information about Kozlov's archaeological discoveries filled not only Russian but also foreign journals and newspapers, including Chinese, especially Harbin's ones. For this reason scientific circles of Beijing were informed about Kozlov's works and were looking forward to meeting him.

That years saw the birth and a true rise of Chinese archaeology. That's why Kozlov had such a warm reception at Beijing University. Another influencing factor was that the University happened to be one of the centers of political life in Chinese capital, and the lecture by Soviet traveler became a subject not only of scientific, but of political interest too. As a token of sincere acknowledgement and gratitude P. Kozlov was offered to leave his signature in the book of honorary visitors of the University.

Kozlov's visit was a notable event also for other foreigners living in Beijing: British embassy organized a special dinner to honor him. There P. Kozlov met Swedish geologist and archaeologist, then councilor for Chinese government Johan Gunar Andersson (1874-1960), one of the most eminent specialists in Chinese late Stone Age. They had a unique chance to share the results of their work.

The same reception on May 20, 1925 was prepared for P. Kozlov by Soviet embassy. Several Chinese scientists, governmental officials and foreign diplomats were invited. P. Kozlov expressed his extreme gratitude to Chinese officials for their support of his previous expeditions as well as his hopes for further help for the present one. He highlighted that Chinese government had done Russian geography a great favor by giving permission to research on the country's territory.

Chinese press reported about Kozlov's staying in Beijing, but, as it often happens, the information was not accurate. So, in November 1925 to Chinese embassy in Moscow a letter from Zhang Yuanji (1867-1959) —philologist, researcher and later president of a famous Chinese publishing house Shangwu Yinshuguan, specialist in editing and publishing ancient manuscripts, was delivered. He, basing on newsletter materials, wished to check information given there and to know if P. Kozlov had really found ancient Chinese manuscripts in Noyon-Uul and if it was possible to get permission of Soviet government and copy them.

Soviet foreign office ordered the Russian Academy of Sciences to help Chinese researcher and considering possible perspectives of cultural and scientific co-operation between the USSR and China to deliver him the copies. I suppose that was the letter of nevertheless, in Noyon-Uul finds of P. Kozlov there were no manuscripts, and possible cooperation had not been developed.

In August 1926, having finished the expedition in Ulaanbaator P. Kozlov met famous

Chinese politician and later scientist Yu Yuaren (1878-1964) on his way back from Moscow to Beijing. Kozlov showed him Noyon Uul barrows that interested Yu Yuaren so much that he even assisted in interpreting hieroglyphs and dating of several objects.

However sad it was, P. Kozlov had no more contacts with his Chinese colleagues. Nevertheless it seems that the visit to Beijing in 1925 was very productive for him. P. Kozlov estimated the state of Chinese archaeology and presented to Chinese scientific community the main results of his own work, having built a serious basis for further contacts. Besides Beijing that P. Kozlov visited for the first time, impressed him a lot. He liked everything: the architecture, the music, the people, the everyday life, Chinese cuisine and a warm reception.

In his Beijing lectures P. Kozlov repeatedly emphasized his hopes for future collaborations with Chinese scientific community in the field of geography and ethnography of Central Asia.

Unfortunately dramatic political situation in China and then the whole Far East, reinforcement of authoritarian regime in the USSR did not allow close contacts between scientists of two countries in 1920s-1930s. P. Kozlov's wish began to realize only 25 years later, after 1949. Then, as you know, there were new difficulties in Russian-Chinese relations for a long time. Cardinal changes in political conjecture and new influences in international, and, subsequently, in Russian-Chinese relations during the last decade made us pay more attention to particular features of such communications, first of all aimed at alliance in a very broad sense. So the more precious for the historians of science is any information on the first steps in scholarly contacts between China and Russia and on the very atmosphere surrounding them.

In modern China P. Kozlov is known first of all due to numerous studies of his Khara-Khoto materials. The close contacts have been established between the Institute of Oriental Manuscripts of RAS, where Khara-Khoto's manuscripts are stored now, and Chinese scholars especially in Tangut Studies. Recently Kozlov's book *Mongolia and Amdo and the dead town of Khara-Khoto* was published in Chinese[5]. Its translator Dr. Ding Shugin is preparing the second edition of this book. We believe that further and closer cooperation between Russian and Chinese scientist will result in translation of another his book "Mongolia and Kam" (SPb, 1905). Written in lively and vivid style it poses a great scientific and historical interest for a modern Chinese reader.

References

[1] Yusupova T I. P. K. Kozlov—the Eminent Explorer of Central Asia. In: Yusupova T I. St. Petersburg and China: Three Century of Contacts (Sankt-Petersburg—Kitay: tri veka kontaktov). St Petersburg: Evropeysky Dom. 153-166 (in Chinese)

[2] Yusupova T I. P. K. Kozlov Mongolia and Sichuan Expedition (1907-1909): the Discovery of Khara-Khoto. In: Yusupova T I. Russian Expeditions to Central Asia at the Turn of the 20th. Popova I F Ed. St. Petersburg: Slaviya, 2008. 112-129

[3] Kychanov E I. The Tangut Collection of the Institute of Oriental Manuscripts: History and Study. In: Kychanov E I. Russian Expeditions to Central Asia at the Turn of the 20th Century. Popova I F Ed St Petersburg: Slaviya, 2008. 130-147

[4] Kozlov P K. Journals of the Mongolia and Tibet Expedition (1923-1926) (dnevniki Mongolo-Tibetskoy ekspedizii). Yusupova T I, Andreev A I eds. St. Petersburg: Nauka, 2003 (in Russia)

[5] Kozlov P K. Mongolia and Amdo and the Dead Town of Khara-Khoto. Translator Ding Shugin. Lanzhou, 2003 (in Chinese)

Negative Effects of Patent on Technological Development: The Perspective of the Theory of Modern Technological Process

Zhang Gaizhen

(The Institute for the History of Natural Science, CAS)

Interest has increasingly played an important role in the process of modern technology nowadays. The process of modern technology can be portrayed as comprising the following four stages: interest expectation, technological ideas, technological realization, technological application (interest realization).

Technological development has not only been regarded as a process of technology involving a kind of transformation from one stage to another, but also one of a technological diffusion from one technological system to another, either in the same or different industrial enterprises, as well as in international ones.

Some of negative effects will be analyzed in this paper. The examples I am going to use are the treatment and control of AIDS and the case of Tamiflu. Technologies that have failed in interest expectation usually have not been developed. In addition monopoly over patented technology makes the prices of technological products unreasonably higher than its value, therefore limiting its dissemination; Monopoly also makes technological diffusion, either domestically or internationally, become very difficult. Serious adverse results in the development of technology as a whole, and of the small businesses in similar or heterogeneous industries, as well as in underdeveloped and developing countries and other vulnerable group have been observed.

My discussion, to some extent, affirms the merits of patent throughout its history, as it has been found to have many positive effects on technological development such as giving motivation to technological invention, promoting technological dissemination and protecting individually creative abilities. The purpose of my paper is to provide a new perspective in thinking about patent and its negative effects on technological development. Much more work is of course needed in order to make a rounded evaluation of the effects of patent on technology, including an extensive evaluation of the relevant arguments etc. A theoretical tool—namely, "the process of modern technology": developed by the author—will be used to analyse two cases: ① the treatment and control of AIDS; and ② the case of Tamiflu (Oseltamivir

Phosphate).

1. The process of modern technology: developed from the theory of technological process

In their work "On Technology"[1] (1986), Professor Yuan Deyu and Professor Chen Changshu firstly advanced the theory of "Technology as Process"[2]. In his article of 2003, Professor Yuan Deyu wrote: "The process of technological development is practically the process of the development of technological patterns, which include technology developing from a subjective one to an objective one, a potential technology to a practical one."[3] "Subjective technology" refers to technology which is in the process of being conceived or in the stage of designing. "Objective technology" is the original form of technological invention and is the same thing with "potential technology" which exists in modes of symbols, tables and single products. "Invented technology" refers to only a pattern of technology, whose function and value is still potential. Through the socialization, industrialization, and commercialization of technological invention (potential technology), technology gets realized into the actually existing one in its concrete pattern.

Professor Yuan and Professor Chen's theory of "Technology as Process" and patterns of technological development can be described as follows:

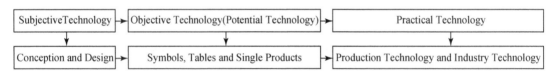

Fig. 1 the theory of technological process

They provided us with a significant and valuable frame of thinking about technology. Being different from other philosophical theories of technology, their theory is a method to go deeply into technologies seen in this context as dynamic activities, not as static results as is usually the case[4]:11. According to their theory, technology develops in stages, and researchers can choose a certain stage or more than one stages to study. Professor Yuan Deyu said their theory was developed in the course of reflecting on static definitions about technology prevailing at that time and finding them limited or inappropriate[5]:11.

There are many new characteristics of modern technology that are different from those of the technology of the past. What does modern technology mean? As to its denotation, there isn't a certain date point when modern technology appeared on the arena of history. It was a gradual process. "There is still no normative definition about the meaning of modern technology"[6], either. Karl Marx (1818-1883) pointed out the phenomenon of alienation as the result of modern technology being part of social system[7]:483, Jurgen Habermas (1929-) underlined sociopathy of modern society through reflecting on modern technology[8]:68. Philosopher Karl Jaspers (1883-1969) thought there were four characteristics of modern technology: "firstly, providing groundwork for automated production; secondly, revolutionary development of energy usage; thirdly, based on modern science; the fourth, strongly supported by cultural values."[9] It was on the latter characteristics of modern technology advanced by Jaspers and on the reflections of Marx's and Harbermas's that my work was based in order to develop further Professors Yuan and Chen's theory of technological process.

According to my own theory of technological process, the process of modern technology goes as follows:

Fig. 2 the process of modern technology

I will use the two examples I have mentioned to explain the theory of the process of modern technology; to expand my demonstration of why interest plays a very important role, right from the first stage of modern technological process, and then onwards, penetrating all of its later stages determining the whole of the technological process; and finally to explain what interest is in part 2.

2. Why interest is a determining factor in modern technological process: the cases of AIDS and Tamiflu

2.1 The case of AIDS

AIDS (Acquired Immune Deficiency Syndrome) is one of the infectious diseases which cause the highest rate of death in the world. Roughly 6000 people become infected with HIV each day, with a total of approximately 2.7 million new infections per year. Worldwide, an estimated 33 million people are living with HIV/AIDS, with 22 million of those individuals living in sub-Saharan Africa[10].

Even if the incidence of HIV infections continues to decline at current rates—for which there is no guarantee—there will still be more than 22 million new infections by 2015. Vaccines have proved to be one of the most powerful and cost-effective of all public health tools, however, up to now, human beings haven't found an effective vaccine to control AIDS, and very few people, companies or organizations invest in a vaccine of AIDS. The poor outcomes of one HIV vaccine trial in 2007 prompted reductions in commercial investment and redirection of government funding toward more basic research[11]. Bill & Melinda Gates Foundation donated 28.7 billion USD to found CAVD (The Collaboration for AIDS Vaccine Discovery) in July 2006[12].

On the other hand, individuals have already found ways to combat AIDS by mixing three kinds of drugs together called drug cocktail treatment (HAART). If AIDS can be treated continuously, it becomes a kind of chronic disease which may not devitalize patients' lives. However, the cost of AIDS treatment is too high, therefore only a few rich people can afford it. To most patients living in Sub-Sahara countries, medical fees of treating AIDS is 2000 times as much as their per capita medical cost, and as much as the whole cost of ten children's primary education. More than 4000 billion dollars are needed to treat patients with AIDS worldwide.

Lots of money has been invested in the treatment of AIDS that can be afforded by wealthy people, already reaching 400 billion dollars, but only 5 billion in vaccine research (before the Gates' Foundation)[13].

There are two key points related in my topic: ①Why cannot a strong social demand for AIDS vaccine attract the same strong interest of investment? ②How can we understand the strong zeal in investing in treatment of rich people? We can indeed find the main factor, here: it is interest.

As to ① above, one reason of such non-investment is the strong contrast of technological difficulty in vaccine-discovering and the high cost of its research. As mentioned above, we have been suffering from the painful failure in 2007. The other important reason is that potential investors worry about whether they can recover the cost and get profits from patients living in poor countries with very low purchasing power. This area, according to Bill Gates is where the market doesn't work[14]. Failure in investing on the vaccine is actually potential investors' failure in interest expectation but also failure in moral principles when, in the course of their competition, they only care about economic profits. We can find the same explanation behind of strong zeal for AIDS treatment ② above, which is taking on in the opposite form. The population of who can afford the costs of AIDS treatment is much smaller than that who need treatment and vaccine. However, the former is a certain group of people having strong purchasing power, so investors can get expected profits from them. Even though investment generally involves risk, in this case the risk was less than that in vaccine investment, too. It was economic interest that forces individuals, companies and organizations to invest the in treatment of AIDS

Investment in AIDS vaccine should make it possible to save more people, to eliminate or limit AIDS, and also save resources spent on treatment, given the low buying power for poor people and difficulties in technology, it will be risk for people to invest in discovering vaccine against AIDS. So, I don't know if philanthropy will be the only way for human beings to deal with this serious problem. Unity is the best way to accelerate the process of discovering vaccine, CAVD which was founded by the Gates Foundation was an effort through philanthropy to unite constructive powers in the world.

2.2 Case of Tamiflu

Tamiflu (Oseltamivir Phosphate), a kind of medicine which was brought to market in

1999 by the Swiss firm Roche Holding AG, and proved to be one of the world's only drugs effective against flu. It was frequently prescribed for flu treatment and flu prevention in adults and children 1 year or older.

As a terrible infectious disease, Bird flu has been spreading globally from 2005. People around the world were facing another menace from infectious diseases. To take precautions against the possibility of its spreading in a large scale, many countries turned to the best choices: Tamiflu storing. However, Roche, whose productive summit for Tamiflu was not forthcoming within one year, held the patent of Tamiflu and has not authorized any organization to produce it. Fortunately, bird flu was controlled in a short time and has not exploded again until now, but it is important for us to reflect on the Tamiflu event[15].

As a relatively mature form of technology, or as practical technology in professor Yuan's theory, or as technological-used stage in the process of modern technology, the dissemination of Tamiflu was controlled by interest: i. e. the economic profits of Roche. It was required by the crisis human beings were faced with from 2005 required that Tamiflus were produced on a large scale while Roche who invented it owned the patent of the medicine. It was a conflict between individual interest and public interest.

2.3 A brief conclusion of part 2

We can conclude from those two cases above that economic interest played a very important role in modern technological process. What does interest mean here? Not always individual's interest; it can be the interest of collectivities, of countries, especially economic profits of countries and transnational enterprises with strong economic and political powers, but not of human beings as a whole, because there is no international organization beyond countries. The interest of humanity as a whole can hardly be taken seriously in relatively normal state nowadays, only when some crisis happened in a way threatening all countries' benefit.

3. Negative effects of patent on technological development: revelations from cases of AIDS and Tamiflu

3.1 How does patent play an important role in the process of modern technology?

One obvious kind of objects falling under patent is new techniques never invented by anybody before the inventor. Different from science, technology has its own special function: this function can be that of industrializing which can bring profits to human beings. One of functions of patent system is to protect profits of technological inventors or patent holders trough protecting ownerships of patented technology.

Since 1474, with the Venice Patent Law, and 1623, with the proclamation of the "monopoly Law" in the United Kingdom, the patent had gradually changed from a first privilege granted by a monarch to a kind of private right set down as a law, which has been valued for its protection of individually creative abilities and for its economic function of promoting technological dissemination.

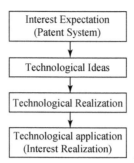

Fig. 3 The role of patent in process of modern technology

The original objectives of patent law were: ① encouraging technological invention by protecting individually creative abilities; ②promoting technological dissemination. However, as it has been discussed by many scholars, patent is becoming gradually a tool for big enterprises and many countries to earn a huge profit. To①②, there were much data to support the ideas and many argumentation to retort it, too.

Next, I will give my description of its negative effects on technological development.

3.2 Conclusion: negative effects of patent on technological development: and considerations for analysing the two cases further

Patent law is to some extent an effective balance between individual and collective benefits (details of patented technology is to be known publicly in a certain year). However, monopolizing over patented technology for patent holders within a certain time has nevertheless caused negative effects in the development of technology.

The meaning of technological development has two aspects: a process of technology from one stage to another stage of transformation; a technological diffusion from one technological system to another, in the same or different industrial enterprises, as well as on an international scale. Technological dissemination corresponds roughly to the latter aspect of technological development. Patent has negative effects on both these two aspects and even becomes a fateful hinder if there is a failure in interest expected.

Let us go back to the two cases examined above. There are three kinds of considerations regarding them: ①In the case of discovering a vaccine against AIDS, without successfully interest expectation here, excepting philanthropy, human beings have no interest to invest in it, therefore we have no inventions here. ②In the treatment of AIDS, when we have an effective method for it, interest expectation has given rise to successful technological invention, however only a few rich persons can afford it, that is to say, the population of using the technology is limited. On the one hand, patent make prices of medicines much more high than their costs, and so poor people can't afford it; On the other hand, it is not legally allowed that poor countries which didn't hold patents to produce the medicines. ③In the case of the Tamiflu, the technology was developed, and many countries could afford and hoped to stock it, but the patent system in this case made the right of its production controlled only by a certain company whose productivity is too limited to meet the need for the medicine.

Tab. 1 Cases in technological stages

Stages \ Cases	Vaccine of AIDS	Treatment of AIDS	Tamiflu
Interest Expectation (ensured by patent)	f	s	s
Technological Ideas	f	s	s
Technological Realization	f	s	s
Technological application (tech-diffusion)	f	some rich people	some people

Notes: f = failure (failing in interest expectation); s = success (success in advancing tech ideas)

In the case of discovering an AIDS vaccine, potential investors failed in the first stage of technological process, technological development being set back at the beginning. In the treatment of AIDS, investors got a successful interest expectation, and the second and third stage of technological process are developed successfully, too. In the fourth stage, affected by patent, the price of medicine was higher than its cost, and so only some rich people can afford the medicine; so according to the definition of technological development, it follows that here technology has not been widely diffused.

In the case of Tamiflu, technological development was failing in the last stage, too, the cause being also the collective interest being protected by patent.

We can therefore develop conclusions as follows by analyzing for these two cases:

(1) Interest-oriented of patent to patent holders make economic interests, not social needs, become the direct driving force for technological innovation, while that technology which does not pass the test of interest expectation usually has not been developed. Patent goes against increasing of quantity of technologies in this way.

(2) Monopoly over patented technology bestowed by patent law makes the prices of technological products unreasonably higher than their value, thus limiting its dissemination and reducing the population of using it.

(3) Monopoly makes technological diffusion become difficult in different areas of a single country, or in different countries and amongst different enterprises.

The most important conclusion has to do with who was affected in the whole process, we can find this easily from the two cases discussed here as well as from other similar ones. Those negatively affected were people with low buying power, small enterprises, developing and underdeveloped countries, and other vulnerable groups who were especially affected. The impact of patent on vulnerable groups have been taken to be obvious, because such negative effects frequently conflict with fundamental human rights, such as the right to life, subsistence, health etc. Such considerations form the bottom line of human being's moral principles.

Acknowledgement

Deeply grateful to professor Byron Kaldis from Hellenic Open University, professor Lu Dalong and my supervisor-professor Dong Guangbi and professor Yuan Jiangyang from the Institute for the History of Natural Science, Chinese Academy of Sciences.

References

[1] 远德玉,陈昌曙. 论技术. 沈阳：辽宁科学技术出版社,1986
[2] 同 [1]
[3] 远德玉. 技术过程论的再思考. 东北大学学报（社会科学版）,2003.11
[4] 同 [3]
[5] 同 [3]
[6] 王健. 现代技术伦理规约. 沈阳：东北大学出版社,2007：29
[7] 〔德〕马克思,〔德〕恩格斯. 马克思恩格斯全集. 第26卷（上册）. 北京：人民出版社,1972
[8] 〔德〕哈贝马斯. 作为"意识形态"的技术和科学. 上海：学林出版社,1999
[9] 〔德〕卡尔·雅斯贝尔斯. 历史的起源与目标. 北京：华夏出版社,1989：114
[10] The CAVD Structure. http：//www.cavd.org/about/Pages/CAVDStructure.aspx〔2009-12-05〕
[11] 同 [10]
[12] 同 [10]
[13] 〔法〕雅克阿塔利. 21世纪词典. 梁志斐等译. 桂林：广西师范大学出版社,2004
[14] 2009 Annual Letter from Bill Gates. http：//www.gatesfoundation.org/annual-letter/Pages/2009-bill-gates-annual-letter.aspx〔2009-12-05〕
[15] 杨立群. 解放日报〔2005-10-28〕

The Proceedings of the 12th International Conference on the History of Science in China

Highlighting the discipline of the history of science and technology in China and promoting its development

The 12th International Conference on the History of Science in China (the 12th ICHSC) was held in Beijing, June 26-30, 2010. The Conference was organized by the Chinese Society for the History of Science and Technology (CSHST), co-organized by the Institute for the History of Natural Science of the Chinese Academy of Sciences (IHNS, CAS) and Tsinghua University, jointly sponsored by China Association for Science and Technology (CAST), CAS, National Science Foundation of China (NSFC), IHNS and the China-Portugal Center for the History of Sciences (CPCHS). The theme of the conference is multi-cultural perspectives of the history of science and technology in China.

More than 150 experts and students in history of science and technology from Bulgaria, China, France, Germany, Greece, Japan, Korea, Portugal, Russia, U. K., USA and other countries and districts attended, and more than 90 papers were presented in the Conference. Prof. Ke Jun, academician of CAS, Prof. Wu Wenjun, academician of CAS, Prof. Joseph Cheng Chen-Yih, University of California San Diego (UCSD), Prof. Li Xueqin, Tsinghua University of China and Prof. Roshdi Rashed of the Centre National de la Recherche Scientifique (CNRS) of France served as the Scientific Advisors (SA), Prof. Liu Dun, President of the Division of History of Science and Technology (DHST), previously director (1997-2005) of IHNS and the former president (2000-2008) of CSHST, chaired the Scientific Committee (SOC) and Prof. Liao Yuqun, Previously Director (2005-2009) of IHNS and the president of CSHST, chaired the organizing committee (LOC) of the conference.

In the morning of 27 June, the Opening Ceremony of the 12th ICHSC, and the 30th Anniversary (1980-2010) of CSHST were grandly held in Siyuan Building of Zhongguancun Science Park of CAS. Prof. Liao Yuqun presided over the meeting and gave the opening speech. Prof. Boris Chendov, Institute of Philosophy at the Bulgarian Academy of Sciences (BAS), Prof. Zhang Baichun, director of IHNS, Prof. Roshdi Rashed and Prof. Joseph Cheng Chen-Yih successively gave their enthusiastic talks. The plenary lecture, entitled "30th Anniversary of the Chinese Society for the History of Science and Technology" delivered by Prof. Liu Dun, has recalled the glorious history, shared the growing experience, highlighted the further development of CSHST, and called all colleagues of the Society to promote researches and education in history of science and technology and make further contribution. Prof. Yuan Jiangyang, member of the council of CSHST and senior researcher of IHNS, gave a lecture on *Report on Advances in the History of Science and Technology 2009-2010* (China Science and Technology Press, 2010) authorized the topics and goals, and discussed some issues on the perfect and improvement of the Report. Mr. Li Guoqiang made a description on the website, register program of CSHST.

In the afternoon of 27 June, the 3rd meeting of the eighth session of the council of CSHST was held in the Conference Hall of Siyuan Building. Members of the Council, Cheng Wei, Deng Mingli, Fang Zaiqing, Feng Lisheng, Gao Xi, Guan Zengjian, Guo Shirong, Hu Huakai, Jiang Zhenhuan, Li Chengzhi, Li Yanping, Liao Yuqun, Lu Dalong, Mei Jianjun, Niu Weixing, Qu Anjing, Ren Yufeng, Wang Fubin, Wang Daming, Wu Guosheng, Hsu Kuang-Tai, Xu Zelin, Yang Jian, Yuan Jiangyang, Zhang Daqing, Chang Hao, Zhang Li, Zhang Zhihui, Zhen Naizhang presented, some directors and the Secretaries-General of Committees of CSHST and Miss Wang Ying, Mrs. Peng Dongling, Secretaries of CSHST, as observers, attended the meeting.

The council meeting successively presided by three vice-president of CSHST, Prof. Mei Jianjun of University of Science and Technology Beijing (USTB), Prof. Hu Huakai of University of Science and Technology of China (USTC) and Prof. Wu Guosheng of Peking University. The directors and presidents of the 16 specialized committees and 2 branches of CSHST, Committee on the History of Mathematics, Physics, Astronomy, Chemistry, Geosciences, Biology, Medicine, Agronomy, Technology, Metallurgy, Architecture, Committee on the Comprehensive History, History of Minority Nationalities, Consult, Science Popularization, Professional Committee on National Records, Education for History of Science and Technology, and Branch of Traditional Craft and of Horology, respectively reported over the past year working situation and the future work plan, and put forward comments and suggestions for the further development of CSHST. The committees and branches have achieved great successes in scientific research and the organization building. Some of the committees and branches have been conducting the national, provincial, departmental and foreign cooperative research projects, and launching and participating in the joint undertaking domestic, international conferences and symposium. On the future prospect, the committees and branches has regarded the professional education as a key development objectives of the society.

Subsequently, Prof. Zhang Li, member of the Standing Council of CSHST and senior researcher of IHNS, reported the achieved progress in the State Council Project, the Collection of Academic Growth Data for the Aged Scientists in China (CAGDAC) under the auspices of CAST. The implement of Collection Project, has been realized as one of the most important decisions made by the Central Committee of the Communist Party of China (CCCPC) and the State Council, having based upon scientific analysis and accurate grasp of the developing situation and tendencies of science and technology in China, also an important action for better promoting the Stratagem of Strong Country with Talent (SSCT), has a very important significance in contracting a harmony social atmosphere of advocating innovation, promoting the growth and enhancement of scientific talents, clearing the academic inheritance context in Chinese scientific community, exploring the developing law of science and technology in China and the growth law of skilled personnel, as well as saving valuable historical document and texts, promoting the formation of the core value system of socialism with Chinese characteristics. Currently, the collection work for 50 aged scientists has been initiated, of which the acquisition team for per scientist, dominantly led by historians and scholars in history of science and technology, while at the same time warmheartedly involved by the scientist's assistant, student or even friends. The council of CSHST appeals that the members actively prepare in the project application, and believes that the collection project, high-effectively organized by CAST and conscientiously conducted by teams, be a quality project of historic, practical and social inspections.

The secretariat of CSHST delivered an annual report on activities organizations and financial situation, and the council of CSHST discussed the preparatory work for the 13th International Conference on the History of Science in China (13th ICHSC) and other issues.

From 28 June to 30 Sep. 2010, the 12th International Conference on the History of Science in China was held in Beijing Friendship Hotel. More than 90 lectures, relating to the conference theme, "Science and technology cross-culture communication and comparative study", "ancient Chinese science technology and medical research", "the ancient world, traditional crafts and non-material cultural heritage" and other related topics, were delivered and called lively discussions. The sections of the Conference were titled as follows: Ancient and Early Modern Mathematics, Ancient and Early Modern Astronomy, West and China, Literature Research, Medicine in Comparative Context, Medicine in Social Context, Cultural Heritage, Philosophy of Science, Historical Research: sample analysis, field survey and retrospect, Case Studies, Communication of Science and Technology: translation and introduction, and Science in Social Context.

Prof. Roshdi Rashed, delivered the first plenary lecture, titled "History of Science: at the beginning of the 21st century", in which he traces back the origins and development of the history of science, analyses the questions of principle in the history of science. He also called on:

> Today more than ever, this self-awareness seems to me necessary if we want the history of science to be constituted as a genuine discipline, instead of being a mere domain of activity. Also today, we must construct a new discipline, as necessary as it is legitimate, simultaneously with the history of science, but independently from it: that of social research on the sciences. Such independence is the guarantee that both the history of science and social research on the sciences may be formed as true disciplines, which deal with the cultural phenomenon of science.

On 1 July, Prof. Roshdi Rashed, Dr. Suzanne Débarbat, historian of Observatoire de Paris, Prof. Manuel S. Pinto, University of Aveiro of Portugal, Prof. Eduard Kolchinsky, director of St. -Petersburg Branch Institute for the History of Science and Technology, Russian Academy of Science (St. -PB, IHST, RAS), Dr. Tatyana Yusupova, executive secretary of St. -PB, IHST, RAS, Prof. Byron Kaldis, Hellenic Open University of Greece, Prof. Wang Bing, senior research fellow of IHNS, Dr. Welf Schnell and other conference participants, accompanied by Miss Zhang Gaizhen, the secretary of the conference, arrived in Xi'an. 2 July, some scholars gave their academic reports for Northwest University Symposium of the Conference. Prof. Qiao Xueguang, vice-president, Prof. Qu Anjing, director of Department of Mathematics, Northwestern University (NWU), had an interview with the scholars, expected the more close exchanges between foreign guests and Chinese scholars in history of science and technology, giving greater support for the development of basic disciplines, in particular for the construction of the history of science and technology in NWU.

On 5 July, Prof. Joseph Cheng Chen-Yih, Prof. Roshdi Rashed, Dr. Suzanne Débarbat, Prof. Byron Kaldis, Prof. Wang Bing, Prof. Eduard Kolchinsky, Dr. Tatyana Yusupova, Prof. Zheng Chengliang, vice-president of Shanghai Jiao Tong University (SJTU), Prof. Guan Zengjian, executive dean of the Humanities College of SJTU and vice-president of CSHST, met with more than 50 Ph. D. candidates and attended the Solstice Colloquium, host by Department for the History and Philosophy of Science of SJTU. On 6 July, Prof. Ye Shuhua, academician of CAS and professor of Shanghai Astronomical Observatory had a interview with Prof. Joseph Cheng Chen-Yih, Dr. Suzanne Débarbat, Prof. Wang Bing, and desired to serve as the academic advisor of the related research project of CSHST. From 1 July to 2 Aug., Prof. Efthymios Nicolaidis, secretary-general of the Executive Council of DHST, realized a short visit in Shanghai.

During the conference, Prof. Eduard Kolchinsky and Dr. Tatyana Yusupova, visited IHNS, and a four years cooperative agreement of between the St. -PB, IHST, RAS and IHNS, CAS of bilateral cooperation framework with a common academic interest was signed by Prof. Eduard Kolchinsky and Prof. Zhang Baichun on 7 July. The signed cooperation topics focus on comparison studies in the development of modern science and technology in Russia and in China, and knowledge dissemination and technology transfer between Russia and China. Prof. Manuel S. Pinto, Mr. João M. G. Barroso, cultural counsellor of the Portugal Embassy in Beijing and director of Portuguese Cultural Centre (PCC-Beijing), and the secretariat of CSHST discussed the proposal of the 14th ICHSC host by Portuguese institution in 2014.

Networks, www. china. com. cn, www. people. com. cn, www. sina. com. cn, www. ynet. com, www. xinmin. cn, www. tianjiawe. com, newspapers, *People's Daily*, *Science Times*, *Beijing Evening News*, journals, *the Chinese Journal for the History of Science and Technology*, *Chinese Heritage*, *Popular Science News*, and other media issued their interviews and reports. Scholars recognized the academic levels of the conference lectures, especially of the plenary lectures, satisfied with the conference organization, and look forward to continuing to participate in such succeed international conferences. They also expressed their heart-warming thanks to LOC of the conference and gave their comments and suggestions to the improvement and perfection the further ICSHC. Prof. Fung Kam-Wing of Hong Kong University (HKU), Prof. S. Manuel Pinto suggested that CSHST organize at the right time setting up the Conference Awards in order to require the greater recognition of scholar's academic achievements.

Now, the International Conference on the History of Science in China (ICHSC), organized by CSHST, has been regarded as one of the most important special-subject international conferences of the high academic level. In 1982, the 1st ICHSC was held in Leuven, Belgium. The series of ICHSC, financially supported by CAS, CAST and other institutions all over the world, has launched an important platform for the exchanges and dialogue between the international peers and the Chinese scholars in history of science and technology.

16-20 January 1996, the 7th ICHSC was host by Shenzhen University of China, and this Conference was realized as a turning point in the development of ICHSC. Prior to this, on 15 March 1995, Prof. Lu Yongxiang, president of the 5th standing council of CSHST and vice-president of CAS, delivered an important speech:

> The Nanshan District of Shenzhen city wills to come up with 200 000 Renminbbi in support of the 7th ICHSC. This indicates that, as the socio-economic development in China, the Chinese community has started to pay more attention to the development of science. This series of international conferences would be continued to every three or four years, we insist on holding the flag. According to our current situation, the conference place and funds will not be a problem, the Chinese Academy of Sciences and National Science Foundation of China should give some support. ①

Since then, ICHSC has embarked on the right track, the 8th ICHSC (Technische Universität Berlin, TUB, 1998, 130 persons), the 9th (City University of Hong Kong, CityU, 2001, 130 persons), the 10th (Harbin Institute of Technology, HIT, 2004, 120 persons) and the 11th (Guangxi University for Nationalities, GUN, 2007, 120 persons) successfully held. Scholars, experts and junior researchers and students, engaged in history of science and technology, have made important contributions to this successful series of convened conference. In January 1996, Prof. Lu Yongxiang served as the president of SOC of the 7th ICHSC. In October 2001, Prof. Lu Yongxiang, president of CAS, Prof. Wu Wenjun, and Prof. Xi Zezong (1927-2008), academician of CAS and historian of astronomy, attended, Prof. Joseph Chen Cheng-Yih and Prof. Liu Dun jointly served as Chairmen of LOC of the 9th ICHSC.

The series of ICHSC and the establishment of other important academic exchange platforms have prepared some necessary conditions for Chinese scholars holding important positions in related international organizations. Prof. Liu Dun, had served as the second vice-president, the

---

① Xi Zezong. Chinese Society for History of Science and Technology: 1980-2000、in Xi Zezong. A New Catalogue of Ancient Novae and Explorations in the History of Science: Self-selected Works of Academician Xi Zezong. Xi'an: Shanxi Normal University Press, 2002: 732

first vice-president and performed other important duties for the scientific community, was successfully elected as the president of the Executive Council of DHST in the 23rd International Congress of History of Science and Technology (ICHST) in August 2009 in Hungary.

Prof. Lu Yongxiang, vice-chairman of the Standing Committee of the National People's Congress (NPC) and president of CAS congratulated on 12 August by letter:

> Prof. Liu Dun: Having known that you are elected as the president of the Executive Council of DHST in the 23rd ICHST in Budapest, I send my sincere congratulation. It has confirmed not only your achievement, but also the Chinese and my Academy's researches in history of science. I wish you make new achievements in leading DHST and research works in the history of science.

In the 22nd and the 23rd ICHST, young Chinese scholar had twice received the nomination of the assessors of the Executive Council of DHST.

The LOC of the 12th ICHSC invited the Member Effectif and Membre Correspondant of Académie Internationale d'Histoire des Sciences (AIHS-IAHS), academicians of CAS and other well-known scholars serve as the scientific advisors, the members of the Executive Council of DHST as the member of SOC of the conference. Scholars from various research disciplines and fields have made in-depth scientific exchanges in the themes of the conference, improved their friendship and expanded the cooperation channels. The conference has certainly strengthened exchanges and cooperation in the researches on the history of science and technology in China (HSTC), extended the academic influence of researches in the history of science and technology in China in the international academic communities and continuously promoted the young Chinese scholars make their due contributions in international organizations. In July 2012, Hellenic Open University of Greece will host the 13th International Conference on the History of Science in China (13th ICHSC), and the 24th International Congress of History of Science, Technology and Medicine (24th ICHSTM) will be held in Manchester, 22nd-28th July 2013.

CSHST will adhere to the principle of "Small scale, high level, great academic, and preferred effectiveness", constantly improves the academic level of ICHSC, leads the development of the discipline of HSTC, promotes research on HSTC growing as one of the global subject areas, maintains ICHSC as an important platform for scientific exchanges and dialogue, effectively expands the new research directions for the discipline of HSTC. CSHST encourages Chinese junior researchers and students to establish international contacts with historians in ICHSC, and preserve the stable, sustained and coordinated development in the education and popularization of history of science and technology.

Edited by:

Zhang Gaizhen, Institute for the History of Natural Science, Chinese Academy of Scineces.

Wang Ying, Lu Dalong, Chinese Society of the History of Science and Technology.

Tong Qingjun, Tsinghua University.